世界绿色建筑
——热环境解决方案

WORLD GREEN BUILDINGS
THERMAL ENVIRONMENTAL ENGINEERING SOLUTIONS

中国建筑文化中心　编

江苏人民出版社

图书在版编目(CIP)数据

世界绿色建筑 : 热环境解决方案 / 中国建筑文化中心编. -- 南京 : 江苏人民出版社, 2012.4
ISBN 978-7-214-08070-7
Ⅰ. ①世… Ⅱ. ①中… Ⅲ. ①建筑热工－案例 Ⅳ. ①TU111
中国版本图书馆CIP数据核字(2012)第058366号

English & Simplified Chinese Editions
Are Exclusively Distributed in India
by Patrika Book Centre

世界绿色建筑——热环境解决方案 中国建筑文化中心 编

策划编辑：胡中琦
责任编辑：刘 焱 曹 蕾
翻 译：张竹村
责任监印：彭李君
美术编辑：周 宇
出版发行：凤凰出版传媒集团
凤凰出版传媒股份有限公司
江苏人民出版社
天津凤凰空间文化传媒有限公司
销售电话：022-87893668
网 址：http://www.ifengspace.cn
集团地址：凤凰出版传媒集团
（南京湖南路1号A楼 邮编：210009）
经 销：全国新华书店
印 刷：利丰雅高印刷（深圳）有限公司
开 本：965 mm×1270 mm 1/16
印 张：20
字 数：350千字
版 次：2012年6月第1版
印 次：2012年6月第1次印刷
书 号：ISBN 978-7-214-08070-7
定 价：299.00元（USD 50.00）

（本书若有印装质量问题，请向发行公司调换）

World Green Buildings——Thermal Environmental Engineering Solutions

前言
Introduction

GREEN

WORLD GREEN BUILDINGS 世界绿色建筑——热环境解决方案
THERMAL ENVIRONMENTAL ENGINEERING *SOLUTIONS*

我国"十二五"规划纲要提出："建筑业要推广绿色建筑、绿色施工，着力用先进建造、材料、信息技术优化结构和服务模式"。

发展绿色建筑对于实现科学可持续发展、建立循环经济模式、建设低碳社会具有重要意义，是实现绿色发展和建设资源节约型、环境友好型社会的必由之路和有力抓手。绿色建筑在我国大规模城镇化进程中，已成为促进产业结构调整、带动产业升级的重要载体。

发展绿色建筑是应对全球气候变化的重要途径，已引起国际社会的广泛关注，并成为世界各国倡导与发展的主流和方向。一些发达国家形成了较为成熟的绿色建筑技术体系、评价机制及推广模式。为了展示世界绿色建筑领域前沿动态，介绍国外最新技术和实践成果，我们编纂了这套"世界绿色建筑"系列丛书，包括《世界绿色建筑——热环境解决方案》《世界绿色建筑——可再生能源应用与建筑一体化》《世界绿色建筑——环境景观规划设计》及《世界绿色建筑——生态城市与住区规划》，计划于2012年内陆续出版。

《世界绿色建筑——热环境解决方案》集中关注如何运用绿色技术、材料，通过系统、优化设计，实现建筑为人类提供健康生活、工作环境的探索实践。编者历经半年多时间，收集了大量国外相关信息资料和项目案例，从20余个国家和地区的200多件来稿和案例中，精选出50个典型项目案例，进行精心整理、编译，呈现给读者。这些案例采用系统的设计方法和适用的材料设备，通过太阳辐射控制、自然通风、节能环保电气设备，被动措施与主动措施综合应用，给出了热环境解决方案，营造出健康舒适的生活、居住环境。这些方案不乏获得了绿色建筑评价认证及国际或区域性奖项的优秀案例，可供建设领域研究人员、建筑师、设计师、工程师、建设管理人员参考借鉴。

China's Twelfth Five-Year Plan put forward that green building and green construction should be popularized in building industry. We should make great efforts to optimize the structure and serving patterns of building industry through advanced construction techniques, new building materials and information technology.

The development of green building is of significant importance in striving for scientific sustainable development, establishing cycle economic model and building low-carbon society, which plays an important role in realizing green development and building resource-saving and environment-friendly society. In the process of the urbanization, green building and sustainable development have been turned into the vectors of industrial restructuring and upgrading.

It is a best way to tackle the global climate change by developing green building, which has aroused worldwide concern and become the mainstream and orientation advocated by the most countries in the world. Mature technical systems, evaluation mechanism and promotion models of green building have been built up in a number of developed countries. With the purpose of presenting excellent achievements of green building and of introducing the latest theory and technology, we compile a series of *World Green Buildings*, including *Thermal Environmental Engineering Solutions*, *Building Integrated Renewable Energy*, *Landscape Planning and Design* and *The Planning of Eco-City and Green Community*, scheduled to be published this year.

World Green Buildings Thermal Environmental Engineering Solutions presents the exploring practices and typical project cases concerning how to realize the aim of providing healthy working and living environment for people by means of systematic design and optimization, usage of green materials and technology.

We have spent several months accumulating over 200 foreign project cases and relevant materials, selected 50 cases among them and compiled carefully, presenting them to readers. Systematic design methods, appropriate building materials and techniques are adopted in these projects, solar radiation control, natural ventilation, energy-saving equipment, passive and active measures applied integrally. These cases provide thermal environmental engineering solutions, show creations of healthy and comfortable living environment. A number of the projects gained national-level green building certifications, international or regional awards. This book serves as technical reference book on green building for researchers, architects, designers, engineers, project managers in construction industry.

中国建筑文化中心
2012年3月
China Architectural Culture Center
March, 2012

公共 Public 006

办公 Office 158

住宅 Residential 274

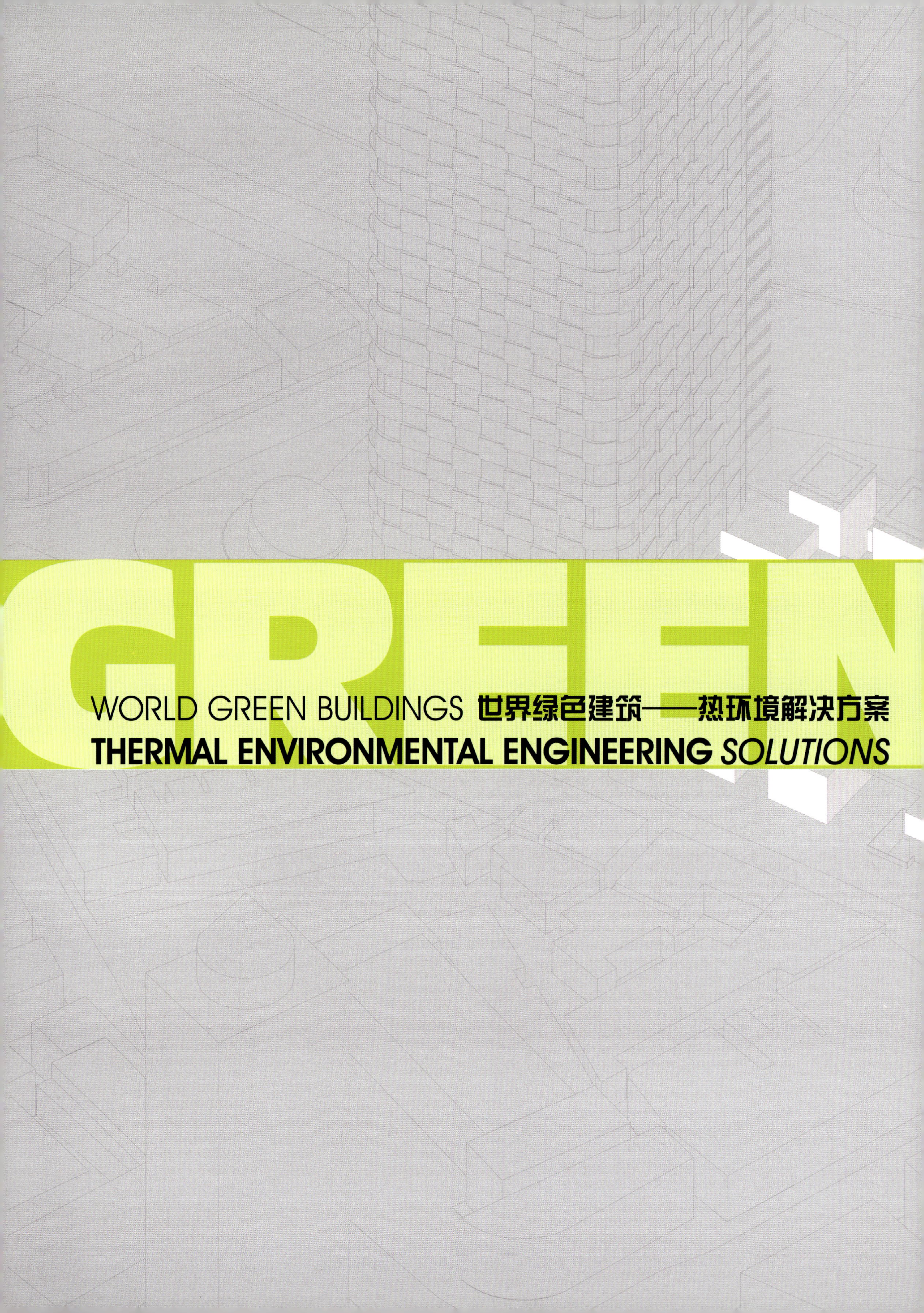
GREEN
WORLD GREEN BUILDINGS 世界绿色建筑——热环境解决方案
THERMAL ENVIRONMENTAL ENGINEERING SOLUTIONS

公共
Public
006

EDUCATION EXECUTIVE AGENCY & TAX OFFICES

教育与税务机关综合管理大楼

2011, UNStudio©Aerophoto E

地点：荷兰格罗宁根
建筑面积：办公48 040 m²，停车场21 000 m²，附属建筑1500 m²
场地面积：31 134 m²
建筑及室内设计：UNStudio
室内设计：Studio Linse
景观设计：Lodewijk Baljon
委托方：Dutch Government Buildings Agency (RGD)
摄影：Ewout Huibers；Ronald Tilleman；Aerophoto Eelde

Location: Groningen, the Netherlands
Building Area: 48,040 m² offices, 21,000 m² parking, 1500 m² pavilion
Building Site: 31,134 m²
Architecture and Interior: UNStudio
Interior: Studio Linse
Landscape: Lodewijk Baljon
Client: Dutch Government Buildings Agency (RGD)
Photography: Ewout Huibers；Ronald Tilleman；Aerophoto Eelde

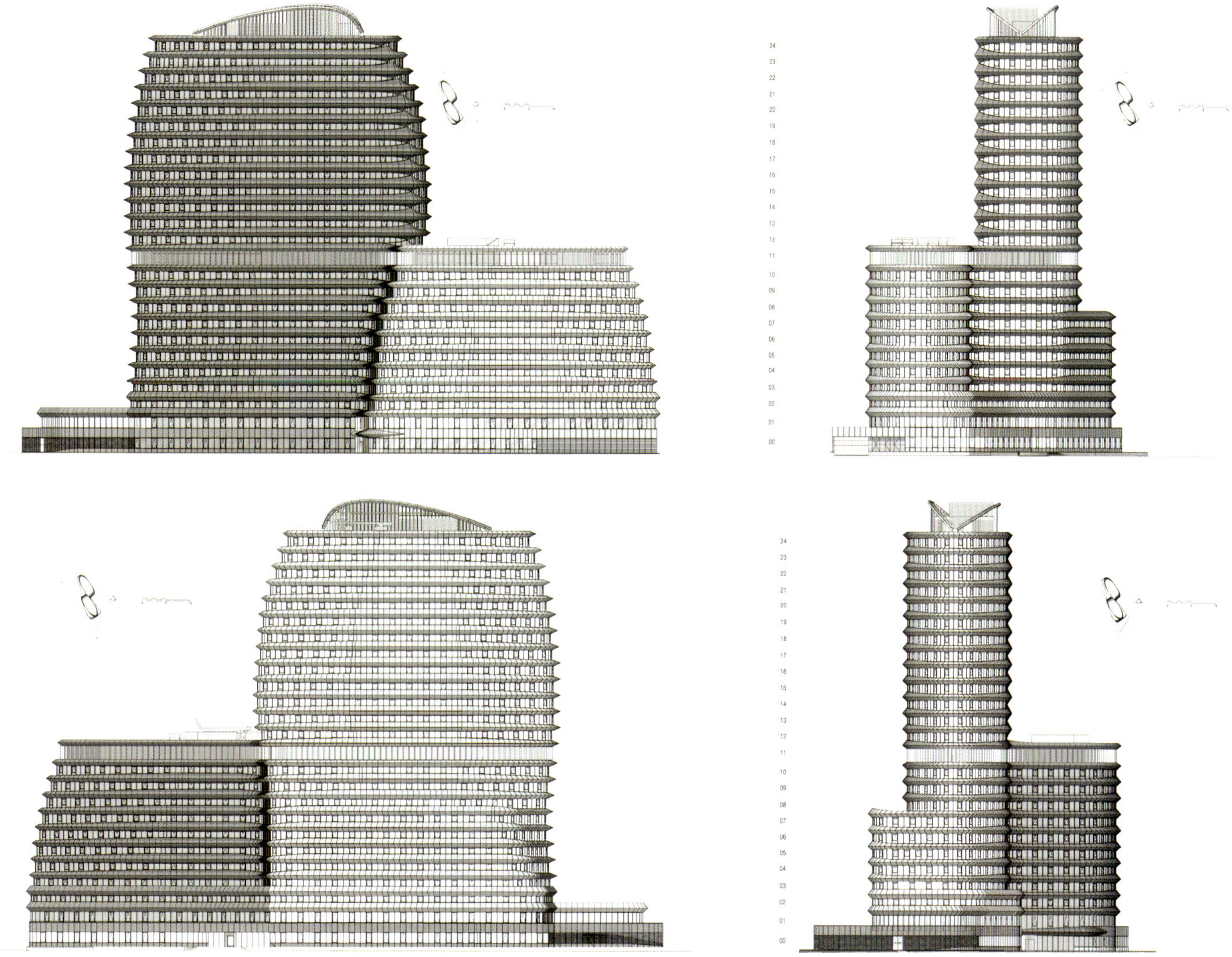

立面图 ELEVATIONS

2011, UNStudio©Aerophoto Eelde

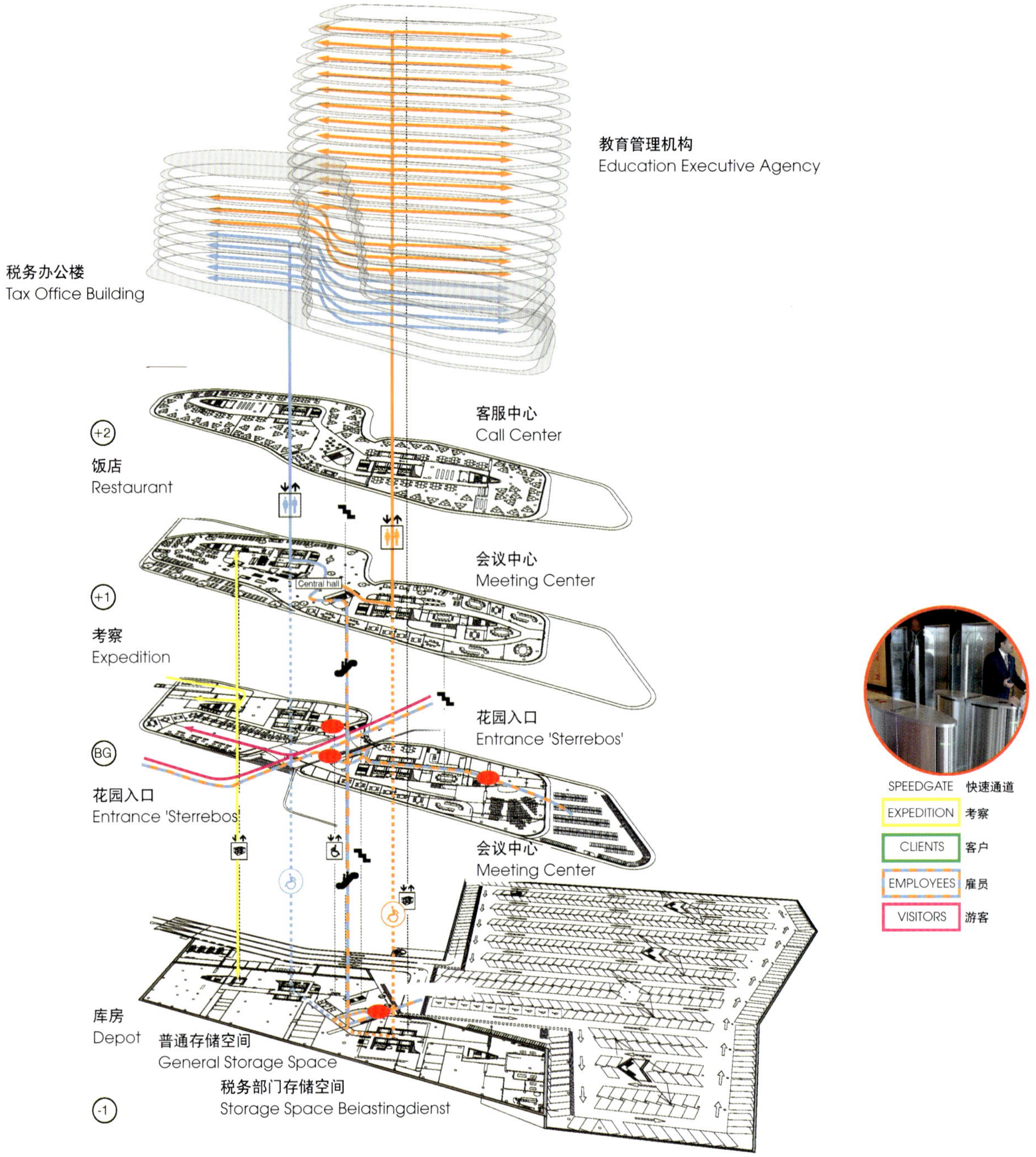

教育管理机构
Education Executive Agency

税务办公楼
Tax Office Building

+2
饭店
Restaurant

客服中心
Call Center

+1
考察
Expedition

会议中心
Meeting Center

Central hall

BG
花园入口
Entrance 'Sterrebos'

花园入口
Entrance 'Sterrebos'

会议中心
Meeting Center

-1
库房
Depot

普通存储空间
General Storage Space

税务部门存储空间
Storage Space Beiastingdienst

SPEEDGATE 快速通道
EXPEDITION 考察
CLIENTS 客户
EMPLOYEES 雇员
VISITORS 游客

分析图 DIAGRAM

2011, UNStudio©Ronald Tilleman

2011, UNStudio©Ronald Tilleman

项目概况

一座92 m高的造型柔和曲线起伏的综合大楼标明了格罗宁根的天际线。UNStudio的Ben van Berkel与DUO² 合作，为政府的两个机构建成的教育管理和税务大楼，是欧洲最成功的可持续设计大型办公楼之一。项目包括国家税务办公室和学生贷款管理局两个公共机构的设计、施工和资金管理。项目拟在第一阶段建设一座办公楼，然后建设一座地下停车场，最后建造城市公共花园和一座商用附属多功能大楼。

建筑热环境解决方案

该项目是欧洲可持续性最好的新建大型写字楼之一。建筑从节约能源、减少耗材以及社会发展和环境保护等多方面出发，利用多种主动或被动措施，实现可持续设计的理念，实现了降低能耗（EPC 0.74），并大大降低材料消耗。建筑物内外形成了良好的生物气候，对人类及动植物都非常有益。

可持续性和节能的目标指导了立面设计，采用了对环境影响最小的技术设备。立面设计结合了遮阳、风控、日光渗透和鳍状元件。横翅阻挡了建筑外部大量的热辐射，减少了对制冷的需求。

建筑的另一项可持续设计是活化混凝土芯与长时间地下储能的结合，这显著地减少了对外部能源的需求。

各工作区独立控温，创造了一个健康、节能的室内微气候，改善员工工作环境也是设计的重点。室内有充足的自然采光，采用可调节加热，独立工作区的通风和新风入口使整个建筑的工作环境舒适优良。

在建筑的第11层，设有高压通风系统，通过设在11楼的主工程轴和立面格栅使自然风进出大楼，从而减少了对人工通风的需求。此外，在后续的施工中，将实现利用数据中心和办公室产生的余热为居室加热。

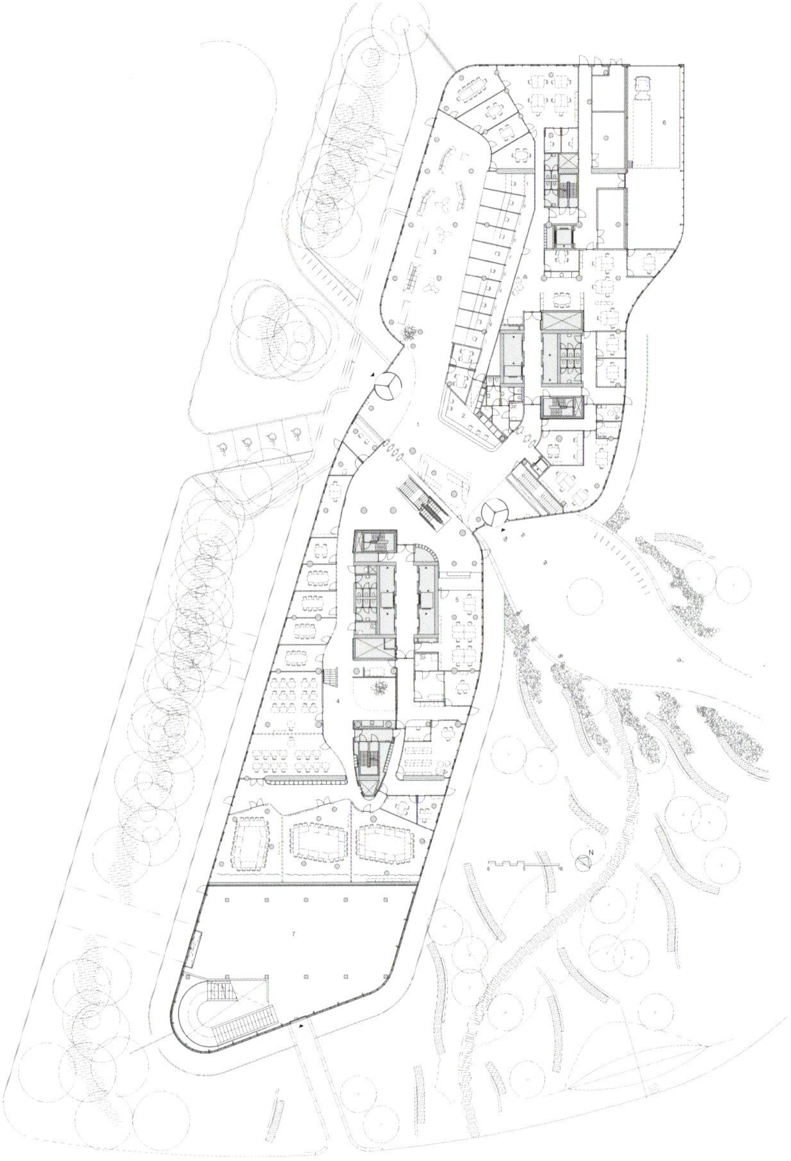

一层平面图 GROUND FLOOR PLAN

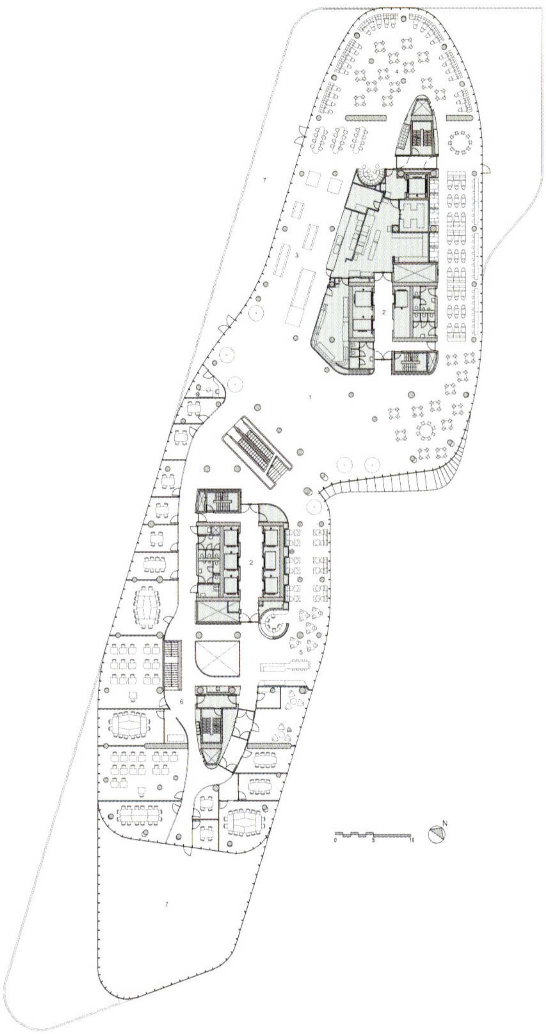

二层平面图 SECOND FLOOR PLAN

2011, UNStudio©Ewout Huibers

2011, UNStudio©Ronald Tilleman

2011, UNStudio©Ronald Tilleman

2011, UNStudio©Ronald Tilleman

2011, UNStudio©Ronald Tilleman

2011, UNStudio©Ronald Tilleman

Program Description

A new, 92 meter tall complex of soft, undulating curves marks the skyline of Groningen. UNStudio/ Ben van Berkel, with consortium DUO², realizes Education Executive Agency & Tax offices, one of the most sustainable large office buildings in Europe for two governmental offices The project includes the design, construction and financing of two public institutions; the national tax offices and the student loan administration. The program includes: an office building in phase A, an underground parking in phase B, a public city garden, and a multifunctional pavilion with commercial functions in phase C.

Building Thermal Environmental Engineering Solutions

The project is one of Europe's most sustainable large new office buildings. The architectural response to this has been to strive for an all-round understanding of the concept of sustainability, including energy and material consumption, as well as social and environmental factors using diverse passive and active environmental and energy efficient solutions. Thus the sustainability manifests itself in reduced energy consumption (EPC 0.74), as well as significantly reduced material consumption. Both inside and outside the architecture generates a bio-climate that is beneficial to both humans and the local flora and fauna.

Program Description

Competition entry for a new Ljubljana administration centre of 60,000 m². The brief proposed several departments to move to the same site but occupy different buildings. The heart of the centre is the main hall where citizens could also arrange all the documents. The site is just on the edge of Ljubljana city centre, by the river and is already occupied by some existing protected buildings. As such, the area represents a chance for unique rearrangement with its identity to become a sort of symbol or landmark of contemporary Ljubljana architecture. The mix of public and private areas inside the program calls for complex organization – both inside and outside. The project took the idea of cylindrical organization from the existing protected garage-building that is in the middle of the site. Three new buildings are designed as cylinders related with loops of spaces in different heights. In the middle of each cylinder the public and entrance areas are organized and offices are arranged around this. The logic of connected and disconnected departments follows the mathematical model of intersections and unions. Two departments that share meeting spaces are connected in various floors with loop-bridges which create a unique space. The loops and circles create a complex landscape with external plazas, squares and bays of greenery, which connect to the river promenade.

Building Thermal Environmental Engineering Solutions

Special care was also given to achieve optimal environmental conditions with minimal energy demands in three steps: reducing the energy and meeting remaining conditions with high-performance building-integrated systems, and sourcing those systems as much as possible with renewable sources.

1. Passive Measures

Control of Sun Radiate. The external facades will feature high performance glazing and an adaptable external shading device to reduce solar gains in summer.

2. Active Measures

Displacement Ventilation and Cooling Slab Air-condition: One of the key ideas of the sustainability concept is to heat and a cool working area like offices and uses the exhaust air of these offices to condition the atrium, which is used as a big buffer space. The concrete slab imbedded pipe system provides cooling without draft problems and in winter comfortable radiant heating.

In summer the majority of cooling for the office spaces will be achieved using the slab system. The decentralized ventilation units vent fresh air into the rooms and serve as peak time cooling in summer. During winter the fresh air will be heated inside the units and distributed into the rooms using the displacement ventilation principle. An overflow to the atrium allows the exhaust air to flow from the offices into the atrium and through louvers in the roof to the outside. As the atrium is a buffer space, which is naturally ventilated and heated/cooled by the exhaust air from the offices it means the temperature in the atrium will swing from 15 °C in winter to ambient conditions at the upper levels in summer (temperature stratification).

Humidity condition. To minimize the peak summer conditions, features inside the atrium like a water wall and a floor cooling system will condition this space in addition. The water wall is used for both cooling and dehumidification purposes in summer, whereas in winter the water will evaporate and humidify the air.

Heat pumps. Special areas inside the atrium will get this "own" climate. This local climate is important to guarantee adequate working conditions. The energy supply system is based on natural resources like a river water heat exchanger to provide cooling in summer as well as heating - in combination with a heat pump - in winter.

Other technical measures

Photovoltaic cells integrated in the atrium roof will help to generate power. Another objective of a sustainable building approach is to minimize the use of energy intensive materials. Instead, locally sourced, sustainable materials will be used.

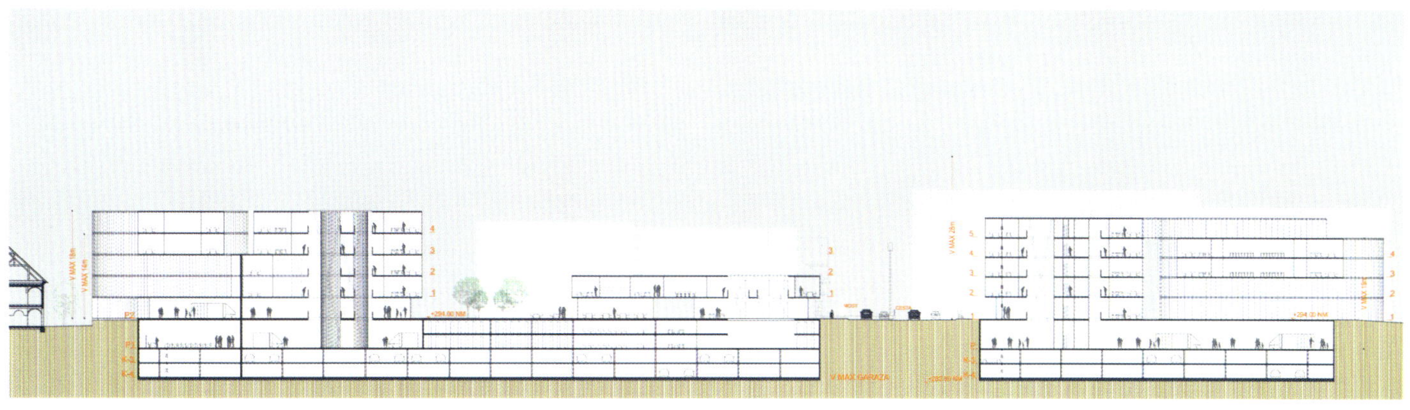

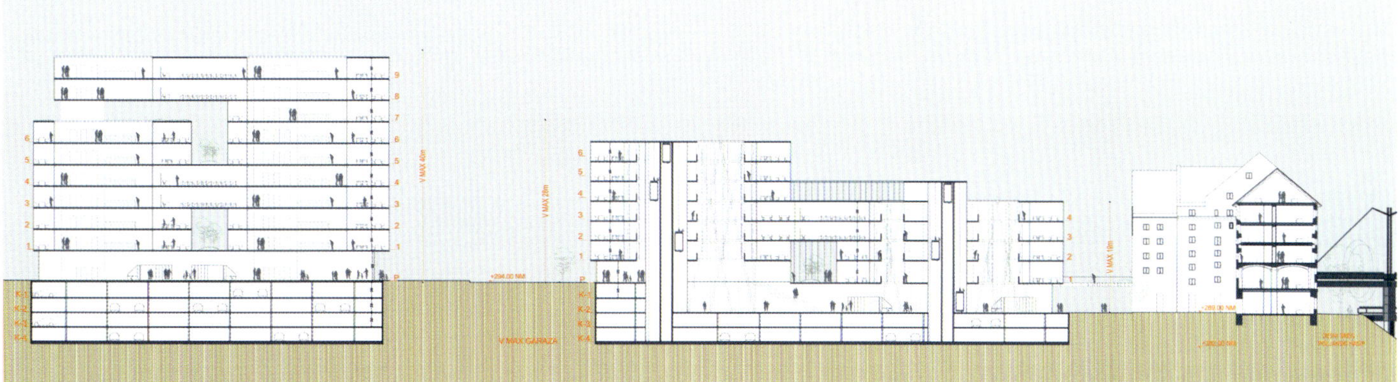

剖面图 SECTIONS

模型图 MODEL DIAGRAMS

- public program - self service terminal 公共区域——自服务终端
- semi public programs - counters 半公共区域——柜台
- controled areas - offices 管制区域——办公室
- reception to controled offices 为限制访问办公室提供的接待处

分析图 DIAGRAM

EMERGENCY AND INFECTIOUS DISEASES UNIT, SKÅNE UNIVERSITY HOSPITAL

斯克纳大学医院急症传染病中心

地点：瑞典马尔默
规模：26 000 m²
建筑设计：C. F. Møller Architects；SAMARK Arkitektur & Design
景观设计：C. F. Møller Architects
所获奖项：2010年马尔默城市设计奖；2010年卡斯帕赛林奖入围；
2006年国际竞标一等奖
委托方：Regionservice Södra Skåne
摄影：Jørgen True

Location: Malmö, Sweden
Size: 26,000 m²
Architect: C. F. Møller Architects; SAMARK Arkitektur & Design
Landscape: C. F. Møller Architects
Selected Awards: 2010 Winner Stadsbyggnadspriset Malmö Stad;
Finalist for Kaspar Salin Priset 2010;
2006 1st Prize in International Competition
Client: Regionservice Södra Skåne
Photography: Jørgen True

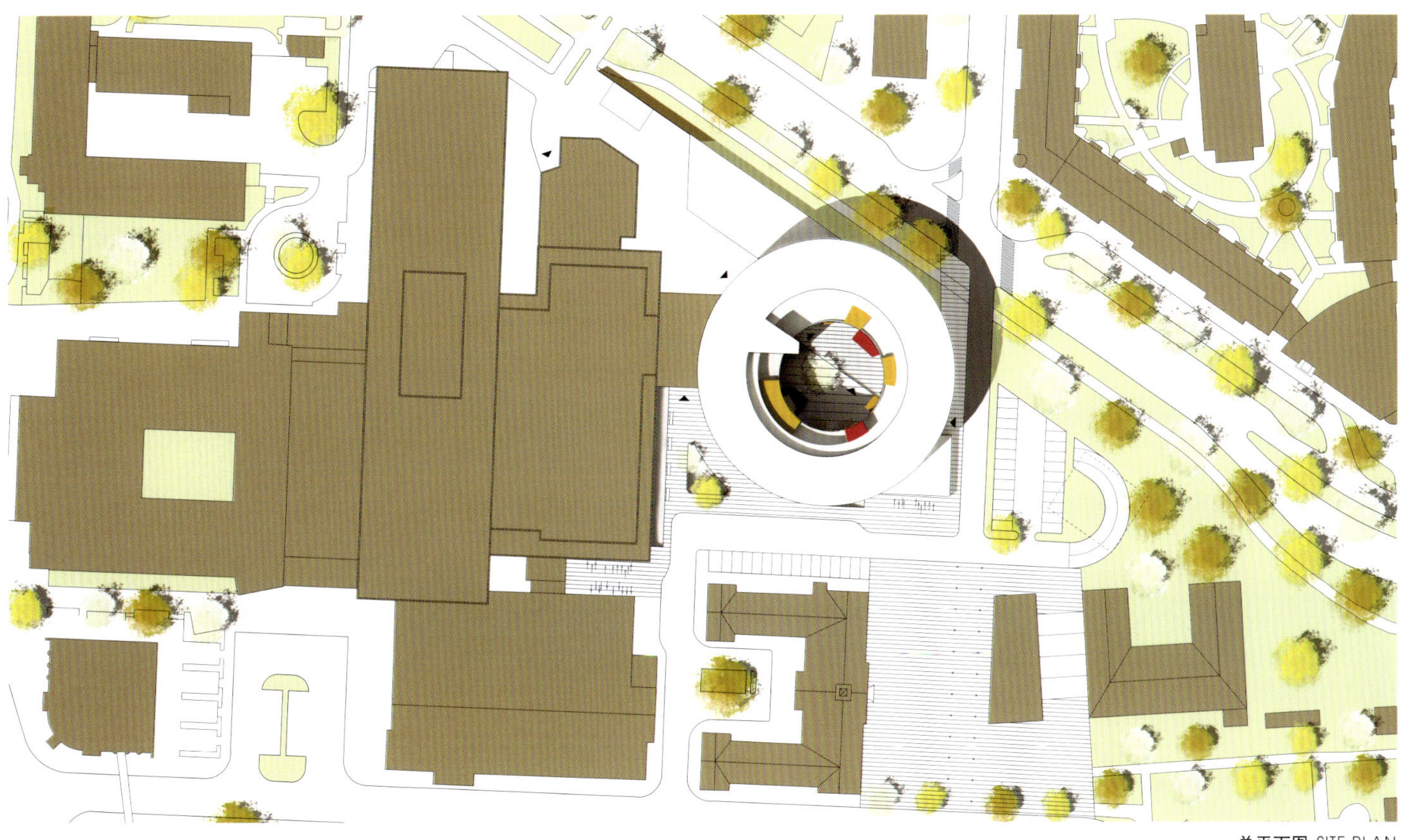

总平面图 SITE PLAN

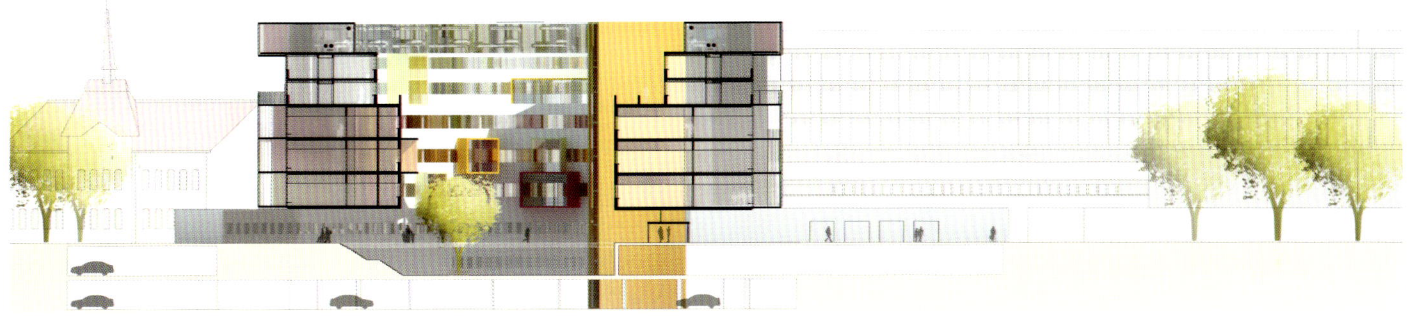

立面图 ELEVATIONS

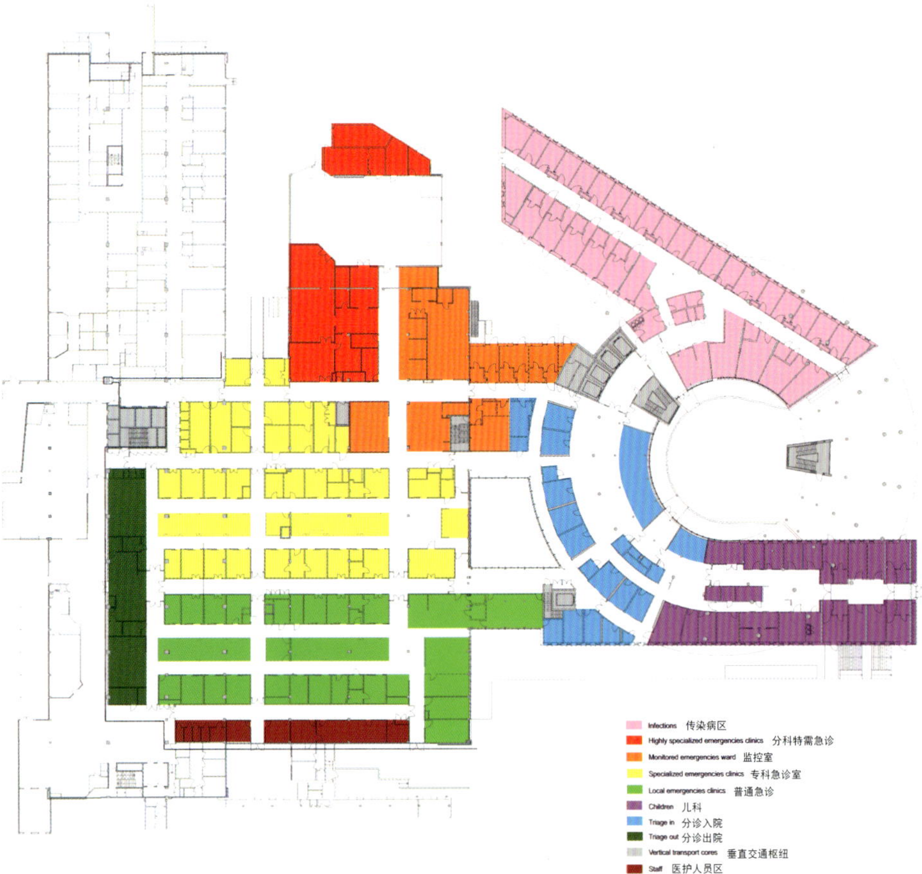

一楼功能分析图 FUNCTIONAL DIAGRAM GROUND FLOOR

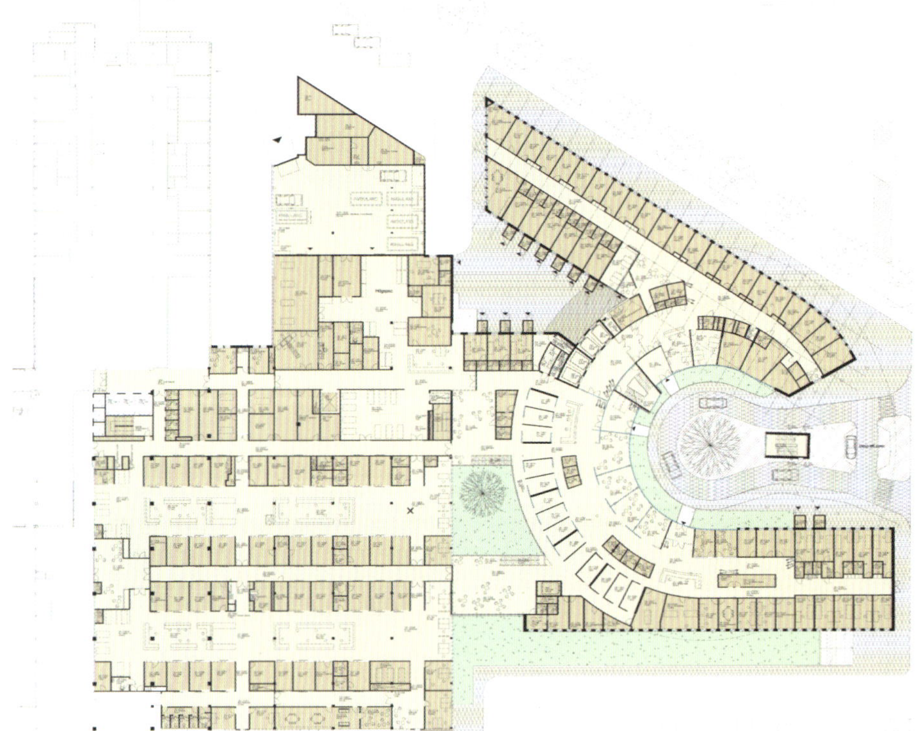

一层平面图 GROUND FLOOR PLAN

位于瑞典马尔默的斯克纳大学医院大楼是一座为重症患者和隔离部门服务的大楼。设计的核心目标在于最大限度地预防感染扩散，并成为当地引人注目的新地标。患者通过从顶层走廊开始的特殊密封舱到达各处。传染病人和医用废弃物有专用的外部电梯，内部有工作人员的专用电梯。在疫情爆发时，每层楼都可以划分成更小的隔离单位。

马尔默隔离中心是瑞典规模最大也是最现代化的传染病中心。鉴于隔离需要，医院使用专业的主动性空气调节设施，同时在建筑围护结构中采用了隔热铝框架、双层玻璃立面以及三层玻璃窗，以达到良好的隔热效果。这座建筑还引入了自然采光，病房沿建筑的外圆周排列，可以享受自然采光。圆柱形的建筑中有一个中央开放中庭，使沿着内圆周分布的房间也能够得到自然日光照射。走廊无法得到日光，但是可以通过玻璃外墙表面的折射、透射等方式间接地从各个房间得到自然采光。

斯克纳大学医院积极响应环保事业，将对环境的影响降到最低。2009年开始，斯克纳大学的医院都使用可再生电能。从2010年开始，SUS致力于实现整体性的综合环境管理系统。

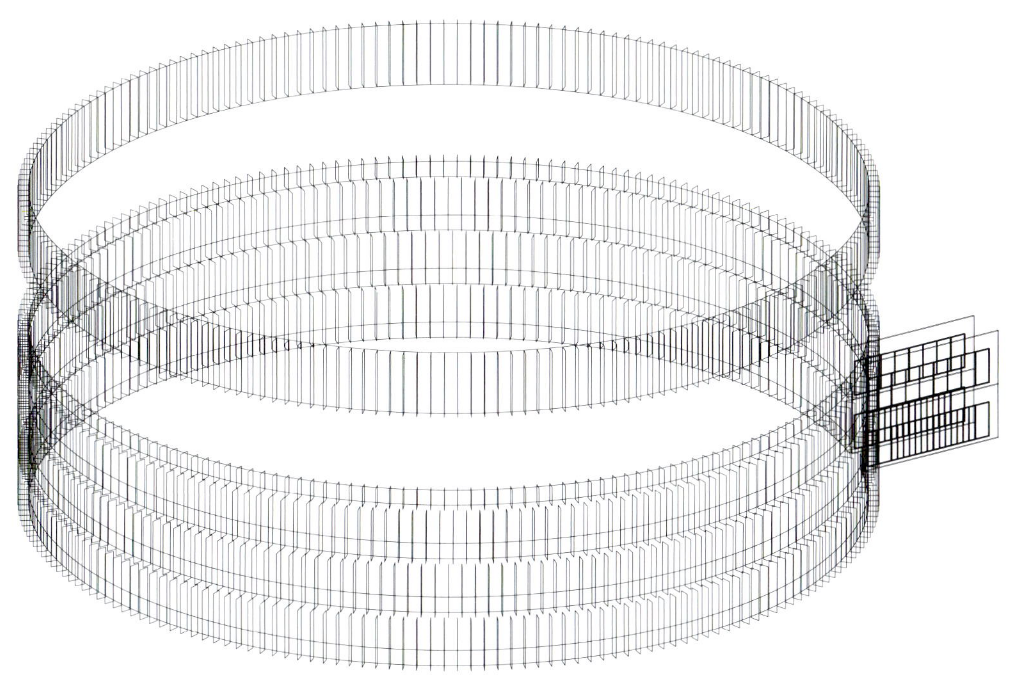

玻璃窗几何分布 GLAZING GEOMETRY

The circular building housing the casualty and isolation departments at Skåne University Hospital (SUS) in Malmö, Sweden – designed to offer the best possible prevention against the spread of infection – is set to become the hospital's striking new landmark.

The Malmö isolation department Sweden's largest and most modern infectious diseases facility. For the need for isolation, specific active air conditioning facilities are used, meanwhile, insulated aluminium frame, double glazing facade and triple glazing windows are used in the build envelope to ensure a high quality of heat insulation. The building invites natural daylight to enter. The wards – which have been laid out around the periphery of the building – benefit from direct natural light. As the cylindrical building has a central, open atrium, the "inner-circle" rooms also benefit from natural light, giving all rooms in the building direct daylight except the corridors, which – through the glazed wall sections – benefit from indirect natural light from the rooms.

Skåne University Hospital is working aggressively on environmental issues. The aim is to have the minimum possible impact on the environment. In Skåne University Hospital, all the hospitals are using renewable electricity since 2009. From 2010, their environment work focus on creating a collective environment management system for SUS as a whole.

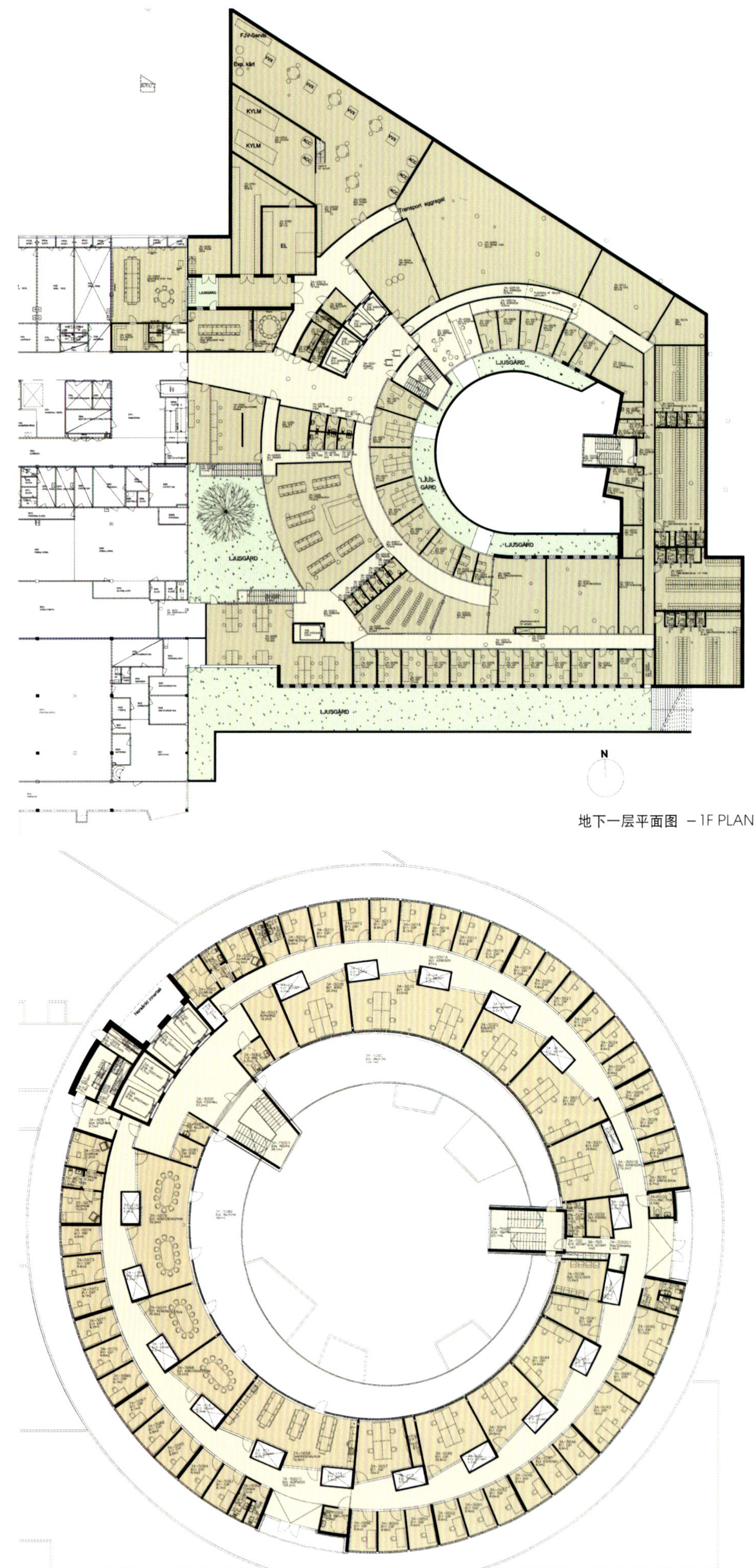

地下一层平面图 –1F PLAN

五至六层办公区平面图 LEVEL05-06 OFFICES PLAN

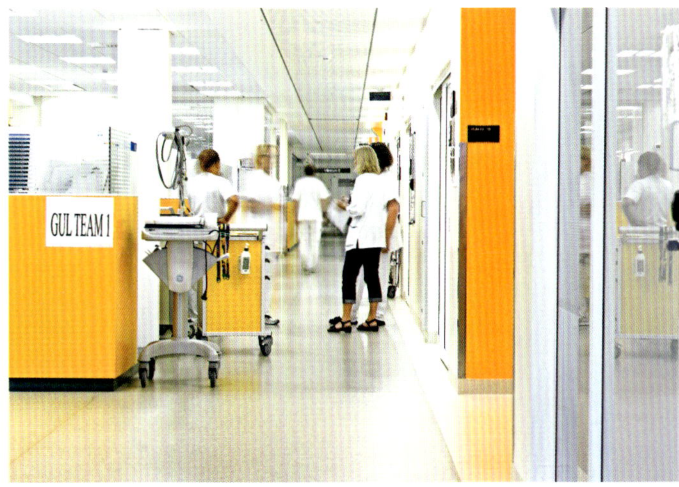
GUL TEAM I

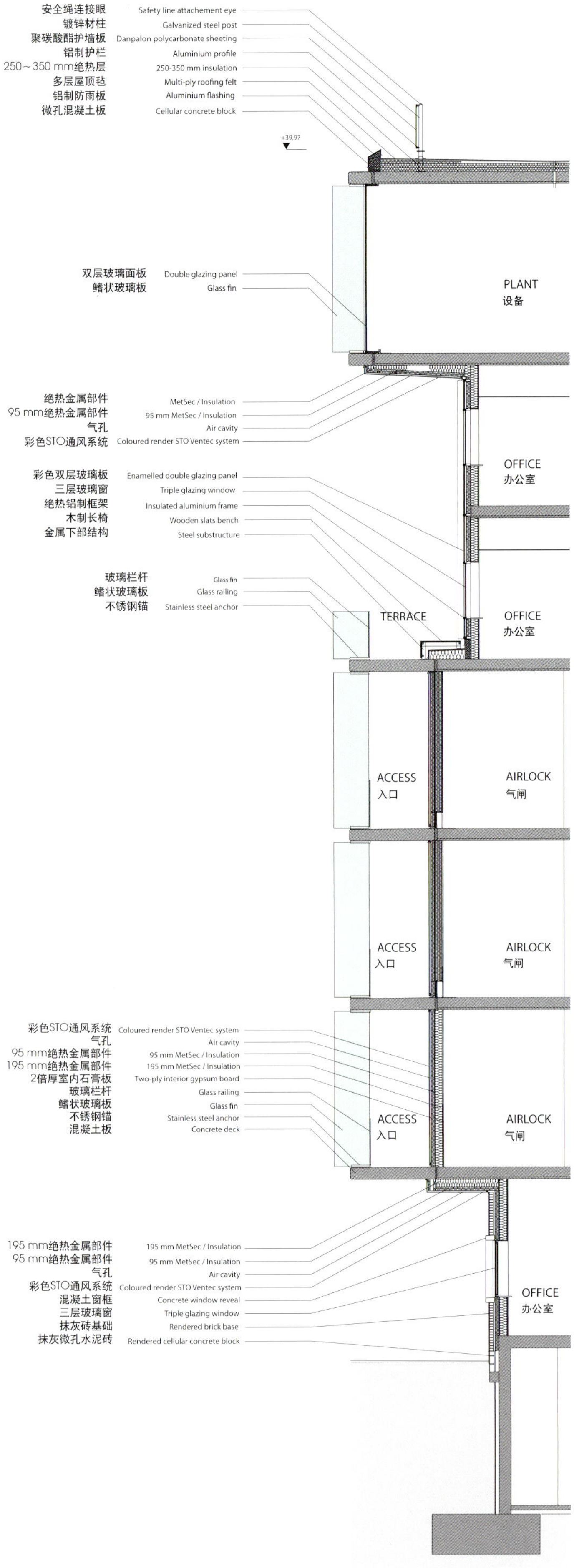

安全绳连接眼 Safety line attachement eye
镀锌材柱 Galvanized steel post
聚碳酸酯护墙板 Danpalon polycarbonate sheeting
铝制护栏 Aluminium profile
250～350 mm绝热层 250-350 mm insulation
多层屋顶毡 Multi-ply roofing felt
铝制防雨板 Aluminium flashing
微孔混凝土板 Cellular concrete block
+39,97
双层玻璃面板 Double glazing panel
鳍状玻璃板 Glass fin
PLANT
设备
绝热金属部件 MetSec / Insulation
95 mm绝热金属部件 95 mm MetSec / Insulation
气孔 Air cavity
彩色STO通风系统 Coloured render STO Ventec system
OFFICE
办公室
彩色双层玻璃板 Enamelled double glazing panel
三层玻璃窗 Triple glazing window
绝热铝制框架 Insulated aluminium frame
木制长椅 Wooden slats bench
金属下部结构 Steel substructure
玻璃栏杆 Glass fin
鳍状玻璃板 Glass railing
不锈钢锚 Stainless steel anchor
TERRACE
OFFICE
办公室
ACCESS
入口
AIRLOCK
气闸
ACCESS
入口
AIRLOCK
气闸
彩色STO通风系统 Coloured render STO Ventec system
气孔 Air cavity
95 mm绝热金属部件 95 mm MetSec / Insulation
195 mm绝热金属部件 195 mm MetSec / Insulation
2倍厚室内石膏板 Two-ply interior gypsum board
玻璃栏杆 Glass railing
鳍状玻璃板 Glass fin
不锈钢锚 Stainless steel anchor
混凝土板 Concrete deck
ACCESS
入口
AIRLOCK
气闸
195 mm绝热金属部件 195 mm MetSec / Insulation
95 mm绝热金属部件 95 mm MetSec / Insulation
气孔 Air cavity
彩色STO通风系统 Coloured render STO Ventec system
混凝土窗框 Concrete window reveal
三层玻璃窗 Triple glazing window
抹灰砖基础 Rendered brick base
抹灰微孔水泥砖 Rendered cellular concrete block
OFFICE
办公室

Frosted glass
Clear glass
Frosted glass
Shelving and drawers

节点图 DETAILS

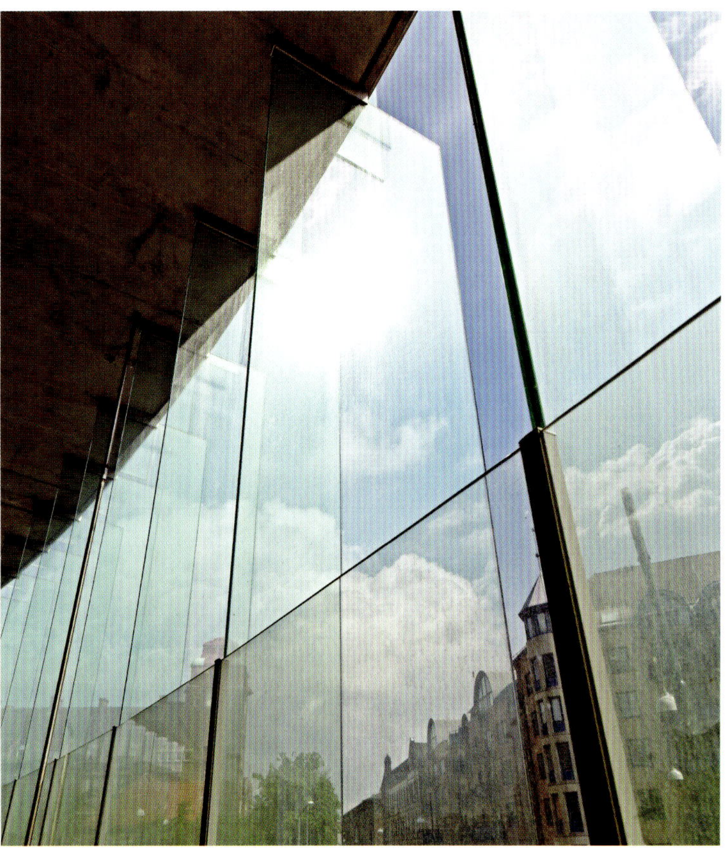

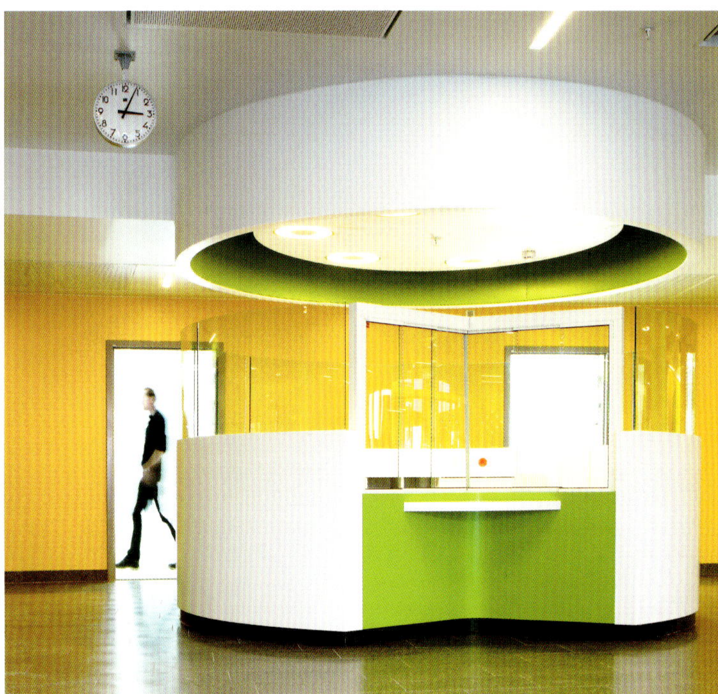

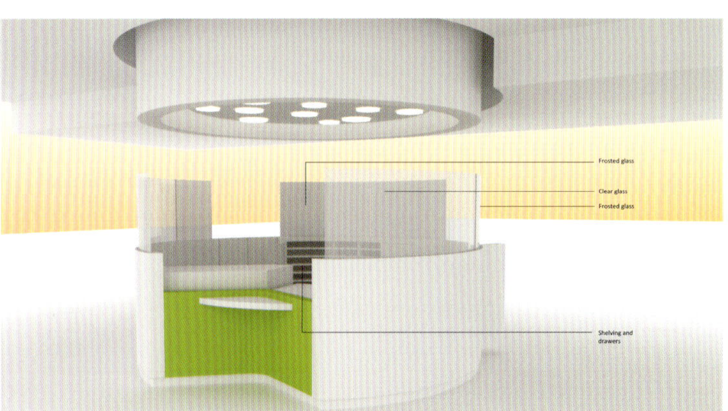

36 Infektionsklinik

彩色双层玻璃板
绝热铝材框架
Guldman GH3 上限
升降机
三层玻璃窗
玻璃栏杆
鳍状玻璃板
不锈钢锚
混凝土板
PATIENT ROOM
病房 入口
ACCESS
入口
Enamelled double glazing panel
Insulated aluminium frame
Guldman GH3 Ceiling hoist
Triple glazing window
Glass railing
Glass fin
Stainless steel anchor
Concrete deck
PATIENT ROOM
病房 入口
ACCESS
入口
Pre-cast concrete landing
Pre-cast concrete stair
inted 10 mmx40 mm steel balustrade
Laminated glass balustrade
Stainless steel tubular handrail
Pre-fabricated steel stair
Pre-fabricated steel landing
RHS 200 steel section
Double glazing facade
Insulated aluminium frame
Coloured render STO Ventec system
195 mm MetSec / Insulation
95 mm MetSec / Insulation
Hard panel for artwork mounting
Recessed lighting fixture
Light sculpture "La Familia"
Wooden deck/sitting platform
Steel IPE section
Granite Paving
预浇铸混凝土平台
预浇铸混凝土楼梯
10 mm×40 mm白色金属
护栏
玻璃板护栏
不锈钢管扶手
预制金属平台
RHS200金属部件
双层玻璃幕墙
绝热铝材框架
STAIRCASE
楼梯
UNDERPASS
通道
彩色STO通风系统
195 mm绝热金属部件
95 mm绝热金属部件
装饰面板
凹进灯光设施
灯具雕塑 "La Familia "
木制桌椅平台
金属IPE设备
花岗石铺地
PLANT
设备
VENTILATION
通风

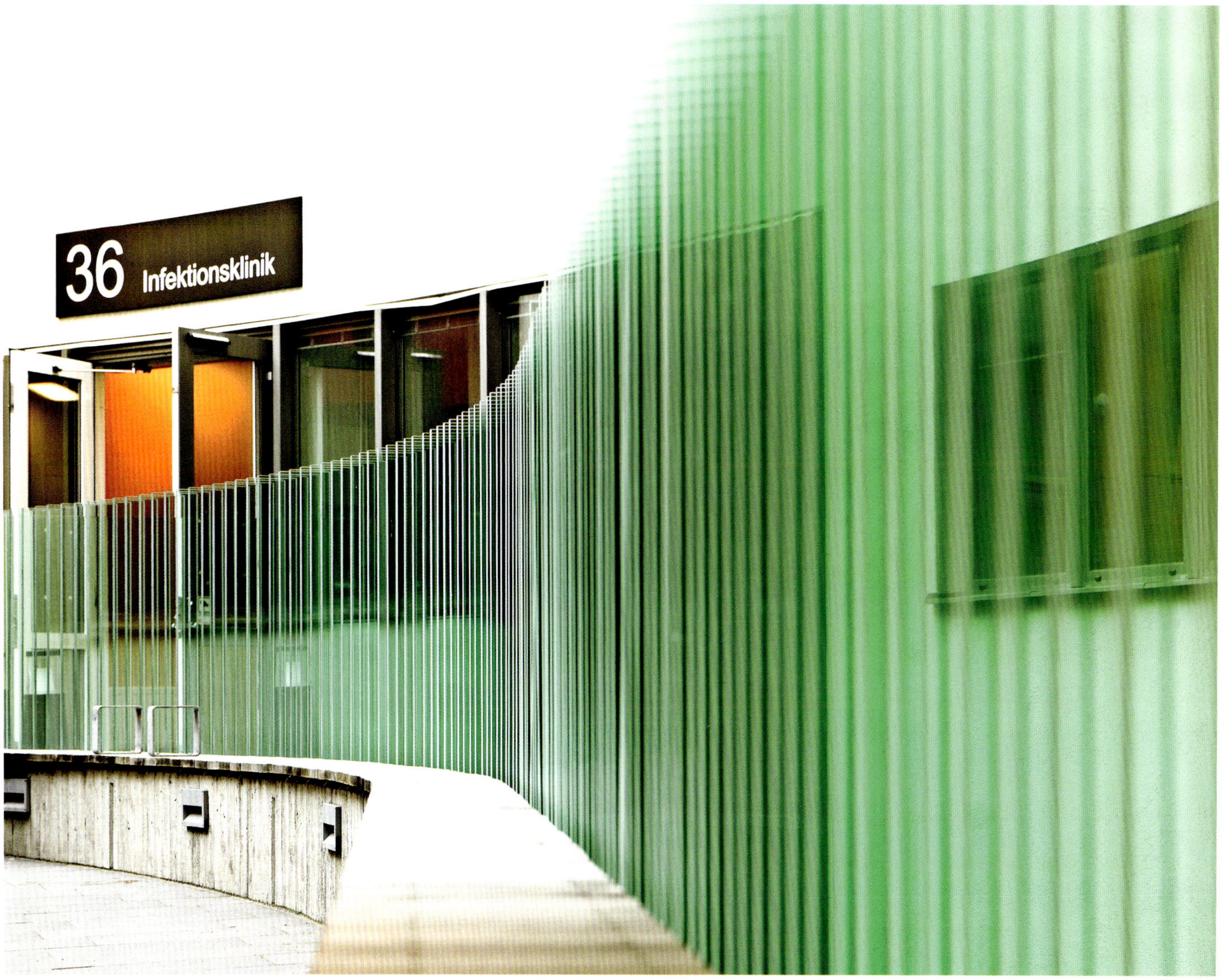

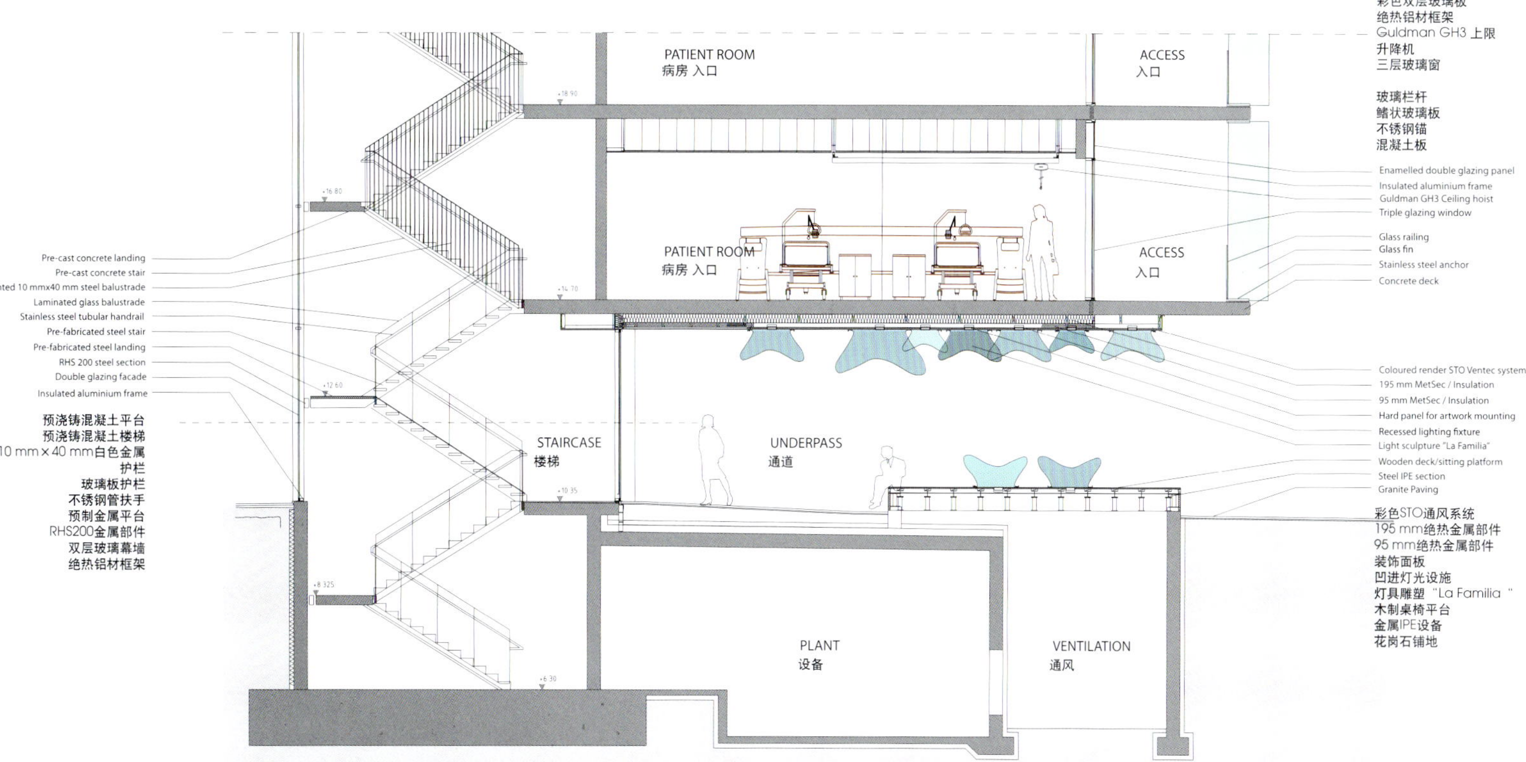

节点图 DETAILS

GRANDE LIBRARY OF QUEBEC
魁北克图书馆

© James Dow / Patkau Architects

地点：加拿大魁北克蒙特利尔
建筑设计：Patkau；Croft Pelletier；Menkès Shooner Dagenais Architectes Associés
景观设计：Scheme Consultants
委托方：Bibliothèque et Archives nationales du Québec
摄影：James Dow；Patkau Architects

Location: Montreal, Quebec, Canada
Architect: Patkau; Croft Pelletier; Menkès Shooner Dagenais Architectes Associés
Landscape: Scheme Consultants
Client: Bibliothèque et Archives Nationales du Québec
Photography: James Dow; Patkau Architects

World Green Buildings—— Thermal Environmental Engineering Solutions

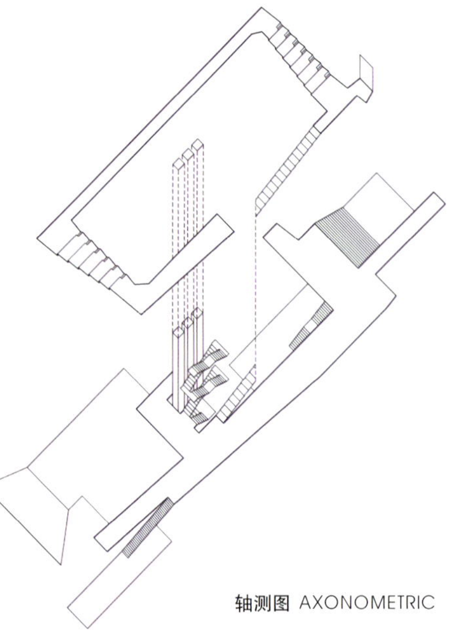

轴测图 AXONOMETRIC

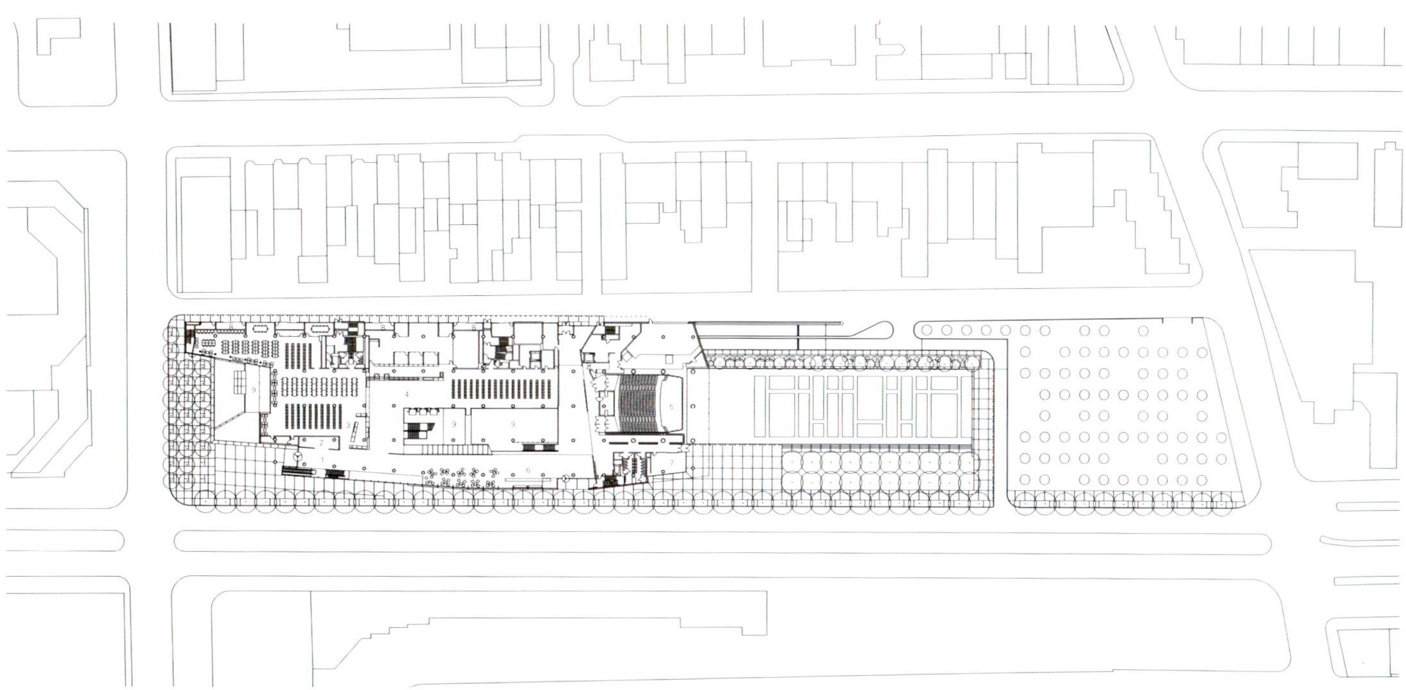

一层平面图 GROUND FLOOR PLAN

魁北克图书馆包括公共图书馆、儿童图书馆、魁北克历史档案藏品区和图书馆外管理区域共四个不同功能的公共空间。图书馆还是蒙特利尔交通系统的重要枢纽。项目将街道和地铁系统融为一体。

下沉的天井为低层空间提供光照，调节光照和辐射。在公共图书馆，藏品集中摆放，边缘是阅读区，能够获得足够的阳光和良好的视野，并使阳光可以射入建筑内部，在冬季，太阳热辐射有助于建筑的保温。

玻璃幕墙及铜质材料结合的建筑外壳将图书馆统一为一个整体。外立面虚实结合，有些区域通透开敞，有些区域私密、封闭，在夏季有效防止通过日晒获得过多热量。立面的设计使建筑在城市中独树一帜。

The Grande Library of Quebec contains four major components: a general library, a children's library, the collection of historic documents pertaining to Quebec, and an assortment of public spaces outside the library control zone. The library is joined to a major intersection in the Montreal metro system. The project knits the street and the subway system spaces together to engage the energies of each.

A sunken court provides daylight to below-grade spaces, adjusting the daylight and radiation. In the general library, the collection is centrally located, with reading spaces at its edges that have access to views and daylight. The daylight therefore can enter the inner space of the building providing a warm heat-up radiate in winter.

A glass and copper building envelope represents the library as a whole. Opaque in some places, diaphanous in others. The arrangement of grids can prevent over-loaded daylight radiation. The facade offers enticing glimpses of the library to the city.

© James Dow / Patkau Architects

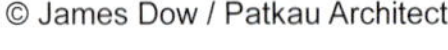

© James Dow / Patkau Architects

© James Dow / Patkau Architects

© James Dow / Patkau Architects

© James Dow / Patkau Architects

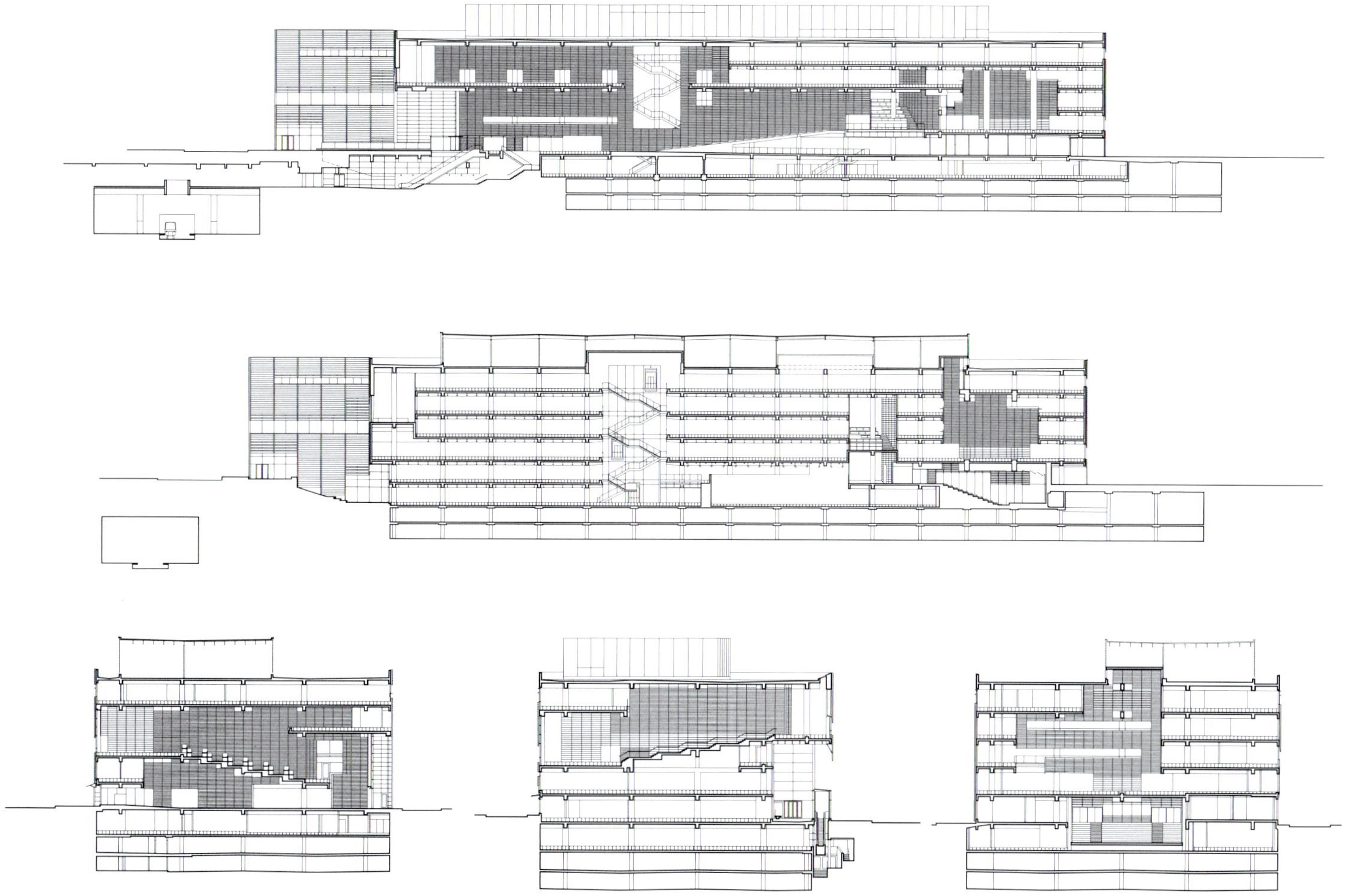

剖面图 SECTIONS

© James Dow / Patkau Architects

GRAND THEATRE OF HEFEI
合肥大剧院

地点：中国合肥
建筑面积：60 000 m²
生态节能设计：五合国际
所获奖项：国家建筑节能试点示范工程

Location: Hefei, China
Building Area: 60,000 m²
Sustainable Design: Werkhart International
Selected Awards: National Building Energy-saving Exemplary Project by MOHRD

项目概况

合肥大剧院被住房城乡建设部评为国家建筑节能试点示范工程，在设计、施工过程中严格按照国家公共建筑节能设计标准进行，其节能技术的应用在国内剧院建设中处于领先水平。

建筑热环境解决方案

1.被动措施

智能遮阳系统。合肥地处夏热冬冷地区，因此整个生态节能系统的设计都以当地的气候特点为依据，尽量利用其气候优势。合肥地区太阳辐射强度大，根据项目的地理位置，采用南向和西向设置智能太阳追踪可调机翼板式外遮阳板设施。通过这项技术，能够有效遮挡80%的太阳辐射。既提高了室内舒适度，又有效避免了高辐射带来的室内升温所导致的能量消耗。

高性能玻璃的应用。合肥大剧院外墙采用双真空玻璃幕墙，以达到保温隔热节能的效果。

高热工性能建筑围护结构。屋盖采用加厚岩棉金属体系，起到良好的保温隔热隔声效果。

自然通风与机械通风结合。采用机械通风预冷系统、置换通风方式。

2.主动措施

废热回收系统。舞台余热回收系统将舞台、灯光产生的大量热能回收，用于剧场空气除湿，以提高空气品质，增加舒适度。

优化供热、通风与空调系统。观众厅采用座椅诱导送风系统；舞台采用独立分层空调系统。划分合理的空调区域，可灵活调节及启停；大堂采用地板辐射采暖。

项目最引人注目的是，大剧院与天鹅湖联姻。大剧院内部的空调系统采用水源热泵系统，利用天鹅湖水天然的蓄水蓄热功能，系统机组同时可以实现供冷和供热，由于其较高的运行效率，比常规冷热源系统节能20%～40%。在冬季，直接利用冰冷湖水这一自然冷源作为辅助冷源，大幅降低了能耗。采用冰蓄冷技术，以冰蓄槽为介质，充分利用夜间低谷电“蓄峰平谷”蓄能，以降低运行成本。

其他可持续设计措施

1.太阳能光伏系统

将太阳光能有效转化为电能，以满足大剧院景观照明需求。

2.太阳能热水系统

结合建筑的大型屋面铺设太阳能积热器，白天制热，晚上使用，为演员、观众和服务人员提供太阳能热水服务，充分利用了太阳能。

3.光导管系统

采用光导管系统，导入自然光，以解决地下车库和阴暗房间白天的工作照明，节约用电。其节能技术的应用在国内剧院建设中处于领先水平。

4. 大温差低温冷冻水系统

5. 优化控制系统与运行方式

Program Description

The Grand Theatre of He Fei gained National Building Energy-saving Exemplary Project by MOHRD. The project team followed the Design Standard for Energy Efficiency of Public buildings strictly, and becomes one of the top theatres of China in the filed of sustainability and energy conservation.

Building Thermal Environmental Engineering Solutions

1. Passive Measures

Intelligent solar protection system. HeFei locates in the hot summer and cold winter zone, therefore the design of ecological energy saving system takes advantage of local climate characteristics. The solar radiation intensity is high in HeFei. The designers proposaled the intelligent sunlight tracing adjustable sun visor in airfoil style setting in the southward and westward directions. 80% of solar radiation can be shaded efficiently thanks to this method, making the interior environment more comfortable meanwhile avoiding the energy consumption due to the high radiation.

High performance glass. The exterior wall of Grand Theatre adopts double vacuum glass curtain wall, which contributes to heat preservation, heat insulation and energy conservation.

Building envelope of high thermal efficiency. The roof was paved with thick rock wool metal system, and received satisfying effects on sound insulation, heat preservation and insulation.

Natural ventilation combined with mechanical ventilation. The building introduced mechanical ventilation pre-cooling system and permutation ventilation.

2. Active Measures

Waste heat recovery system. The stage waste heat recovery system reclaims the heat energy of great amount produced by the stage and lighting system to dehumidify the interior air and make the room more comfortable.

To optimize the heating system, ventilation and air-condition system. The audience halls are equipped with seat-inducing blowers, while the stage is set with independent stratified air conditioning. The air condition areas are divided reasonably, which can be adjusted when necessary. The lobby uses radiant floor heating.

And the most remarkable keypoint of the project is that the combination between the theatre and the Swan Lake. The air conditioning system are adopted with water-source heat pump, taking advantage of the heat-harbour function of the water in the lake. The heat pump units actualizes heat supply and cold supply in the same time, which can save energy comparing with general heat/ cold source systems by 20%-40%. At the same time, the cold water in the lake can be used as natural cold source directly, reducing the energy consumption greatly. The building is introduced with ice storage system, using ice storage tank as medium. The system stores up low ebb electricity energy during night time make the best of peak clipping and valley filling to reduce costing.

Other Sustainable Design Measures

1. Solar Energy Photovoltaic System

Turnning solar energy into electric energy, the solar energy photovoltaic system supply electricity for landscape illumination.

2. Solar Hot Water System

Heat accumulating implements are paved on the huge roof of the theater, producing heat water in the day and using it in night time. The system provides the audience, players, and service staff with hot water.

3. Light-Pipe System

Light-pipe system introduces natural light to resolve the demand of working illumination of underground garage and shadow room in the day time. The technology has reached the advanced level in the filed of theater constructing in the country.

4. System of Low Temperature Freezing Water in Great Temperature Difference

5.Optimization Control System and Relative Running Functions

THE DEE AND CHARLES WYLY THEATER
迪和查尔斯·威利剧院

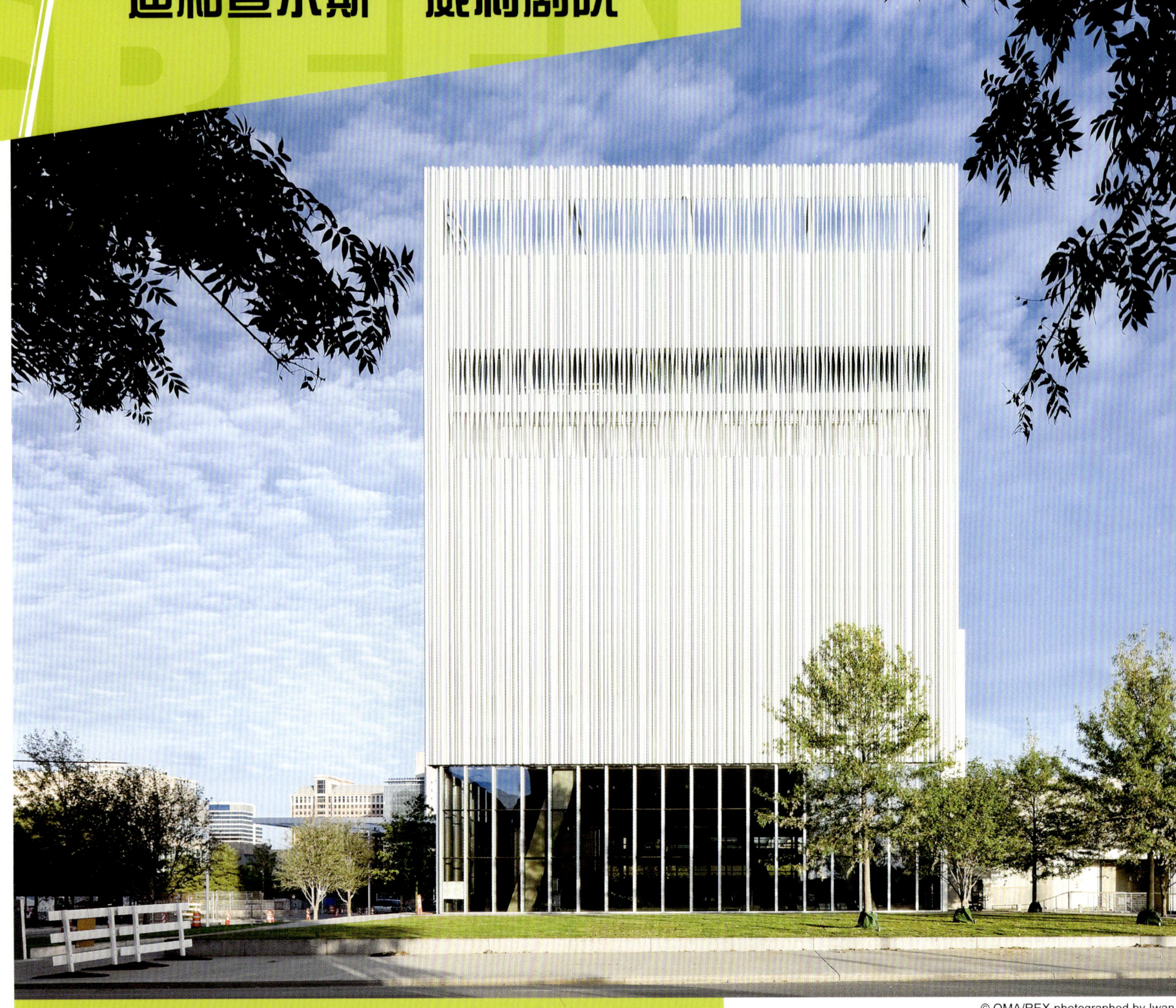

地点：美国得克萨斯州达拉斯
建筑设计：REX / OMA
委托方：Dallas Center for the Performing Arts
摄影：Iwan Baan

Location: Dallas, Texas, USA
Architect: REX / OMA
Client: Dallas Center for the Performing Arts
Photography: Iwan Baan

© OMA/REX photographed by Iwan Baan

与传统剧院的功能区域布局方式不同，由REX/OMA设计的迪和查尔斯·威利剧院共11层，布局紧凑，纵向发展。剧院大厅完全由玻璃幕墙合围而成，独占一层，而化妆间、排练室、办公室等区域分布于二至十一层，地下一层为停车场。这样，整个剧院的附属区域对核心区域（剧院大厅）形成的是一种竖直、立体的包围。

Wyly剧院包含了许多可持续设计要素。例如，在大堂安装的辐射管道，冷却水梁，灵活使用的荧光灯照明灯具，以及在休息室、更衣室、办公场所使用的照明及调光装置。冷梁技术与其他相关技术相比能够提供最大的舒适度，因此被选中。建筑装备了一种被动冷梁扩散器用来为设备组织气流。扩散器不需要送风连接，而是利用水盘管和自然对流冷却。封闭机柜的下部有整体循环空气出口。剧院还装配了天花板空气扩散器。剧院的很多地方都安装了这种设备。扩散器可提供360°的空气扩散模式，并在加热和冷却设备中得到很好的应用，它在变风量空调系统中也表现出良好的性能。

© OMA/REX photographed by Iwan Baan

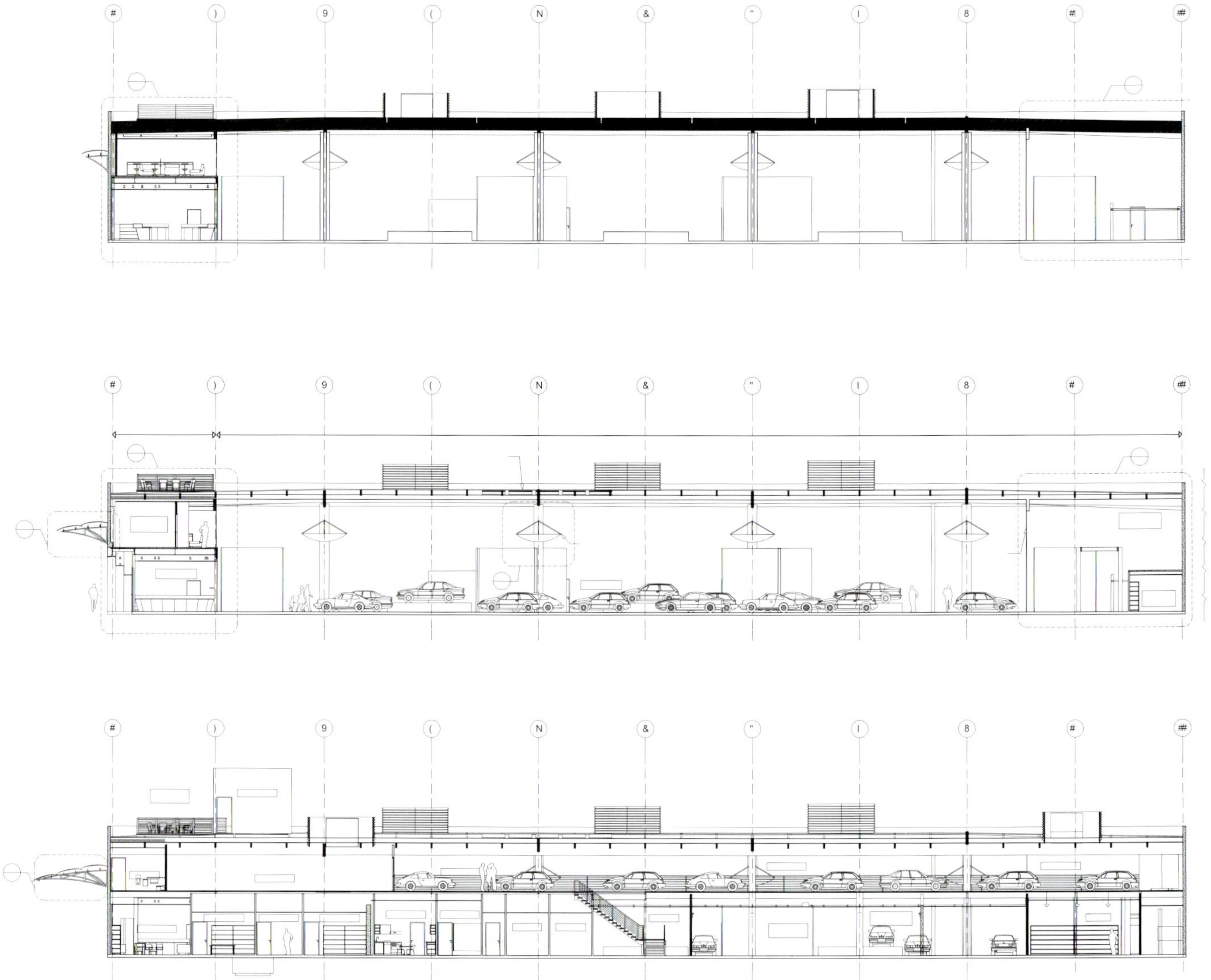

剖面图 SECTIONS

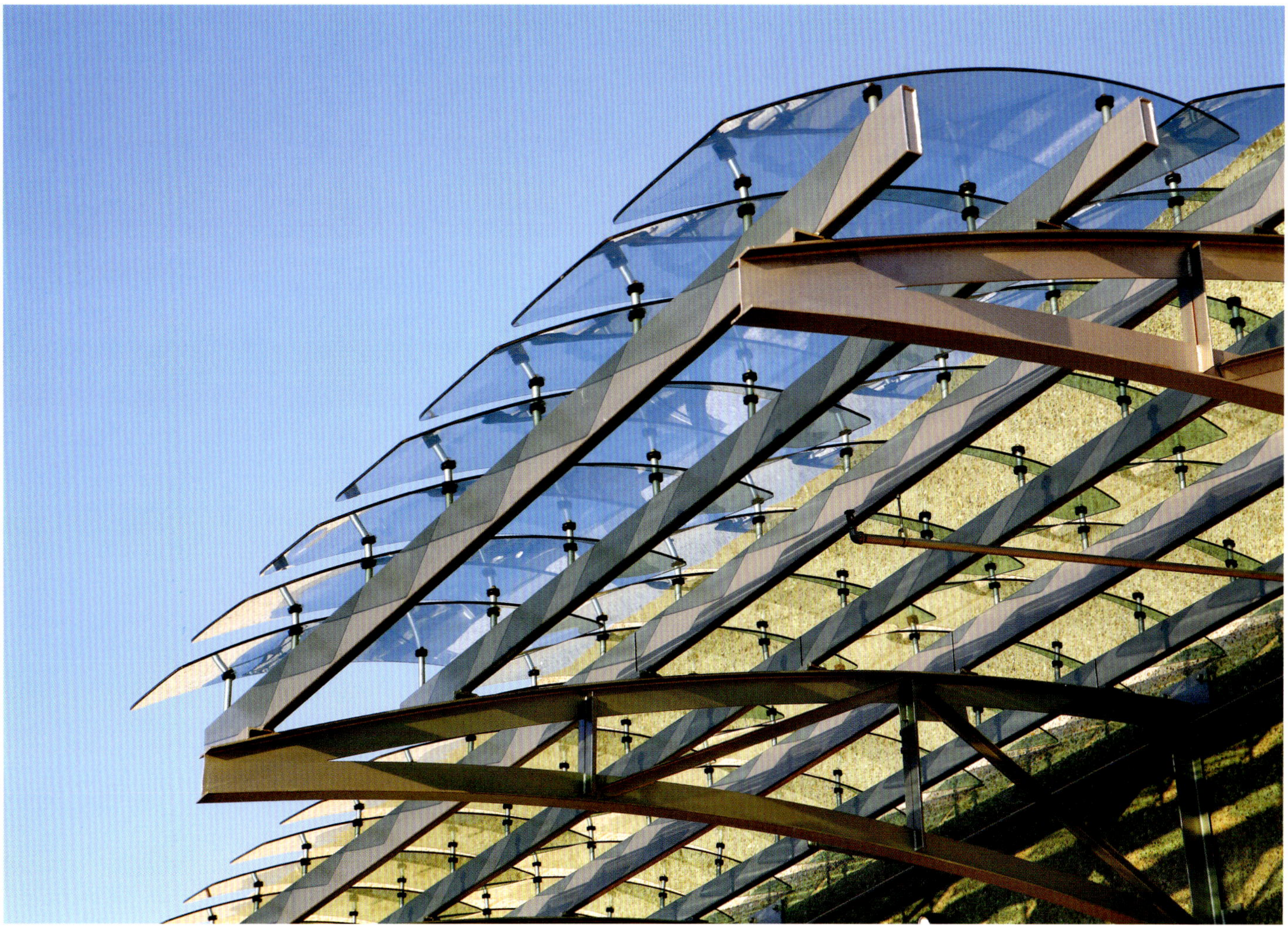

Mullin汽车收藏博物馆是对一座面积为4181 m²的混凝土仓库进行的整体设计。将被建成为全球最大型的收藏20世纪30年代布加迪和法国赛车的博物馆。内部包括一个剧场、礼品店、私人会所，办公室，以及汽车展示区。建筑采用了多种可持续绿色建筑设计措施，创造良好的热环境。设置有日照及自然通风监控器，太阳能光电屋顶，绿色屋顶，电梯机房，风力涡轮机，大型挡风玻璃构成的室外顶棚。

The Mullin Collection Automotive Museum is a complete design of an existing 4181 m², typical tilt-up concrete warehouse, which will be built into a automotive museum for the world's largest collection of Bugatti and French race cars of the 1930s. Interiors include a theater, gift store, private club, archives, offices, as well as display areas for cars and automotive artifacts. The building's many sustainable green building features creating a comfortable thermal environment, include monitors for daylight and natural ventilation, a solar photovoltaic roof, a green roof with a new elevator penthouse and wind turbines, and a large exterior entry canopy made of windshields.

WASTE TREATMENT PLANT-AMAGERFORBRAENDING

AMAGERFORBRAENDING 废品处理处

地点：丹麦哥本哈根
建筑设计：BIG-Bjarke Ingels Group
建筑面积：95 000 m²
委托方：Amagerforbraending

Location: Copenhagen, Denmark
Architect: BIG-Bjarke Ingels Group
Building Area: 95,000 m²
Client: Amagerforbraending

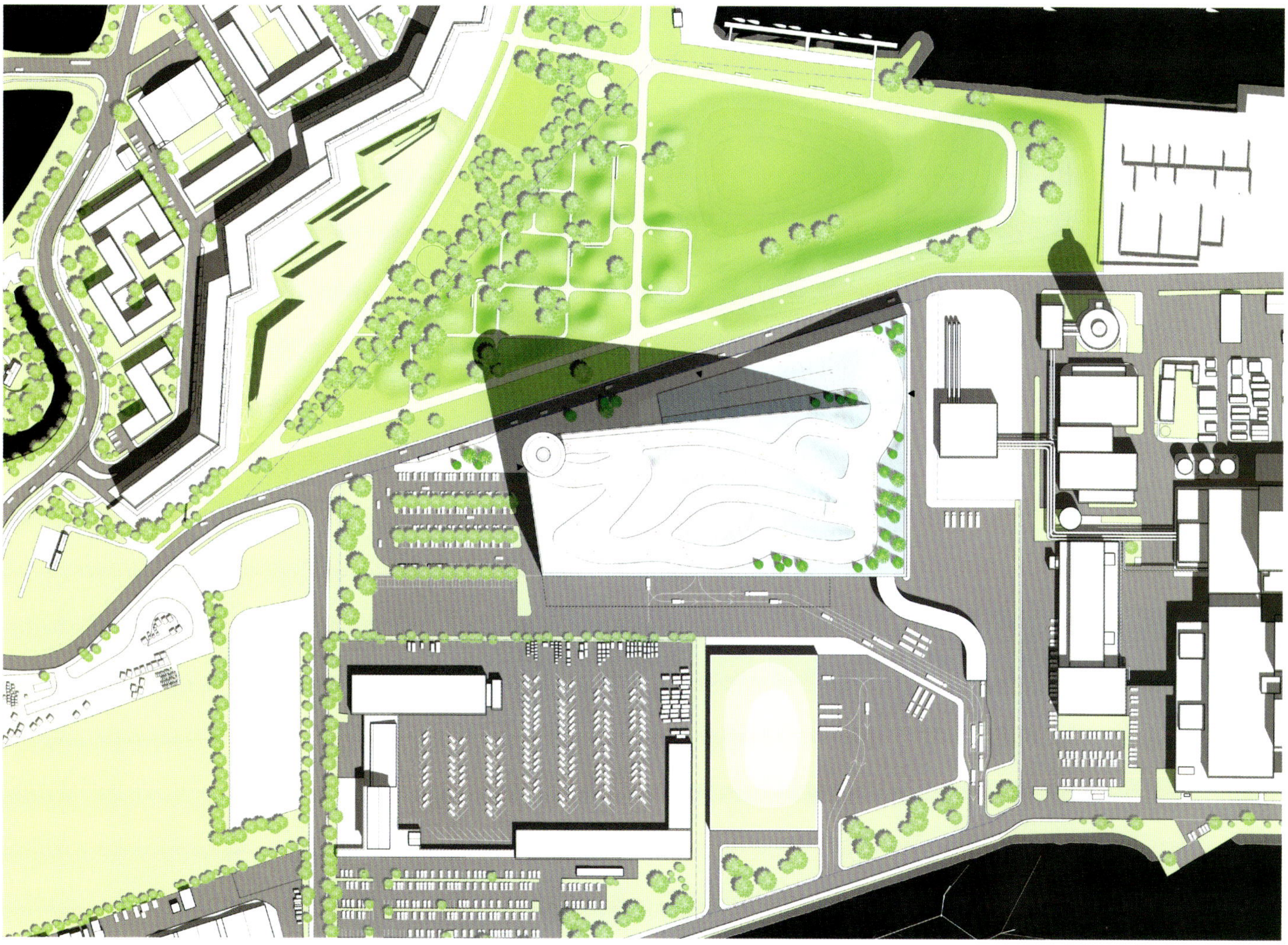

总平面图 SITE PLAN

鸟瞰图 AERIAL VIEW

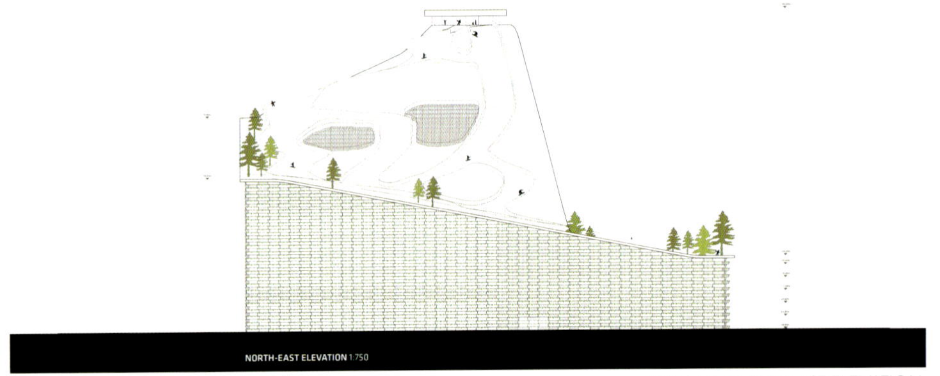

东北立面 NORTHEAST ELEVATION

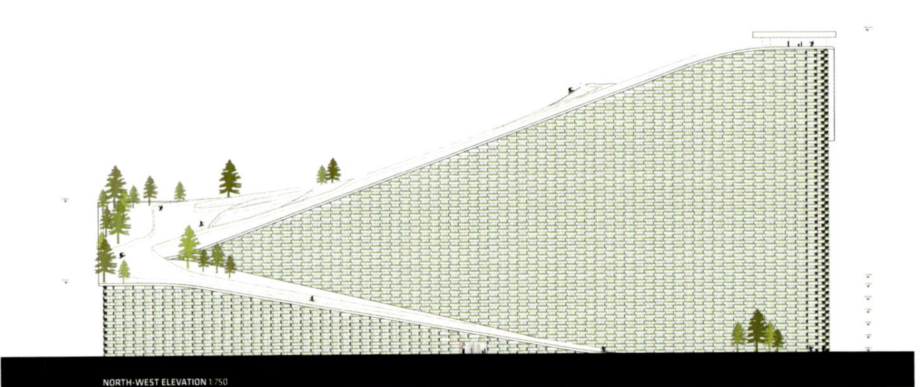

西北立面 NORTHWEST ELEVATION

项目概况

位于市中心附近的一个工业区的废物－能源转换工厂将成为废物管理和能源生产的典范，同时成为哥本哈根城市的标志性建筑。项目预算35亿丹麦克朗，是丹麦单笔款额最大的环保订单，将取代附近一座40年历史的旧工厂，将废物处理和环境保护的最新技术相结合。BIG建筑师事务所赢得了国际竞标的胜利。项目预计2016年建成。

BIG在这一设计中加入了工厂屋顶滑雪道，将工业建筑与公共娱乐设施相结合，为建筑带来了生机与活力，也加强了和城市之间的联系，将建筑本身变为人们游览、聚集之地。为了增强公众对利用可持续能源的意识，方案针对工厂烟囱作了一些调整，使得烟囱可以向天空释放烟圈。当每一吨化石燃料中的CO_2排放到空中时，烟囱就会释放出标志性的烟圈，它将作为与公众交流的信号，用这种具象化的方式时刻提醒人们能源消耗的影响。夜晚，一束热跟踪激光会将一个展示CO_2实际排放量的饼状图投射到烟圈上。

建筑热环境解决方案

从远处看，整个建筑都包裹在立体绿化的植物模块中，立体绿化可以调节温度湿度并起到良好的保温隔热作用。建筑的立面系统在办公区引入了大量自然光，同时将哥本哈根的城市风光作为背景。光架使自然光射入室内深处，浅肋板利于办公室自然采光。根据堆栈效应中庭空间自然通风得以实现。在地下车库有新风进口，冷空气由此被吸入建筑，并适当预热。多余的热空气膨胀上升，并从建筑高层的天窗溢出。

Program Description

Located in an industrial area near the city center the new Waste-to-Energy plant will be an exemplary model in the field of waste management and energy production, as well as an architectural landmark in the cityscape of Copenhagen. The project is the single largest environmental initiative in Denmark with a budget of 3.5 Billion DKK, and replaces the adjacent 40 year old Amagerforbraending plant, integrating the latest technologies in waste treatment and environmental performance. BIG wins the international competition 1 prize. The building is planned to welcome vistors in 2016.

The roof of the new Amagerforbraending is turned into a 31,000 m² ski slope of varying skill levels for the citizens of Copenhagen, its neighboring municipalities and visitors, mobilizing the architecture and redefining the relationship between the waste plant and the city by expanding the existing recreational activities in the surrounding area into a new breed of waste-to-energy plant. All of this while the smokestack is modified to puff smoke rings of 30 m in diameter whenever 1 ton of fossil CO_2 is released. These smoke rings which are the brainchild of Germany-based art studio realities: united will form due to the condensation of water in the flue gases as they slowly rise and cool, serving as a gentle reminder of the impact of consumption and a measuring stick that will allow the common Copenhagener to grasp the CO_2 emission in a straightforward way - turning the smokestack traditionally the symbol of the industrial era into a symbol for the future. At night, heat tracking lights are used to position lasers on the smoke rings into glowing artworks.

Building Thermal Environmental Engineering Solutions

From a distance, the entire building is wrapped in a vertical green facade formed by planter modules stacked like bricks turning it into a mountain from afar. Vertical green facade helps to accommodate temperature and humidity, and of high efficiency in heat preservation and heat insulation. The facade system allows for generous daylighting in staff areas as well as view back to Copenhagen. Light shelves bring daylight deep into interior spaces, and shallow floor plates are ideal for daylight of office spaces. According to the stack effect, natural ventilation of atrium space is realized. There are fresh air inlets underground in the parking area. Cold fresh air are sorbed into the building while the warm excrescent air goes up and expeled out through the louvers of the atrium.

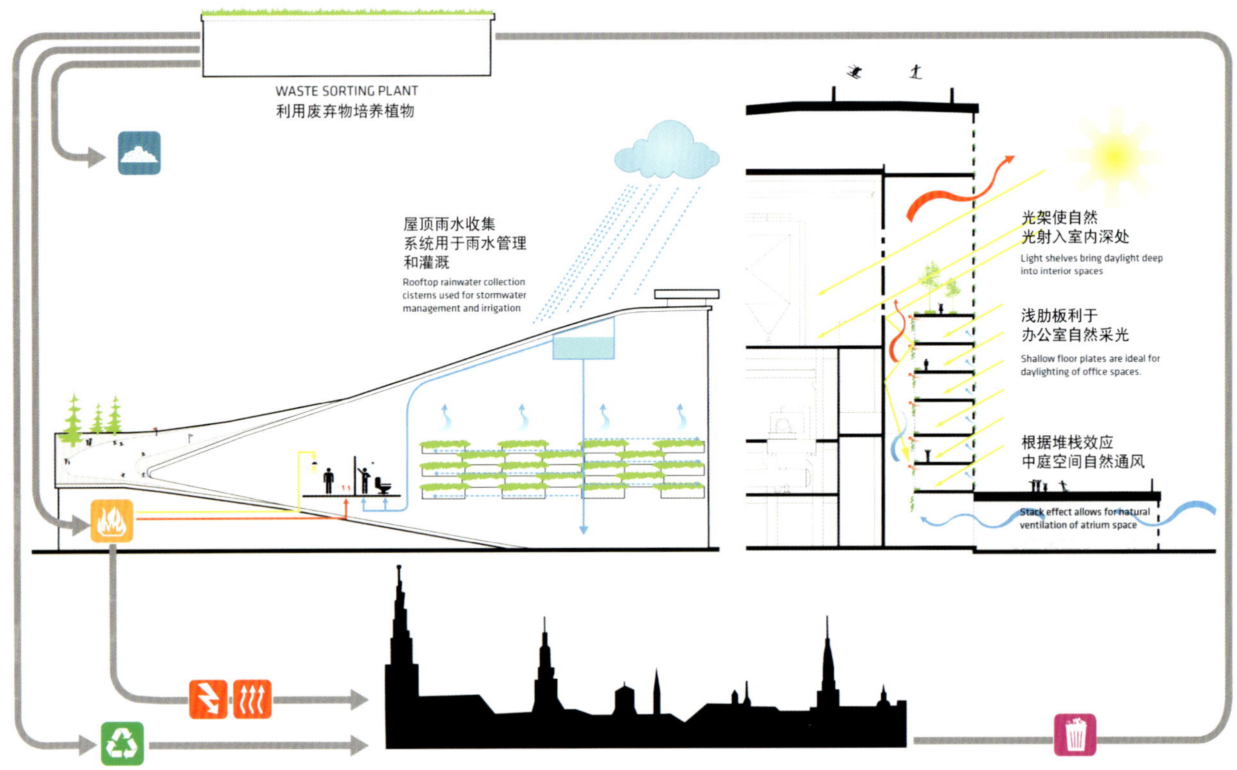

分析图 DIAGRAM

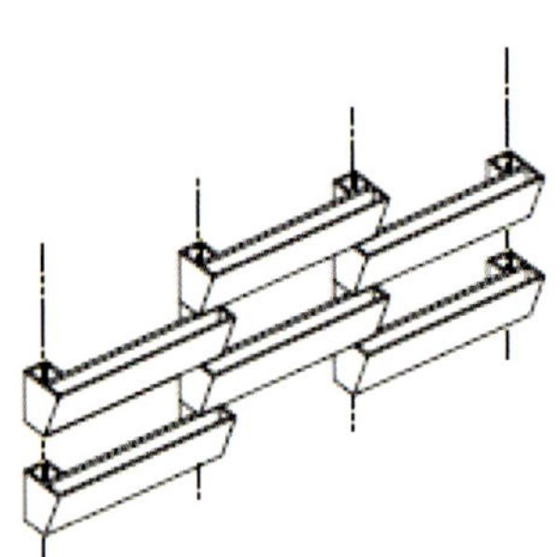

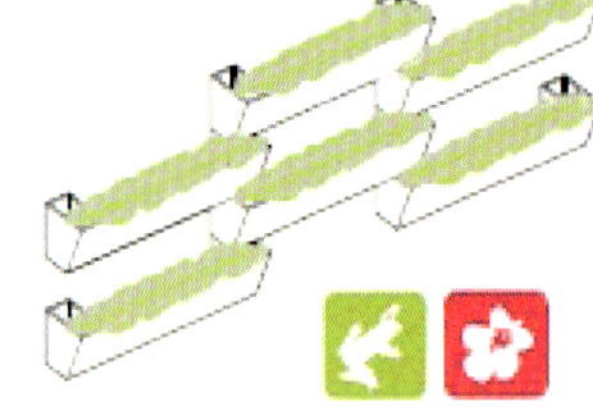

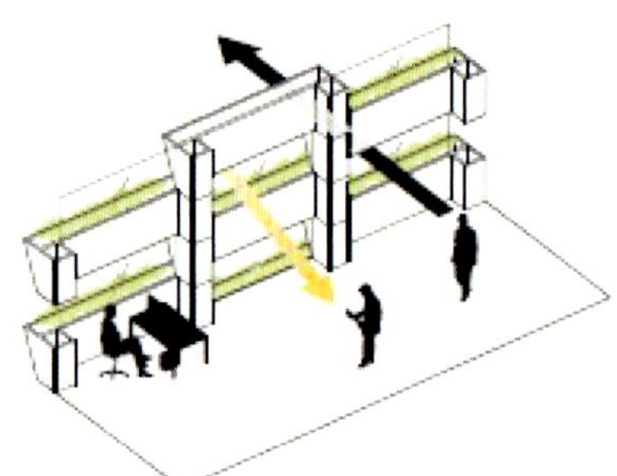

CLADDING MOOULE

The vertical facades are form by planter modules stacked like bricks. They can be flexibly configured and support required services distribution and maintenance routes behind the facade.

覆层模块
立面由像砖块一样堆砌的植物模块组成。它们可以灵活安放，并在立面后提供必要的辅助设备分布空间和养护通路。

ECOLOGY

Green facades support an increase in habitats and biodiversity and are a potential stormwater management solution. Studies show an improvement in air quality from reductions in levels of CO, NO_2, $PM_{2.5}$, and SO_2.

生态关系
绿色外立面为动植物提供生存空间，促进生物多样性，有效解决雨水排放问题。可以改善空气质量，降低有害气体浓度。

WORK ENVIRONMENT

The facade system allows for generous daylighting.
In staff areas as well as views back to Copenhagen.

工作环境
外立面系统缓和射入工作区的阳光强度，室内还可观赏到哥本哈根的风景。

分析图 DIAGRAM

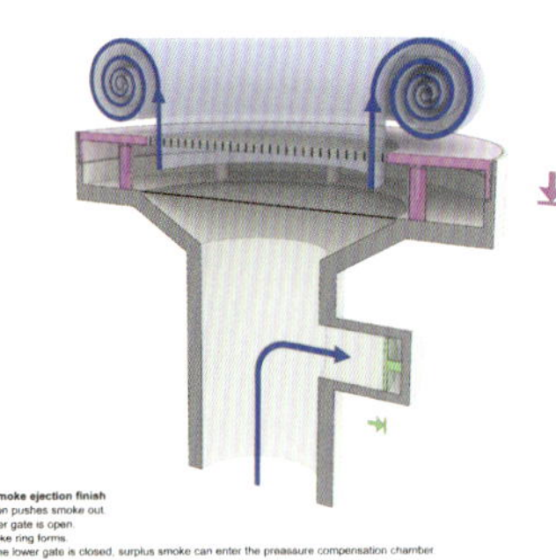

1. Smoke ejection finish
Piston pushes smoke out.
Upper gate is open.
Smoke ring forms.
As the lower gate is closed, surplus smoke can enter the preassure compensation chamber.

1.喷气装置
活塞推出气体；
高层阀门开启；
形成环状烟圈；
低层阀门关闭，废气可以进入压力补偿室。

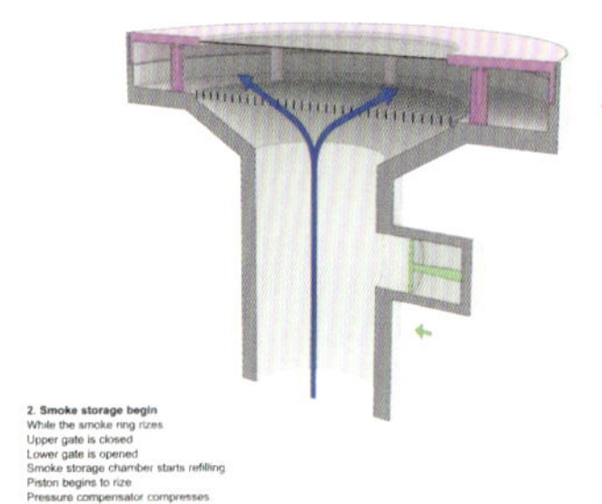

2. Smoke storage begin
While the smoke ring rizes
Upper gate is closed
Lower gate is opened
Smoke storage chamber starts refilling
Piston begins to rize
Pressure compensator compresses

2.开始蓄积气体
当环状烟圈升起时，
高层阀门关闭，
低层阀门打开；
储气室开始再次充气。
活塞开始上升。
压力补偿装置收缩。

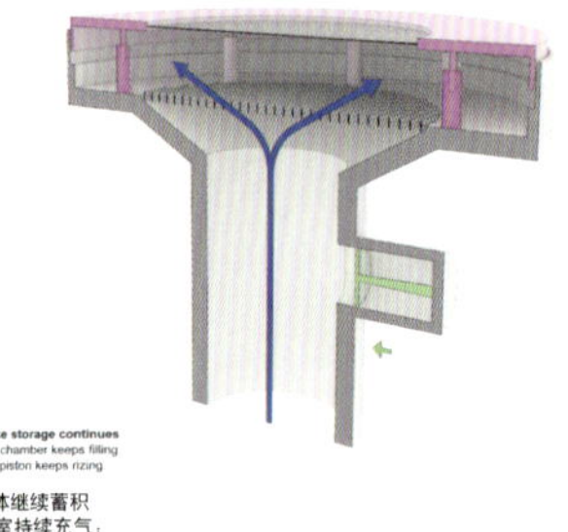

3. Smoke storage continues
Storage chamber keeps filling
And the piston keeps rizing

3.气体继续蓄积
储气室持续充气；
同时活塞持续上升。

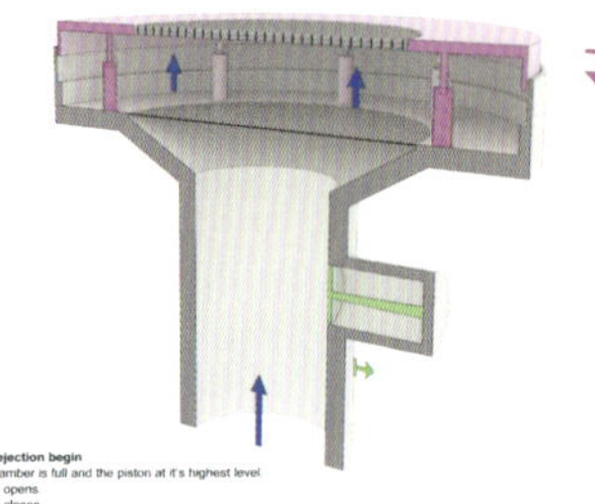

4. Smoke ejection begin
Storage chamber is full and the piston at it's highest level
Upper gate opens.
Lower gate closes.
Piston starts moving downwards to push smoke out.
The preassure compensation chamber is ready to expand.
The complete cycle takes about 30 seconds.

4.气体开始排出
储气室充满空气，活塞到达最高位置；
高层阀门开启，
低层阀门关闭。
活塞开始向下运行以便将气体排出；
压力补偿室准备好膨胀。
整个过程大概用时30 s。

分析图 DIAGRAM

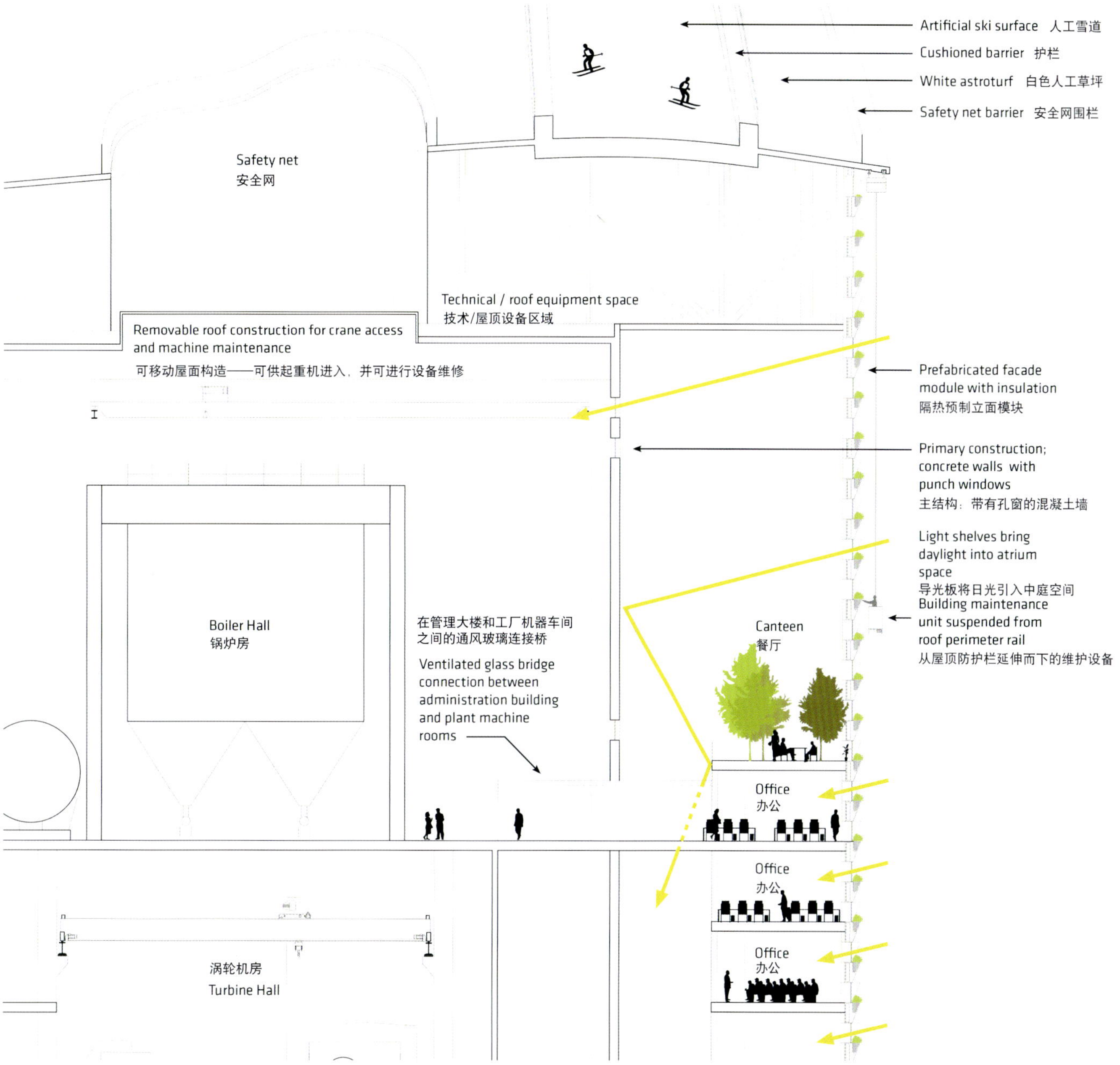

Artificial ski surface 人工雪道

Cushioned barrier 护栏

White astroturf 白色人工草坪

Safety net barrier 安全网围栏

Safety net
安全网

Technical / roof equipment space
技术/屋顶设备区域

Removable roof construction for crane access and machine maintenance
可移动屋面构造——可供起重机进入，并可进行设备维修

Prefabricated facade module with insulation
隔热预制立面模块

Primary construction; concrete walls with punch windows
主结构：带有孔窗的混凝土墙

Light shelves bring daylight into atrium space
导光板将日光引入中庭空间

Building maintenance unit suspended from roof perimeter rail
从屋顶防护栏延伸而下的维护设备

Boiler Hall
锅炉房

在管理大楼和工厂机器车间之间的通风玻璃连接桥

Ventilated glass bridge connection between administration building and plant machine rooms

Canteen
餐厅

Office
办公

Office
办公

Office
办公

涡轮机房
Turbine Hall

DETAIL SECTION OF RELATIONSHIP BETWEEN ADMINISTRATIVE BUILDING, MAIN PLANT VOLUME, AND ROOF
剖面详图——管理大楼、工厂车间及屋顶的关系

NEW AREZZO COURTHOUSE

新阿雷佐法院大楼

地点：意大利阿雷佐
建筑设计：Manfredi Nicoletti
规模：23 000 m²（新翼楼5000 m²；既有建筑改造18 000 m²）
摄影：Simone Levi – Studio Nicoletti Associati

Location: Arezzo, Italy
Architect: Manfredi Nicoletti
Size: 23,000 m² (New Wing 5000 m²; Restorated Old Building 18,000 m²)
Photography: Simone Levi – Studio Nicoletti Associati

World Green Buildings——Thermal Environmental Engineering Solutions

项目概况

大楼位于美迪斯城堡旁边的历史公园里面。这个新的大楼包含了法院的主要部分以及总统公正广场大厅，同时它也和一个新古典主义大楼相连。新翼楼的设计象征了它周围的环境：公园茂密的植被和城市中世纪般的结构。新楼和它周围的环境都被一边的带状贝壳墙保护，而在另一边环境对广场、花园和领地开放。同时新翼楼的北面被一个内凹的黑色花岗石板墙封闭。这些石板的表面都经由火烧而变成一种优雅黯淡的深灰色。

建筑热环境解决方案

南面为形态波动的透明不锈钢遮阳立面。这个立面借鉴许多生物（包括叶子）的皮肤特性，生成了几何体被包裹的形态。复杂的曲线形态只能通过直线型元素建造。因此，一个银色的生态气候感应叶片通过一个发光的阴影遮蔽了室内空间，同时还在与周围新古典主义建筑元素不冲突的情况下，将公园景色融入其中。

在室内空间里，镜子般锃亮的黑色花岗石地板的强烈反射同绿色玻璃的建筑外皮、被狭窄的枫叶状立面分隔的水平灰色隔声板形成鲜明对比。三层高的入口大厅从屋顶接受阳光，跟地板效果由发光黑色花岗石主导一样，大厅空间效果主要由垂直电梯厢控制。

在北面，有个壳状结构，它被黑色花岗石和由加固型混凝土结构衍生的线条覆盖。混凝土结构是一个椭圆底部的倾斜锥体的一部分。在结构之下的片和面的不可比较性创造出一种尺度惊人的效果。这样形成了一个可通风的立面，同时提供了寒冷冬季风时有效的保护，还允许了夏季的自然通风。

外表面的遮阳避免了过量吸收阳光带来的问题——室内过热以及对制冷设备要求的增加。这也是大楼需要在南面有双层表皮的原因。这样的措施既保证了自然通风又提供了有效的遮阳。钢结构的防晒南立面固定在了背光的室内遮挡物上，这是一种生态气候调控的叶状遮挡物。这个立面同时也是由不同角度的柱子钢结构支撑。除了这些结构，建筑还有一个不同角度看去都可见的整体玻璃表面。

扭曲表面的重叠创造了一个可变形的波浪形透明防晒立面，这个表面是通过一个被包裹的几何体生成形态的，这个几何体仅仅由直线型元素建造。

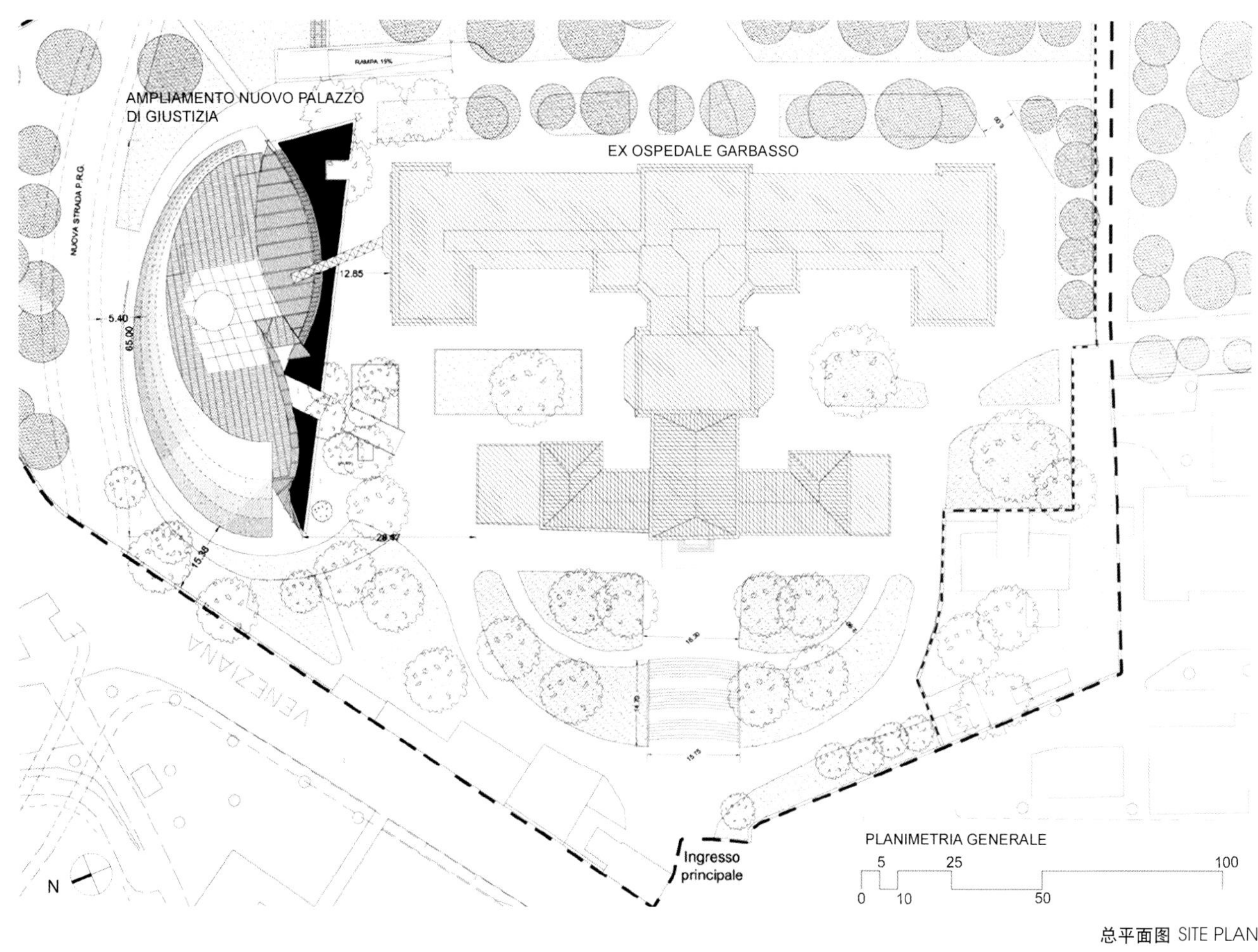

总平面图 SITE PLAN

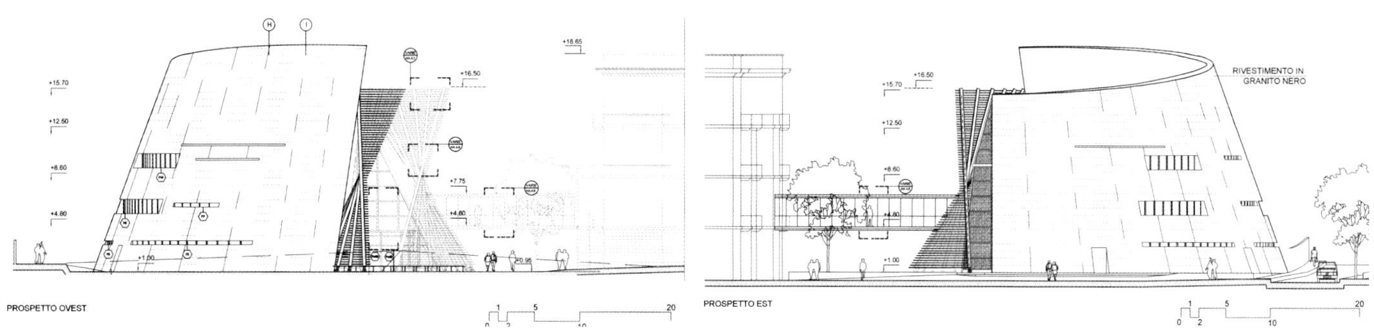

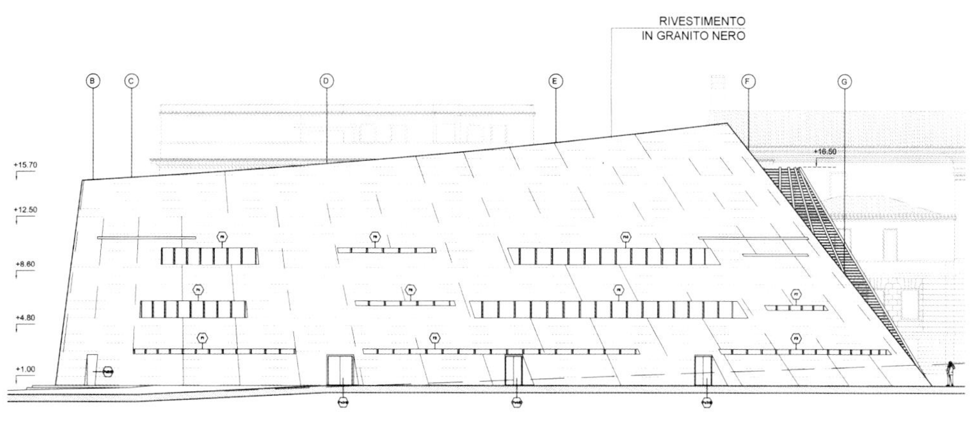

立面图 ELEVATION

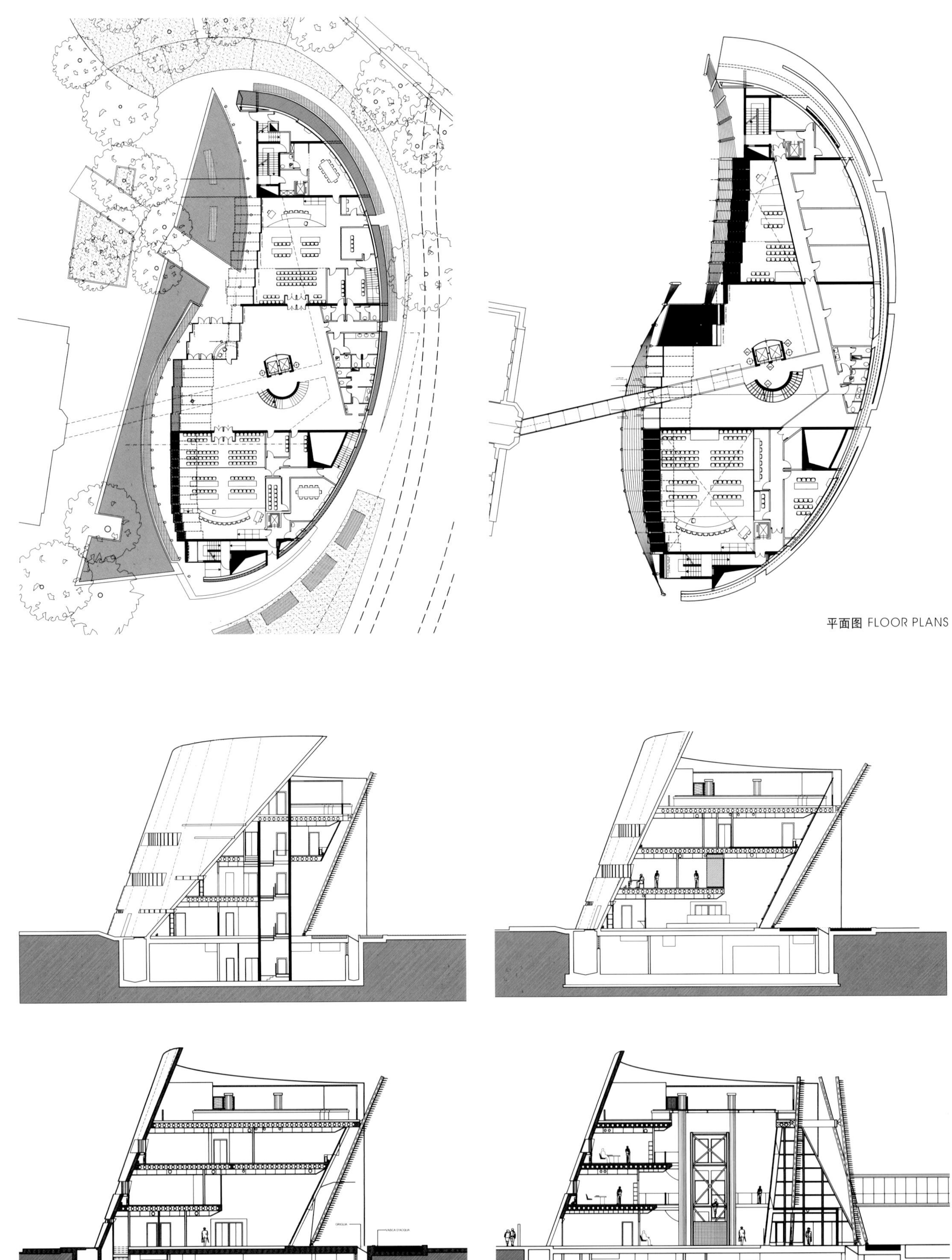

平面图 FLOOR PLANS

剖面图 SECTIONS

Program Description

Near the Medici's Fortress, inside an historical park, the new building houses the main courtrooms and the Hall of Justice President headquarter and is linked to a neoclassical building, once a large hospital, to be restored and used for offices and archives. The design of the New Wing, symbolises its environment: the rich vegetation of the park and the city's medieval structure, which was protected by a belt of conch walls on one side and, on the other, open to the squares, gardens and the territory. Also the New Wing is enclosed on the North side by a coved wall of black granite slabs whose surface is flamed, as to acquire a very elegant matted dark-grey tone.

Building Thermal Environmental Engineering Solutions

To the South, an undulating transparent sunscreen facade of stainless steel is shaped according to a geometrically warped geometry which is characteristics of the skin of many living creatures, including leaves. Those complex curvilinear forms can be built using only rectilinear elements. Thus, a silver bioclimatic foliage protects the interior spaces by a luminous shadow and blends with the park without clashing with the neoclassical nearby architectural elements.

In the interior spaces, the astonishing reflection of the mirror-like polished black granite floor is contrasting with the greenish glazed envelop and the horizontal acoustic grey panel separated by narrow maple fascias. The three level high entrance hall receives light from the roof and is dominated by the vertical elevator case clad alike the floor by shining black granite panels.

To the north there's a shell covered with sheets of black granite sets along the generative lines of the reinforced concrete conoidal structure, part of a leaning cone on an elliptic base. The non coplanarity between the sheets and surfaces below them create a startling scales effect.

This forms a ventilated facade and allows to provide efficient protection from the cold winter winds while allowing natural ventilation during the summer.

External solar shading is crucial to prevent solar gains from causing overheating and increasing cooling requirements. That's why the buildings has to the south a double skin; this approach allows to provide efficient shading from the sun while allowing natural ventilation.

The sunscreen facade the south with steel ripping forms a bioclimatic leaf shelter in the interiors against sunlight held up by weft of pillars of different inclination in steel. Besides there's a total glazed facade with different inclinations.

The over-layering of the twisting surfaces creates a diaphanous undulating transparent sunscreen facade, shaped according to a geometrically warped geometry and built using only rectilinear elements.

GARDEN CITY K66

K66花园城

建筑设计：OFIS Architects
所获奖项：LEED金质认证；2010年应邀参加国际竞标获二等奖
委托方：S.T. Hammer d. o. o.

Architect: OFIS Architects
Selected Awards: LEED Gold Certified; Invited International Competition - 2010_second Prize
Client: S. T. Hammer d. o. o.

World Green Buildings——Thermal Environmental Engineering Solutions

立面图 ELEVATIONS

总平面图 SITE PLAN

项目概况

项目包含两座大楼，多种功能分布其中。建筑底层为公共空间和办公室，高层为公寓。大楼的高度控制在城市建筑限高之内，并形成了阶梯式的花园。底层的公共空间相互隔离，并形成了一个开放广场，与室外广场和办公空间相互呼应。建筑内空间开敞明亮，布局合理，视野良好，与周边的Strenia广场、公园和道路有机结合。该项目获得LEED金质认证。基本的可持续设计原则通过建筑形体设计实现。

建筑热环境解决方案

1.公寓和阶梯露台

阶梯式露台的设计，使公寓最大限度地得到自然采光、光照和自然对流通风。每套公寓都面向2～3个方向设置门窗，并有遮阳设备，提供充分的遮阳设施，并创造了私人空间。植被丰富的宽大露台将在夏季创造凉爽的效果。还将在露台上设置可拆卸的顶棚作为遮阴避雨之用。在低层面积稍小一点的公寓中设有凉廊，也可在较小的空间中获得同样的效果。这些房间还装有外部遮阳设备。

2.办公

室内庭园的周围设置了办公空间和医疗中心。这样的设计，使房间可以获得自然光线和通透的自然风，提供优良的工作环境。建筑师没有进行深度设计，垂直芯型的空间使居住者可灵活布置、组合空间。

3.外部空间

一楼的广场通风良好，空间开敞，自然采光，朝向良好，方便人流通行。此处进行了绿化，并设有水景，创造了舒爽宜人的工作环境。绿色屋顶、花园和阶梯露台使大楼与外部环境融为一体，为居住者、工作人员和来访者提供了友好绿色的环境。

4.复合空间

生活区域、办公空间及服务场所的紧凑排布和搭配组合最大限度地减少了交通需求。紧凑的体积可以带来较低的能源消耗，减少能源浪费。外部遮阳和花园之上的可拆卸屋顶使建筑最有效地实现了防晒保护。

其他可持续设计措施

为了节约资源，将选择无需经常浇水的植物，并配备雨水收集和重新利用系统、节水厕所及淋浴等。在部分楼层的屋顶和表面安装光电池板。由于建筑的外形较为独特，因此外墙材料可以相对简单。大楼的底层使用玻璃外墙，选用内部着色的透明玻璃。玻璃稍加着色，其间装有层状金属网。外墙的栅格为1.00 m，而且办公空间和公寓都采用了模块化设计。公寓的外墙设有大型全景推拉窗，并配有外部遮阳设备（银铝穿孔金属百叶窗）。阶梯露台铺装木质地板，木材和天然石材也将应用于公寓当中。

Program Description

The project proposes two layered volumes embracing the rich mix of different programs and distributed in: base of the volume as public programs and offices, top floors as apartments. The tower strip steps up and down according to the urban height limits and form terraces-gardens. Public Base volumes are separated and create open squares - dialog of external plaza with public programs and offices. Spaces are communicative, bright, fluid, and easily accessible and offer nice views and connections with surroundings: Strenia square, Park and roads. The program gains LEED Gold Certified. Basic principles of sustainability are generating volume of the building.

Building Thermal Environmental Engineering Solutions

1. Apartments and Terraces

Terraces are creating apartments with maximum daylight, sunlight and cross natural ventilation. Each apartment has openings to 2 or 3 sides and external shading device, which gives most sufficient sun-protection and creates privacy. Large terraces with rich vegetation will create cool effect in summer. They will have removable textile roof shading for shadow and rain protection. Smaller apartment units in lower floors have loggias that create similar effect in smaller scale. Also these units have external shading.

2. Offices

Offices and medical centre are generated around internal courtyard/park. That provides spaces that have natural light and natural cross ventilation and provide friendly and eco working environment. Plans are not deep. Distribution of vertical cores allows different arrangements of plans and flexibility.

3. External Spaces

Plaza on level +0 is well ventilated, opened, with natural light and allows pedestrian flows and good orientation.It is combined with green and water features which creates cool and friendly environment.Green roofs, gardens and terraces produce direct contact to external environment to all; residents, employees and visitors and create friendly and green quality.

4. Compact Volume and Program Mix.

Compact volume and program mix of living/working/service minimizes transport needs. Compact volume produces low energy consumptions and reduces energy waste. External shading and textile removable roofs above gardens produce maximum sun-protection.

Other Sustainable Measure

Vegetation that needs no extra watering will be chosen. The planner propose rain water collecting and re-usage and elements such as low-water usage toilets and showers; and propose photo-voltaic panels on some roofs and surfaces. Since the volume creates unique form the facade materials can be simple. Base of the building has glazed facade – this is combination of transparent glass with internal shading. The glass is light colored with laminated metal mesh in between. The facade raster is 1.00 m and is modular for both - offices and apartments. Apartment's facade have large panoramic sliding windows with external shading (silver alu perforated metal blinds).

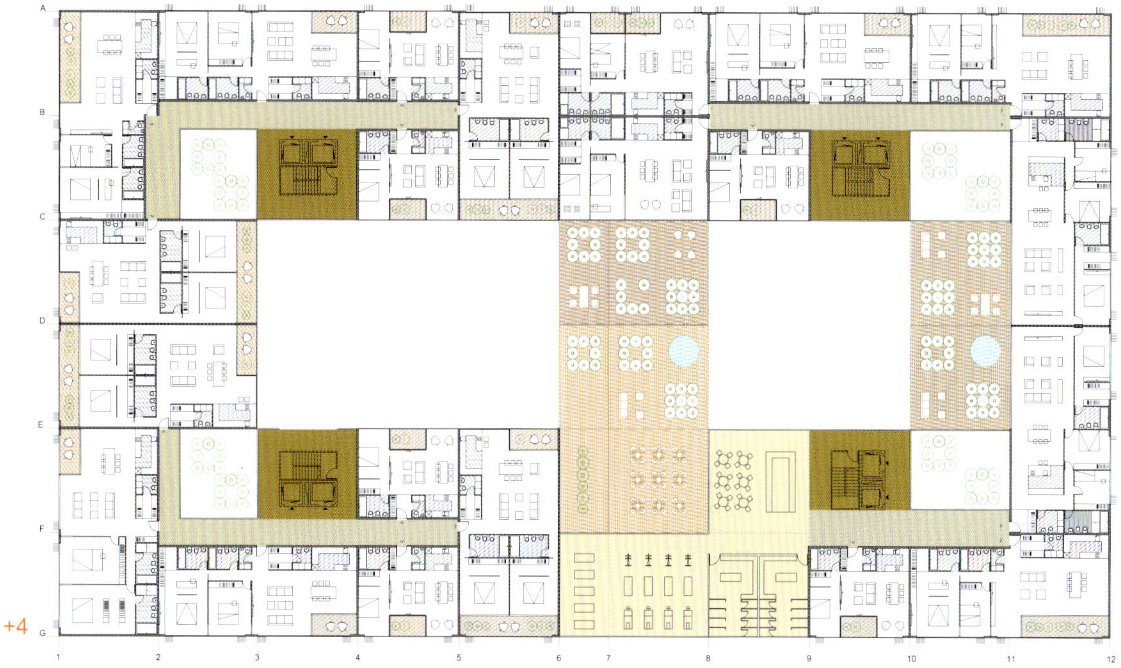
+4
A
B
C
D
E
F
G
1
2
3
4
5
6
7
8
9
10
11
12

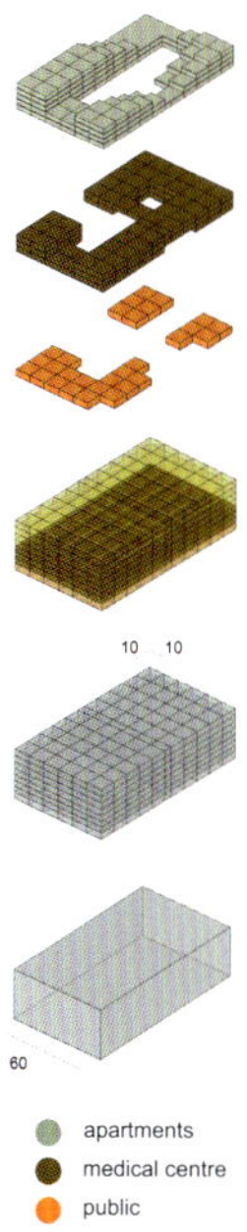
10 10
60
apartments
medical centre
public

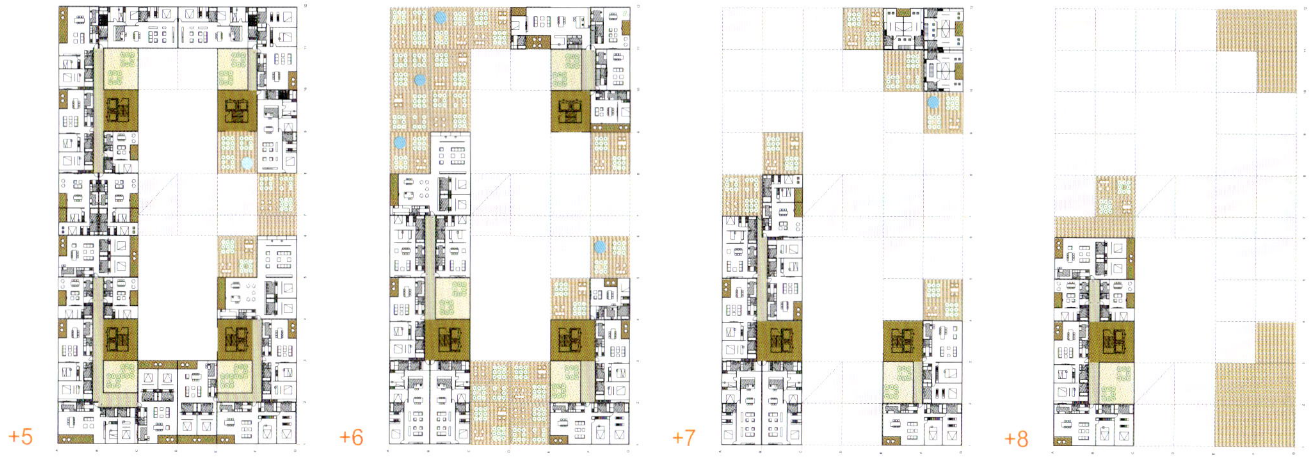
+5
+6
+7
+8

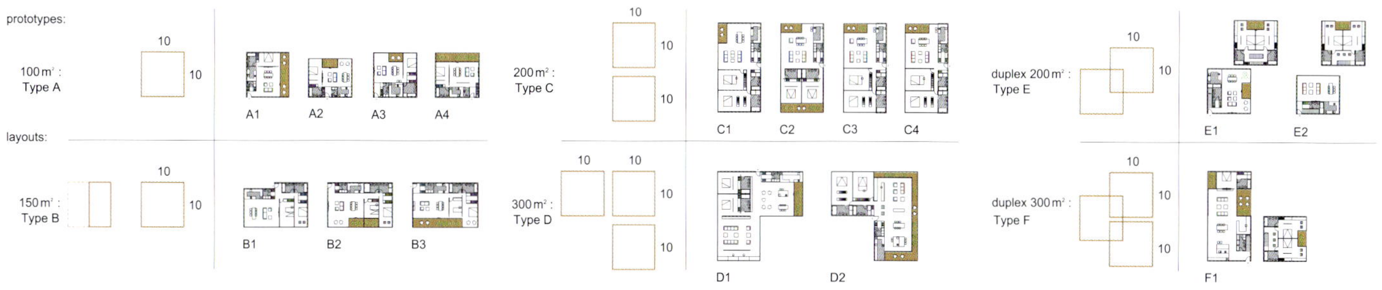
prototypes:
layouts:
100m² : Type A
10
10
A1
A2
A3
A4
150m² : Type B
B1
B2
B3
200m² : Type C
C1
C2
C3
C4
300m² : Type D
D1
D2
duplex 200m² : Type E
E1
E2
duplex 300m² : Type F
F1

模块系统分析图 MODULAR SYSTEM DIAGRAM

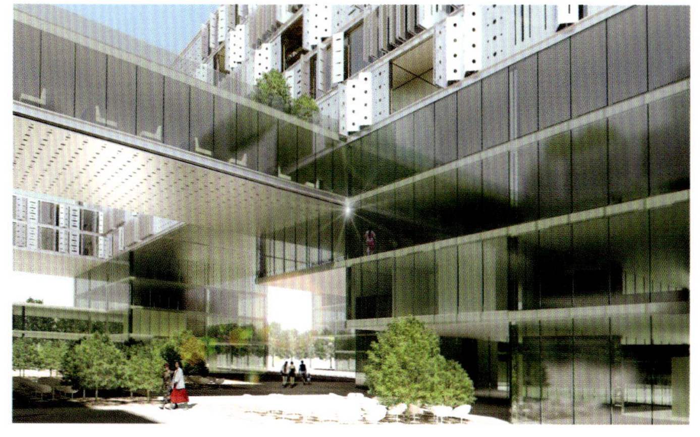

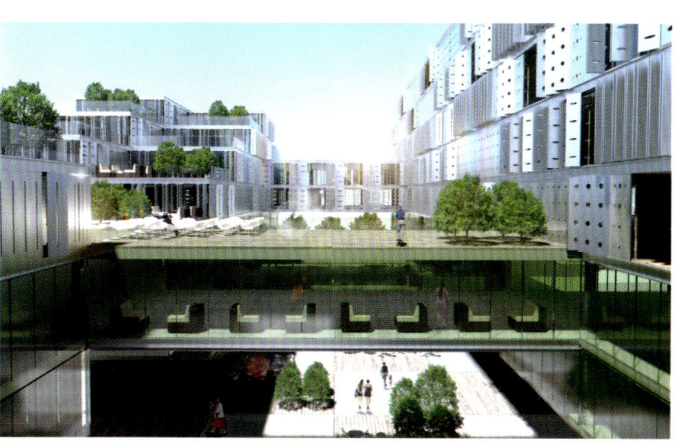

CAMPUS CENTER
校园中心大楼

地点：美国佛罗里达州迈阿密
占地面积：232 258 m²
建筑设计：Oppenheim Architecture + Design

Location: Miami, Florida, USA
Area: 232,258 m²
Architect: Oppenheim Architecture + Design

总平面图 SITE PLAN

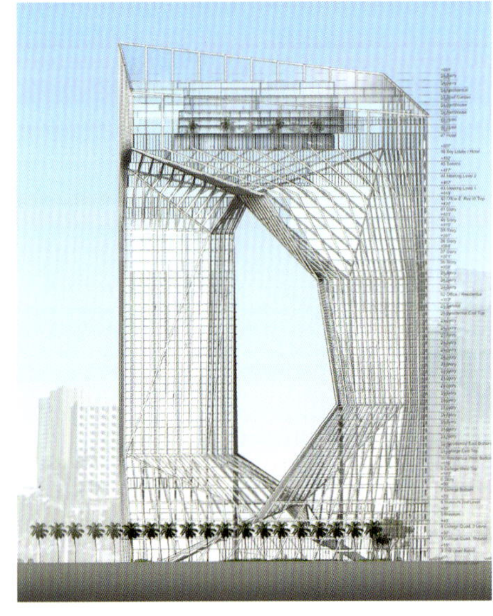
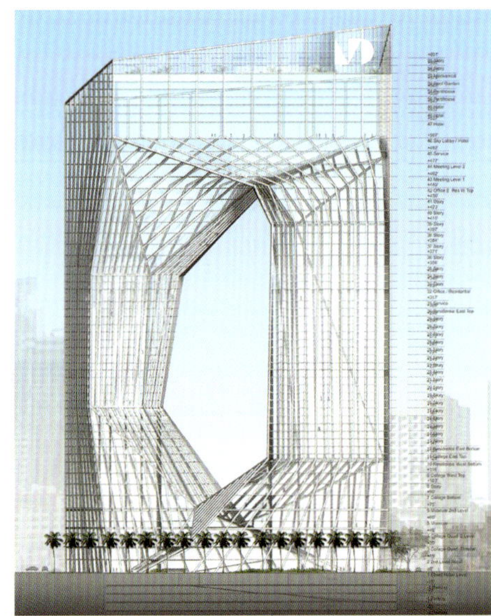

立面图 ELEVATIONS

项目概况

Campus Center大楼的设计大胆而新颖。通过进一步观察，就会发现其复杂的形体是由基本的元素和简洁的构架结合而成。方案中利用通过LEED认证的结构，组建起一个具有底面和顶面的大门，两边由塔楼支撑，在底面和空中提供了大量的外部公共空间。项目将使园区形象大为提升，并带动周边区域的兴旺。

项目主体是一个多功能大楼，除了为学校提供23 226 m²的居住空间之外，还将包括13 601 m²规模的两层临街公共空间，在三楼将设置一个3809 m²的露天"艺术方庭"项目还将包括面积23 235 m²的办公空间，5574 m²会议空间，3716 m²运动中心，以及一座300个客房、面积24 935 m²的中型宾馆。停车场将全部建在地下，不对地上建筑的使用造成影响。地下车场面积39 789 m²，提供1700个车位。

大楼顶层，是建筑的机械系统，包括风力涡轮机和太阳能热水收集装置，为三联产系统提供动力，并增加整个项目的可持续性。建筑外部为钢结构，营造了一个壮观的建筑外骨骼系统，同时使建筑避免设置大量的剪力墙结构。

建筑热环境解决方案

1.高效玻璃

在建筑的南部和西部，暴露在阳光下的时间最长，获得的热量最多。对此建筑师将采用熔块玻璃和LOW-E膜。质地轻薄，用肉眼几乎观察不到的低辐射涂层是覆盖在窗子玻璃上表面的金属或金属氧化层，主要通过降低热辐射吸收量来降低*U*值。多层玻璃的热传导机制是热辐射从温度较高的表面向热量较低的表面传导。用LOW-E材料覆盖的玻璃表面将大大降低此类热传导现象，并从整体上减少由窗口处获得热量。LOW-E涂层对于可见光是透明的，但隔离红外线，不同类型的LOW-E材料性能不同，可控制进光量。因此有助于空间的热量保存，这就使窗子的效能进一步提升。从室内获得的热量可以保存在屋中，有助于冬季室内保暖；同时，太阳辐射出的红外线被挡在室外，有助于夏季保持室内凉爽。

2.绿色屋顶

绿色屋顶将设置在大楼的人行平台，发挥高效的隔热作用，防止城市"热岛效应"。

3.三联产技术

三联产是指利用单一的热源（例如太阳能或燃料），同时产生机械能（通常转化为电能）、热能并制冷的技术。在运行过程中产生的副产品——余热也将用来发电，这样就提高了系统的整体效率。在夏季，建筑对热能的需求大大减少，但是在CHP过程中散发的热量可以通过一个吸热冷却装置转化为冷气。三联产也可以表述为CCHP(制冷、产热和产能结合)。

4.通风与光照

建筑的主体中空，旨在使空气和光照能够贯穿大楼。良好的视线、自然光及交叉循环通风对于迈阿密著名的自由大厦非常重要。

5.太阳能热水板

太阳能热水板是一种利用太阳能对液体进行加热的热水器，并将热能传递给一个热量存储装置。例如，对于家庭而言，可以加热饮用水，并存储在热水水箱中。平板太阳热能收集装置经常放置于支架上，有一个装有液体管道的吸热平面。具有特质黑色表皮的集热器将太阳能转化为热能，管道中的液体将热量携带至储热设备。经过加热的液体被泵抽到热交换器，通常为存储容器中的螺旋圈或者外部热交换设备。在此液体完成放热降温，并进入新一轮加热循环过程。液体循环流动的动力可由机械泵提供，泵体能源可由光电电池提供。或者在安装条件允许的情况下利用温差环流系统。

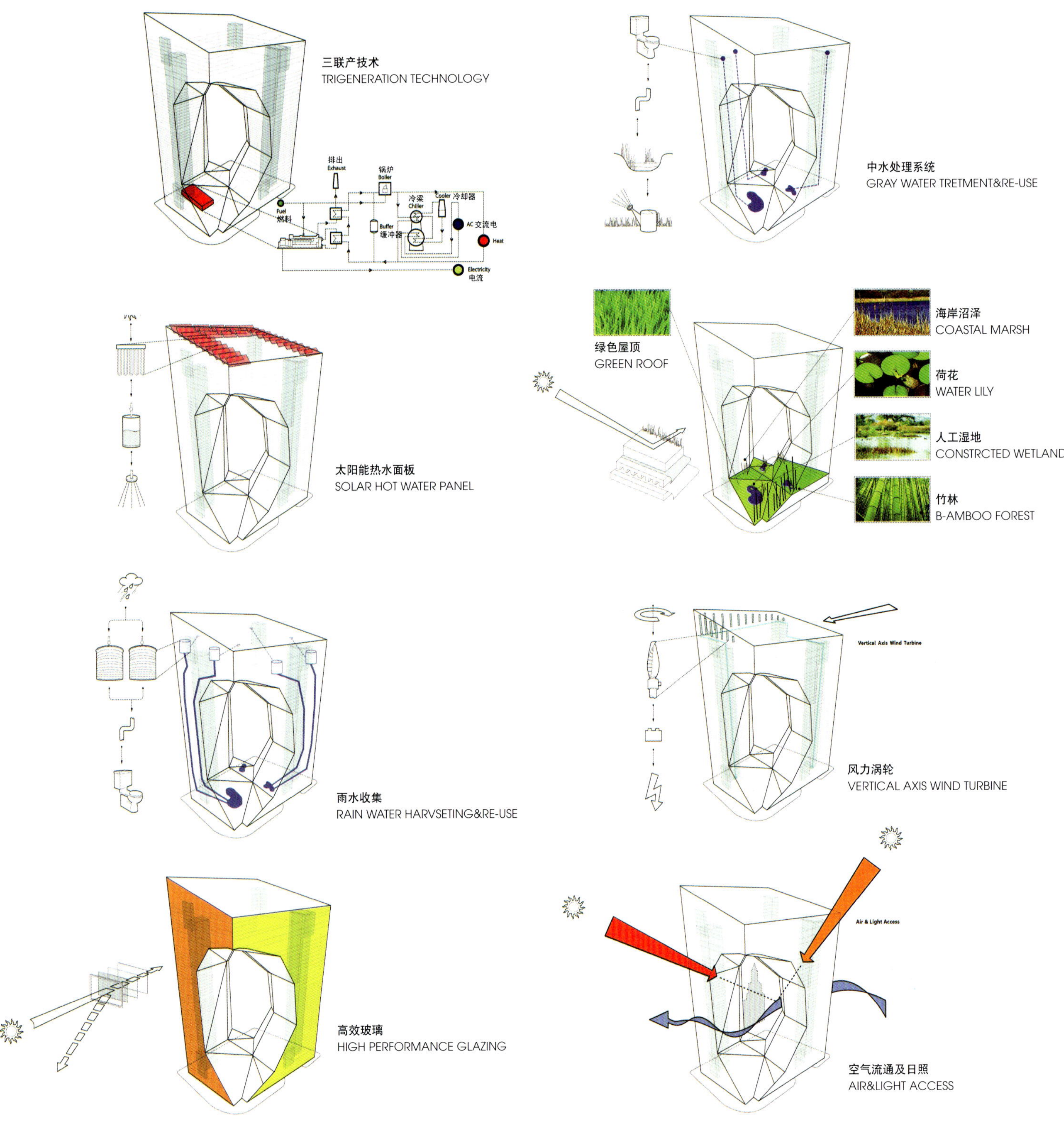

分析图 DIAGRAM

Project Description

The dramatic design set forth for Campus Center is visually daring and bold; yet upon further introspection, inherently elemental and concise in its organization of the complex programmatic mix. The proposed LEED certified structure is conceived of as a portal comprised of a base and a top supported by two towers that allow large exterior public spaces at ground and in sky. The project is to serve as a catalyst-exasperating significant enhancement to the campus experience and image through a local revitalization of the surrounding neighborhood.

The proposed project is a mix-use building that, in addition to housing the College's 23,226 m^2 Center, will include 13,601 m^2 of two-level street fronting commercial space wrapping the entire base of the building as well as a 3809 m^2 open-air campus "Arts Quad" on the third level. The project also incorporates a 23,235 m^2 office component; a 5574 m^2 meeting facility, a 3716 m^2 athletic center, a 74,188 m^2 residential component with 1142 small studio, and one-bedroom rental units; and a 300 key, 24,935 m^2 full service mid-range hotel. The parking garage is proposed to be completely underground, innocuous to the above ground functions. The underground parking structure accommodates 1700 vehicles in 39,789 m^2.

Located above the Penthouses are the building's mechanical areas with farms of wind turbines and solar hot water collectors that provide the tri-generation power system and add to the overall sustainability of the project. The buildings exterior is a pure expression of structure where veins of steel create an elegant exoskeletal system increasing building efficiency while eliminating the need for massive shear walls aside from the core.

Building Thermal Environmental Engineering Solutions

1. High Performance Glazing

On the southern and western elevations, where sun exposure in quality and quantity are strongest, both frit glazing and LOW-E film will be utilized. Low emittance coatings are microscopically thin, virtually invisible, metal or metallic oxide layers deposited on a window or skylight glazing surface primarily to reduce the U-factor by suppressing radiative heat flow. The principal mechanism of heat transfer in multilayer glazing is thermal radiation from warm surfaces to cooler surfaces. Coating a glass surface with a LOW-E mittance material reflects a significant amount of this radiant heat, thus lowering the total heat flow through the window. LOW-E coatings are transparent to visible light, and opaque to infrared radiation. Different types of LOW-E coatings have been designed to allow for high solar gain, moderate solar gain, or low solar gain. These coatings reflect radiant infrared energy, thus tending to keep radiant heat on the same side of the glass from which it originated. This often results in more efficient windows because: radiant heat originating from indoors is reflected back inside, thus keeping heat inside in the winter, and infrared radiation from the sun is reflected away, keeping it cooler inside in the summer.

2. Green Roof

Green roofs will be used throughout the elevated pedestrian level to provide excellent themal insulation and avoid the heat island effect which is so prevalent in downtown locations prone to high temperatures.

3. Trigeneration Technology

Trigeneration implies the simultaneous production of mechanical power (often converted to electricity), heat and cooling from a single heat source such as solar energy or fuel. As with cogeneration, the "waste heat" byproduct that results from power generation is harnessed, thus increasing the overall efficiency of the system. In summer, the heat demand is much lower but the lost heat of the CHP (Combined heat and power) process can be transformed into cooling energy by an absorption chiller. Trigeneration is sometimes refered to as CCHP (combined cooling heating and power).

4. Air & Light Access

The massing of the proposed building, with a large middle opening, is aimed at facilitating the free movement of air and light throughout the site. Of particular importance is visual, daylighting, and cross circulation wind access to Miami's historic Freedom Tower.

5. Solar Hot Water Panel

A solar water heater that uses the sun's energy to heat a fluid, which is used to transfer the heat to a heat storage vessel. In the home, for example, potable water would be heated and then stored in a hot water tank. Flat-plate solar-thermal collectors are usually placed on the root and have an absorber plate to which fluid circulation tubes are attached. The absorber, usually coated with a dark selective surface, assures the conversion of the sun's radiation into heat, while fluid circulating through the tubes carries the heat away where it can be used or stored. The heated fluid is pumped to a heat exchanger, which is a coil in the storage vessel or an external heat exchanger where it gives off its heat and is then arculated back to the panel to be reheated. Fluid circulation can be assisted by means of a mechanical pump (which itself could be powered by photovoltaic cell), or (where mounting conditions allow) by allowing convection to arculate the fluid to the storage vessel mounted higher in the circuit, also known as a thermosiphon.

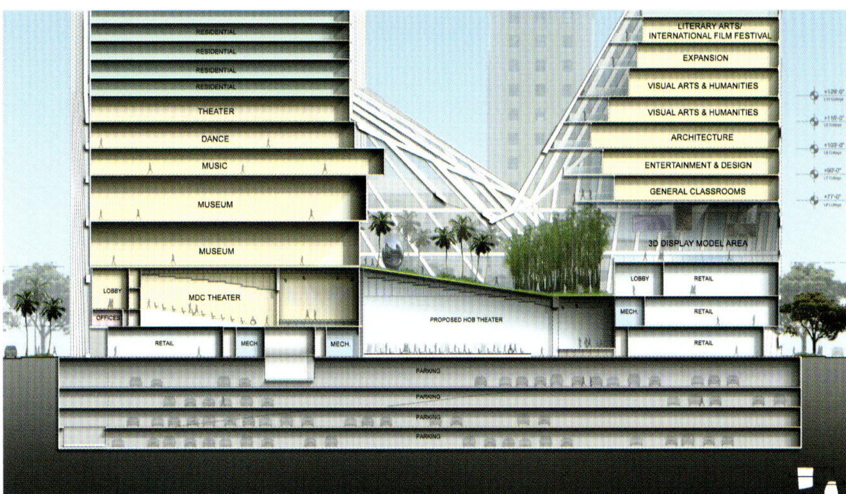

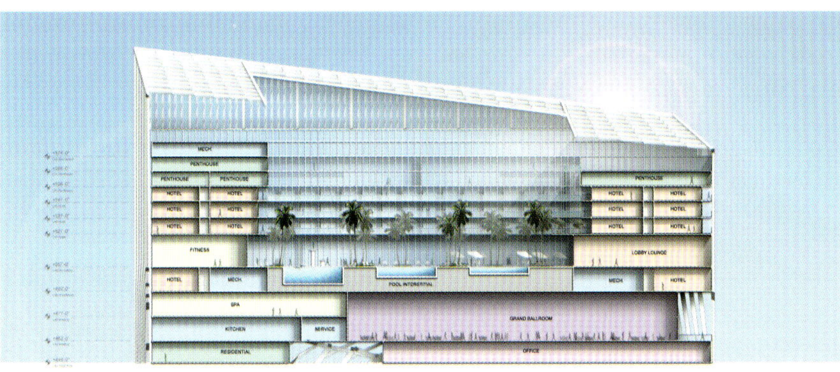

剖面图 SECTIONS

MARINA + BEACH TOWERS

玛琳娜海滩大楼

地点：阿拉伯联合酋长国
占地面积：246 086 m²
建筑设计：Oppenheim Architecture+Design

Location: U. A. E
Area: 246,086 m²
Architect: Oppenheim Architecture+Design

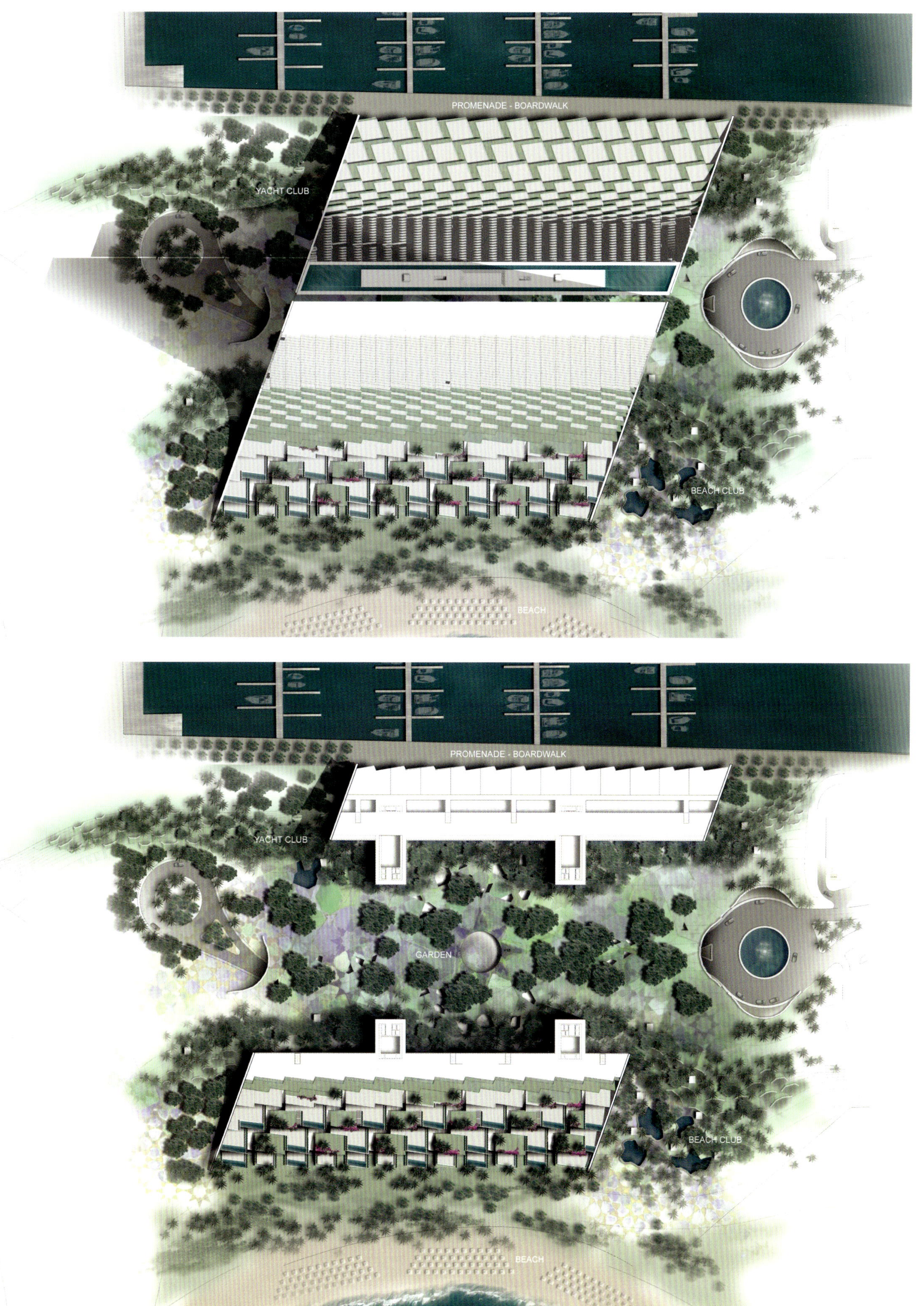

总平面图 SITE PLAN

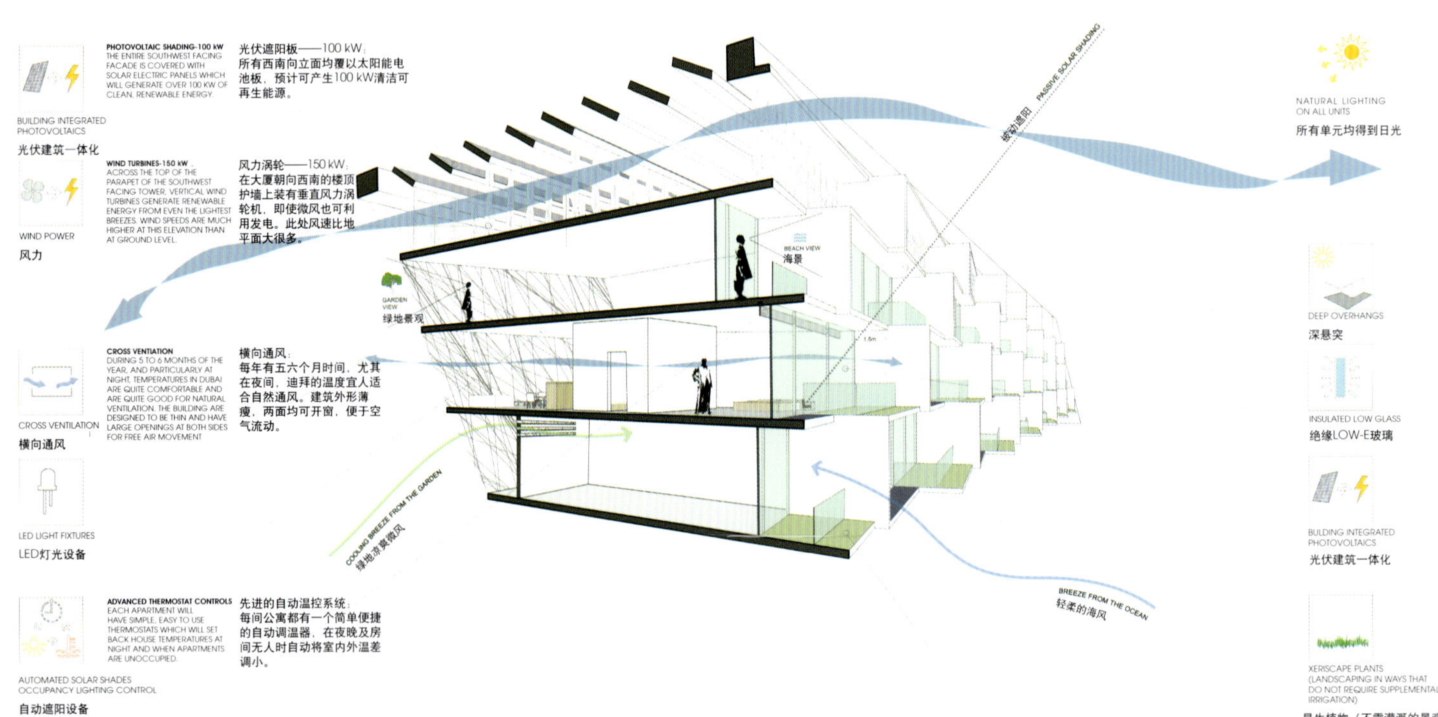

分析图 DIAGRAM

项目概况

基于对基地的最初理解，我们采用了最大化沙滩上建筑使用单位数量的策略，这样就认同了在村庄边缘空间对新建筑类型的引进。设计中除考虑建筑类型之外，还有对于建筑和景观的共同考虑：利用建筑流动性去融合天空和水域空间，这样让更多多样的、超越普通摩天楼功能的居住体验成为可能。因为对建筑设计的一个整体考虑是对于其自身可持续性的关注和对自然资源的合理使用，所以我们开发了能源利用装置。

建筑热环境解决方案

建筑使用单元的模数标准化允许了楼层的不断重复，这样楼层就能通过重复移动来生成织毯似的立面。这个表面既可以呼应阳光的每次细微变化又可以在阳光强烈时对建筑提供关键性保护。采用被动方法为人营造良好居住环境。每年有五六个月时间，尤其在夜间，迪拜的温度宜人，适合自然通风。建筑外形薄瘦，两面均可开窗，便于空气流动。建筑的一侧接受绿地吹来的凉爽微风，另一端有海风送来新鲜空气，实现了自然通风。这里所有的单元都可接收自然光线。安装了被动太阳能遮阳系统，深悬突、高R值立面、隔热LOW-E玻璃广泛应用于建筑，防止大楼吸收过多热量。我们参考了当地区域久经考验的风斗传统，让大楼本身变成了一个让凉风到达遮蔽了阳光区域的装置。

其他可持续设计措施

大楼的结构巧妙利用了太阳能和风能，进而生产大楼所需的部分能量。其他针对大楼自身持续性的整体考虑还包括再利用穿过基地大量流水的各种方法。可持续性主题体现在整个建筑生命周期：有为绿色建筑的想法提供了创新的机会；大楼的生态和可持续性允许了自由使用能源，避免了因大量使用能源产生的负罪心理。

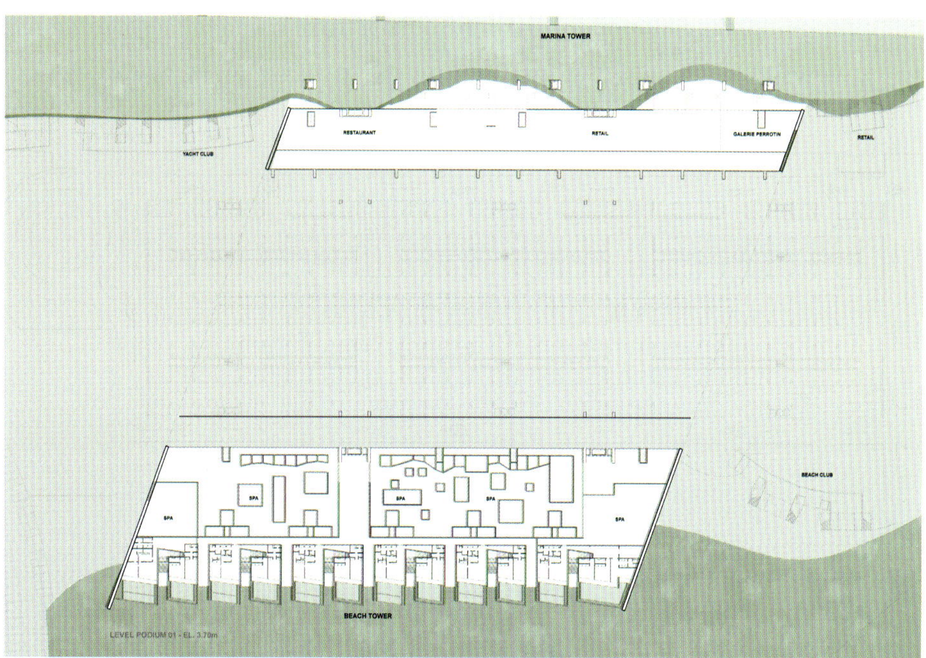

Project Description

Based on our initial reading of the site we took the opportunity to maximize the quantity of units on the beach. This approach has permitted the introduction of new unit types with the spatial adjacencies of a village. What has emerged, in addition to the typologies, is a response that is simultaneously building and landscape; a project whose fluidity merges sky and water. This allows for more varied living experiences beyond the capabilities of the normative tower. An integral aspect of this orchestration has been our focus on self sustainability and an embrace of the natural resources. Herein we have deployed our only apparatus of exploitation.

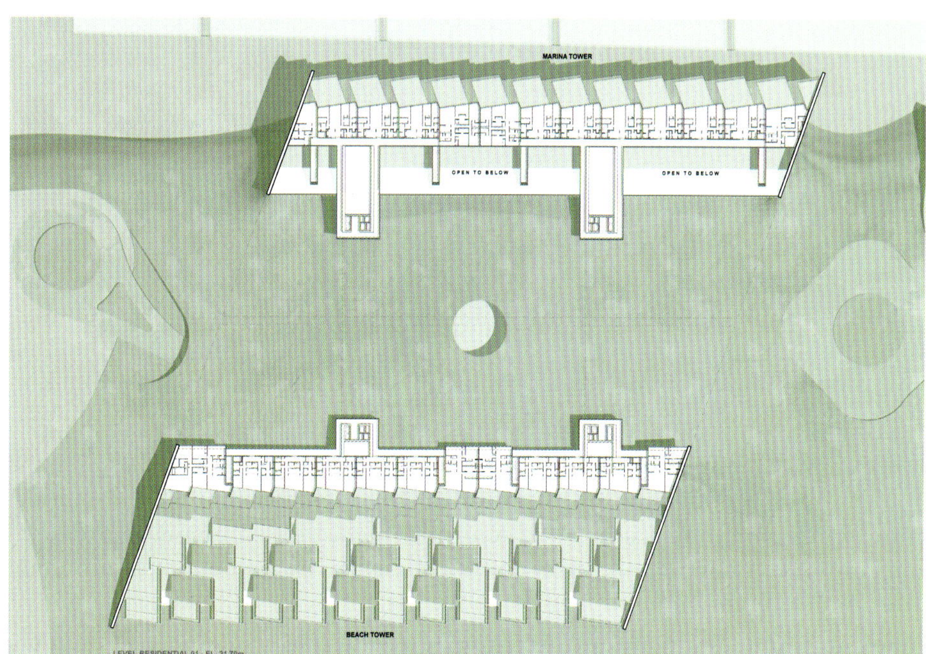

Building Thermal Environmental Engineering Solutions

Basing the unit types on a standard module has allowed its shifted repetition to develop into the woven tapestry of the facade; a surface that responds to every nuanced shift of light, while providing critical protection from the intense sun. Passive methods are used to create a comfortable interior environment. During 5 to 6 months of the year, and particularly at night, temperatures in Dubai are quite comfortable and are quite good for natural ventilation. The buildings are designed to be thin and have large openings at both sides for free IR movement. This way, one side of the building accepts cooling breeze from the garden, while the other side are breezed by the ocean winds. Cross ventilation are realized here. All the units here are designed to receive natural lighting. Passive solar shading are equipped in the building, deep overhangs, high R-value insulation facade and insulated LOW-E glass are used to protect the building from over-heat gain. The building form itself becomes a device for luring cool breezes to spaces shielded from the sun, referencing the regional time-tested tradition of wind catchers.

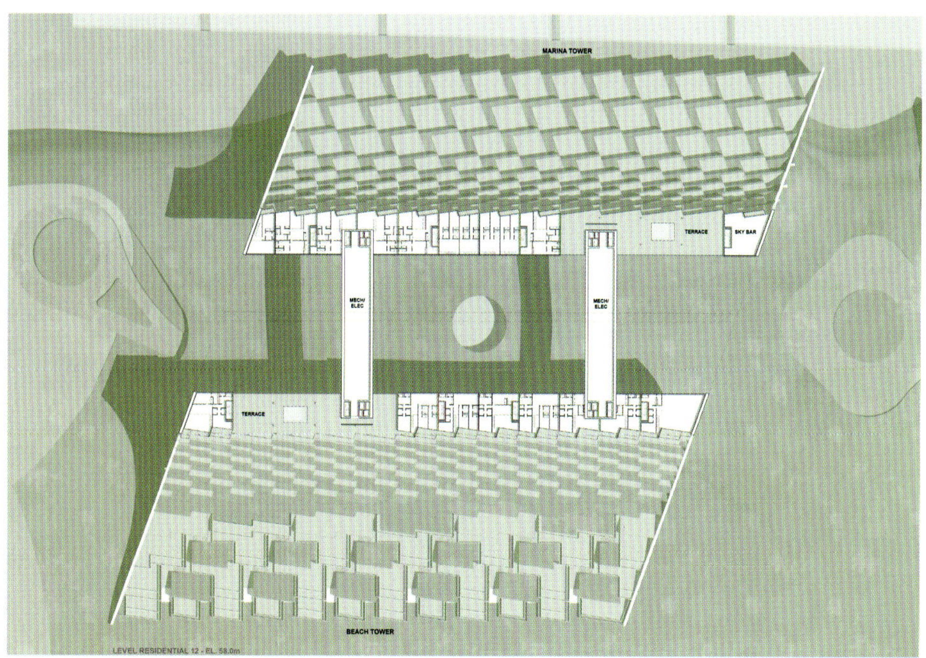

平面图 FLOOR PLANS

Other Sustainable Features

Inconspicuously and opportunistically the structure incorporates solar and wind arrays for the generation of some of this building's required energy. Other integral systems deployed towards complete self sustenance, include various methods for the reuse of the vast amounts of water flowing through the site. Sustainability is imbued throughout the life cycle, where intelligent planning provides innovative opportunities for energy and resource conservation, up-cycling (as opposed to re-cycling), waste, and healthy building initiatives; where ecology and sustainability allow even greater guilt-free luxury, not self denial. The orchestration of all these imperatives has generated an elegance derived from simplicity that addresses all of the needs of development without any of the proliferating exuberance.

COR
COR综合大楼

地点：美国佛罗里达州迈阿密
占地面积：44 593 m²
建筑设计：Oppenheim Architecture+Design

Location: Miami, Florida, USA
Area: 44,593 m²
Architect: Oppenheim Architecture+Design

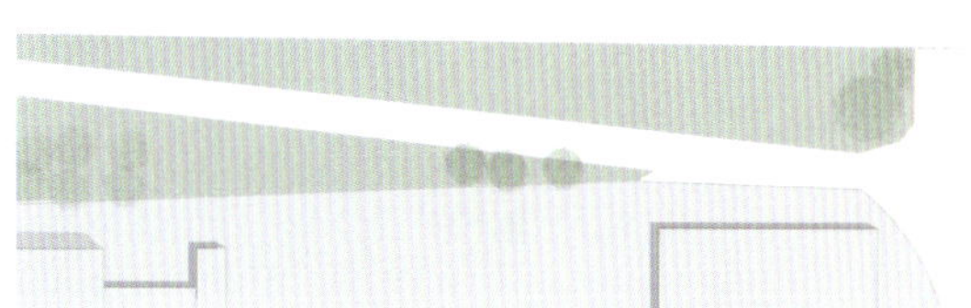

总平面图 SITE PLAN

COR是佛罗里达州迈阿密市的第一个可持续性、功能复合的综合大楼。它代表着建筑、结构工程和生态设计的完美结合。COR高122 m，利用风轮机、光电板和太阳能热水器的最新科技进展，从它周围环境中汲取能量，同时也把这些设施融入到大楼的建筑身份里。一个超高性能的外骨架壳同时提供了大楼结构——隔热块，自然通风的遮阳装置，平台的围合，风轮机的转动以及人们可以在地面聚集的凉廊。COR包含了商业、办公、健身、生活/工作和纯粹的居住空间，它为生活水平的提高提供了一个非常特别且灵活的平台。

Building Scale Strategies

建筑尺度策略

一些设计策略提高了中央大楼系统的性能。可更新的能量，水资源再利用和中央空调系统需要对大楼进行整体的考虑和投资，即使这些策略所带来的好处只能被公寓的个体所有者感受到。绿色屋顶将最有益于建筑的阁楼部分。一些环境策略对于城市和区域环境同样有利。

Some design strategies improve the performance of the central building systems. Renewable energy, water reuse systems and central air systems require whole-building consideration and investment even though benefits will be seen by the condo owners. Green roofs may benefit the penthouse units the most. And some environmental strategies benefit th greater city and regional environmental as a whole.

风能

在空中40层楼的位置，风轮机将从海风中产生可再生能量。这些设在女儿墙上的风轮机是没有污染且来自当地的，他们将产生足够的电力去支持所有中央区域的电力需要。

Wind Energy

40 stories in the air, wind turbines will generate renewable energy from breezes off the ocean. Pollution free and local, these turbines on the parapet will generate enough electricity to supply all central building electrical demands.

绿色屋顶

绿色屋顶为太阳能热吸收，暴风雨水流失和城市热岛效应提供了一个缓冲区。公寓的屋顶花园作为一块被占用的绿地，给公寓单元提供了额外的宜人空间。水池区域也给绿色屋顶带来了同样的益处。

Green Roofs

Green roofs provide a buffer for solar heat gains, stormwater runoff flows, and the urban heat island effect. Rooftop gardens for condo units provide an added amenity as an occupiable green area. Pool areas also provide the same benefits of green roofs.

太阳能热水

家用的太阳能热水器将会是大楼能量需求的一个显著部分。太阳能是迈阿密市的重要能量来源，作为一种可再生能源满足人们的热量需求。

Solar Hot Water

Domestic hot water heating will be a significant portion of the energy demand for this building. The significant solar resource of Miami can be utilized to supply the heating demand from a renewable source.

Bldg 集成光电板

高楼的立面是可以全天连续吸收太阳能的大型外表面。当PV材料替换了涂层材料，他们就能成为可更新能量的重要产生装置，同时因为替换了涂层材料，所以大楼的建造花费也减少了。

Bldg Integrated Photovoltaics

The façades of tall buildings are huge surfaces which absorb solar energy all day, every day. When PVs replace cladding materials, they can be significant generators of renewable energy, while the cost is partly absorbed by that of the cladding material they are substituting.

足够的太阳光遮挡

炎热充满阳光的迈阿密气候让遮阳变得尤为重要，这里遮阳是基于对热舒适度、眩光和能量保存的考虑。任何在迈阿密的绿色建筑必须有一个在建筑的四个立面都可以作出回应的遮阳策略。

Adequate Solar Shading

The hot, sunny Miami climate makes solar shading very important for thermal comfort, glare and energy conservation. Any 'Green Building' in Miami must have a responsive solar shading strategy that varies across all four façades.

玻璃规格

在不同的立面朝向设置不同质地的玻璃，可以减少太阳能吸收，同时可以最大化阳光的透入。南立面和西立面应该有更暗的可反射涂层。所有的立面都应该有一个LOW-E涂层。

Glazing Specifications

Careful selection of glazing quality across different façade orientations can reduce solar gains and maximize daylight penetration. South and West facades should have darker, reflective coatings. All façades should have a LOW-E coating.

灰水收集和再利用

雨水和其他灰水可以被用来灌溉室外的景观。一个标有不可携带标志的简单的水箱和管道可以组成一个雨水的收集系统。

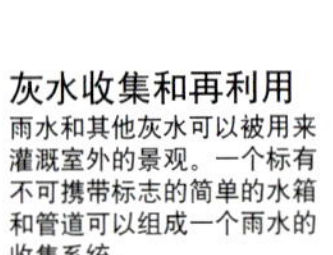

Graywater Collection & Reuse

Rainwater, and other graywater runoff can be used to irrigate exterior landscape features. A simple tank and pipe marked 'non-potable' can make up a rainwater collection system.

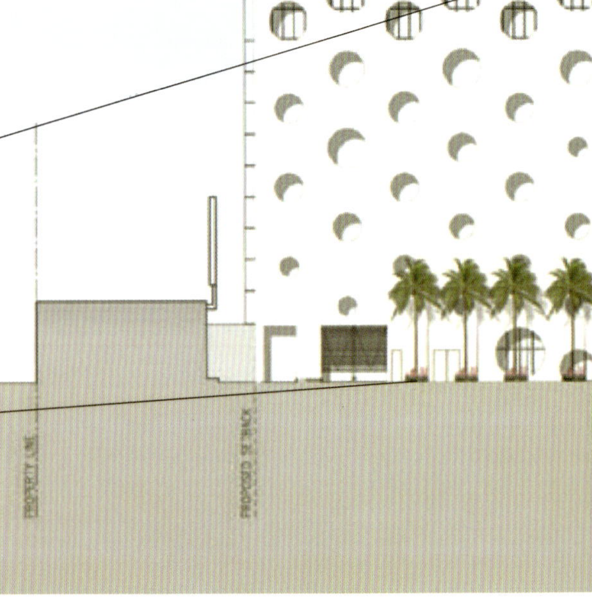

Dwelling Scale Strategies
居住尺度策略

Other design strategies improve the experience of the individuals in their homes. There are a number of elements of a Green Residence. The following items are the basics and absolutely essential to calling a Condo tower 'Green'. Sustainable design includes more than just energy conservation. It must include water saving strategies, careful attention to daylight and airflow design. And, possibly most importantly, a strong connection to the outdoors with views outside and plants indoors are very important to creating an atmosphere of 'environmental' living.

其他的设计策略促进了个体在家庭的居住体验。在一个绿色住宅中有许多关键要素，以下为一些基本和绝对关键的对于绿色公寓的评价标准。可持续性设计包含的不仅仅是能量保存，它必须包括节水措施、细致的日光和空气流通设计。最重要的是将户外景观和室内有机结合，因为这对于创造一个生态的环境氛围都很重要。

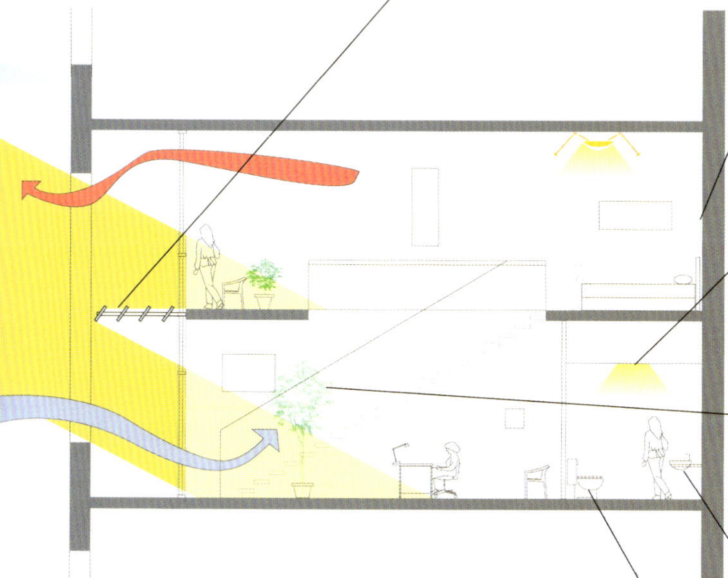

Good Daylight Design

Good daylight in a space does not always mean more daylight. Diffuse, indirect daylight washing across a white ceiling produces the best quality visual experience. In hot climates, like Miami, it is important to balance % glazing, solar shading and daylight penetration.

良好的日光设计

在一个空间里，好的日光设计并不意味着更多的阳光。穿越一个白色天花板的漫射光——间接日光可以产生最好质量的视觉体验。在像迈阿密这样城市的热天，平衡玻璃的遮阳和日光的渗入是非常重要的。

Energy Star Appliances

Energy Star certified appliances should be a requirement for all base-building appliances. A wide range of choices are available for refrigerators, dishwashers, laundry machines, hot water heaters and air conditioners.

星级节能装置

星级节能认证装置应该是对大楼所有基础设施的要求。在广泛的范围中选择冰箱、洗碗机、洗衣机、热水器和空调。

Low VOC Paints & Carpets

Low VOC paints and carpets have been one of the most significant areas of market change from the US Green Building Council's LEED program. A wide range of colors and qualities can now be purchased extensively.

低含量挥发性有机化合物的油漆和涂料

低含量挥发性有机化合物的油漆和涂料已经成为美国绿色建筑协会LEED体系对建材市场改变最大的一个方面。许多不同颜色和质量的油漆和涂料都可以买到。

Energy Efficient Lighting

Lighting technologies have improved dramatically over the past 10 years. Compact fluorescent bulbs fit in all light sockets and put out an increasingly warm light. Compact fluorescents use 10-20% as much energy as incandescents.

节能灯

制灯技术在过去的十年中进步了很多。紧凑型节能灯只用白炽灯10%～20%的能量。

Plants in space

Plants absorb carbon dioxide and release oxygen during photosynthesis. In addition, they help purify the air by trapping airborne pollutants and particles. This helps support a healthy indoor environment, both physically and psychologically.

建筑空间的植被

植物在光合作用中吸收了CO_2同时释放了O_2。另外，它们还通过截留空中传播的污染物和粒子帮助净化空气。从物理和心理两个角度来讲，这样可以有助于提供一个健康的室内环境。

Low Flow Fixtures

Another very basic and cost neutral green building requirement is low flow sinks, showers, toilets and appliances. The green building market today has a wide range available.

低流量的装置

低流量的装置是绿色建筑的另外一个基本的减少花费的要求，比如低流量的水槽、淋浴器、厕所以及其他设备，这些设备现在绿色建筑市场上都非常普及。

Dual Flush Toilets

Dual flush toilets have one button for a 0.8 gallon flush and one for a standard 1.6 gallon flush. Toto, Caroma, Gerber and Ifo Cera all make dual flush toilet models that have been third-party tested.

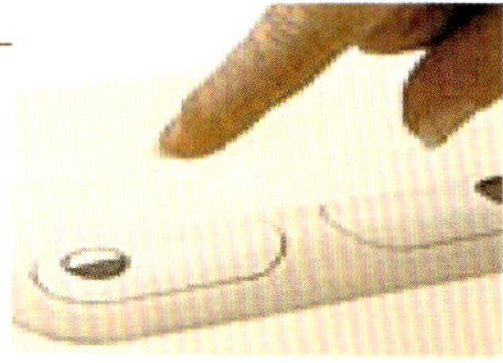

双冲水马桶

双冲水马桶一个按钮提供3 L的冲水量，另外一个提供6 L的标准冲水量。Toto、Caroma、Gerber和Ifo Cera这些品牌都制造双冲水马桶模式，这些模式都已经由第三方测试。

FSC Certified Wood

All wood products, particularly wood floor finishes should be specified as Forest Stewardship Council Certified. Creating a market demand for FSC improves forest habitat and ensures sustainability.

森林管理评价委员会认证木材

所有的木材生产尤其是木地板装饰都应该达到森林管理评价委员会认证标准。为森林管理评价委员会创造的市场标准可以改善森林栖息地质量，保证可持续性。

COR, the first sustainable, mixed-use condominium in Miami, Florida represents a dynamic synergy between architecture, structural engineering and ecology. Rising 122 m above the Design District, COR extracts power from its environment utilizing the latest advancements in wind turbines, photovoltaic's, and solar hot water generation— while integrating them into its architectural identity. A hyper-efficient exoskeleton shell simultaneously provides building structure, thermal mass for insulation, shading for natural cooling, enclosure for terraces, armatures for turbines, and loggias for congregating on the ground. Comprising commercial, office, fitness, live/work, and pure residential spaces—COR provides a uniquely flexible platform for lifestyle enhancement.

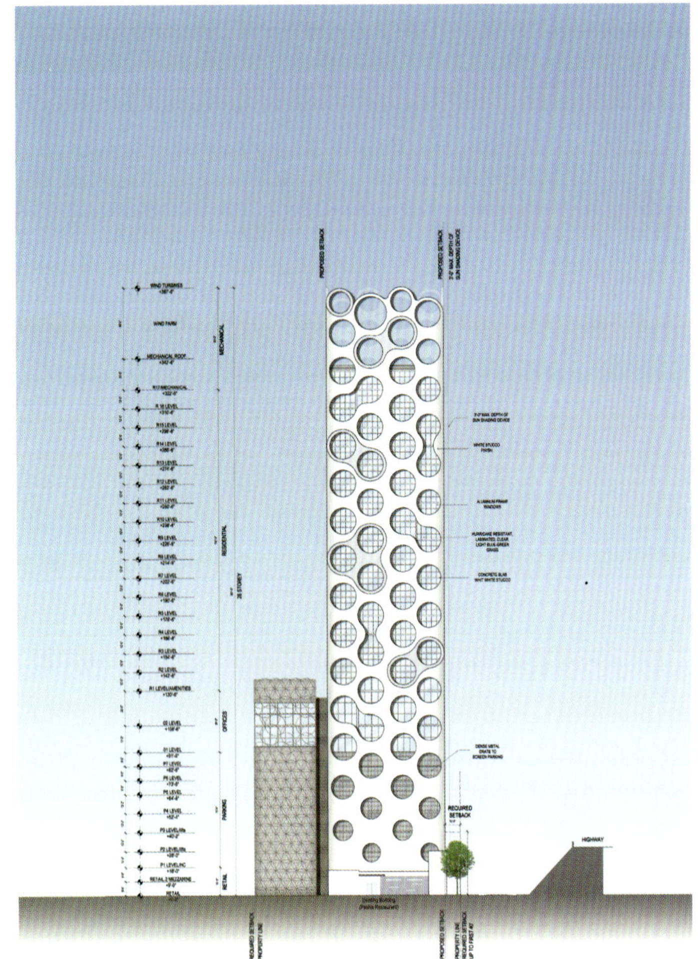

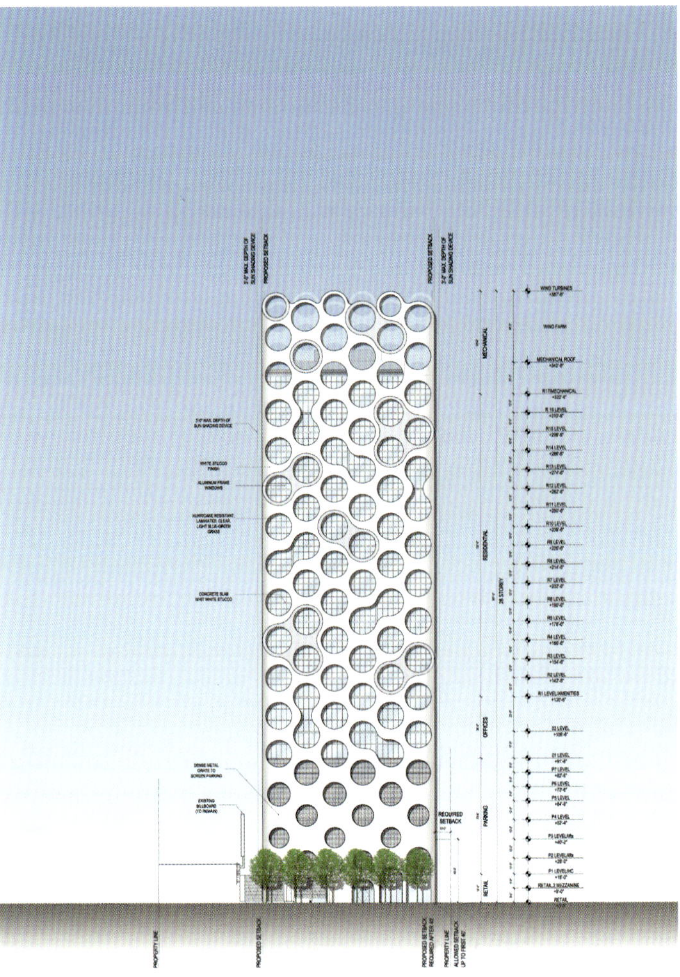

一层平面图 GROUND FLOOR PLAN

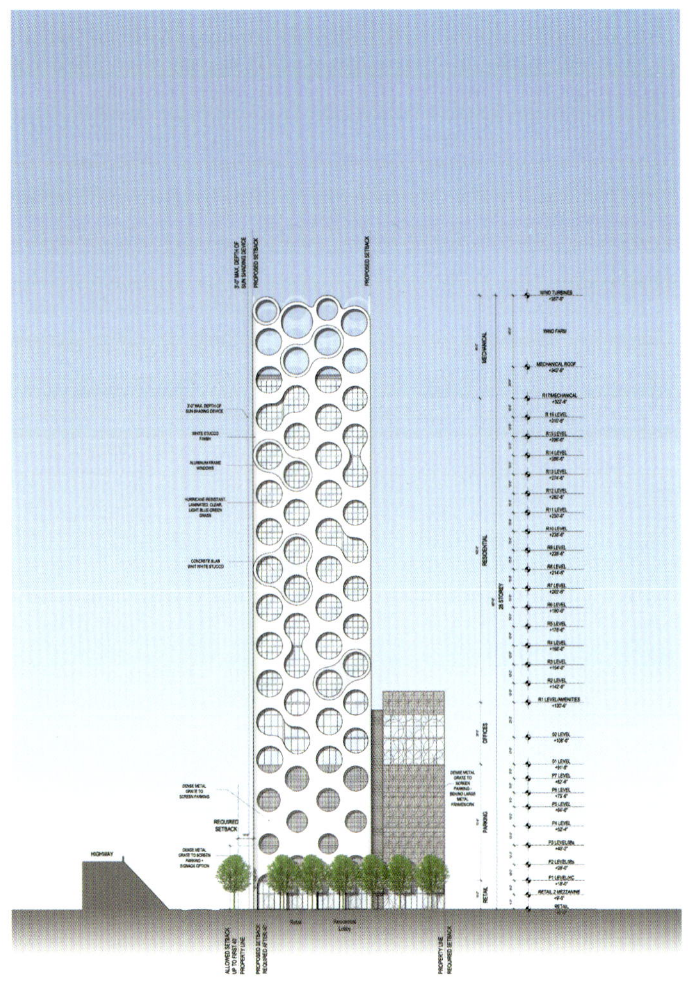

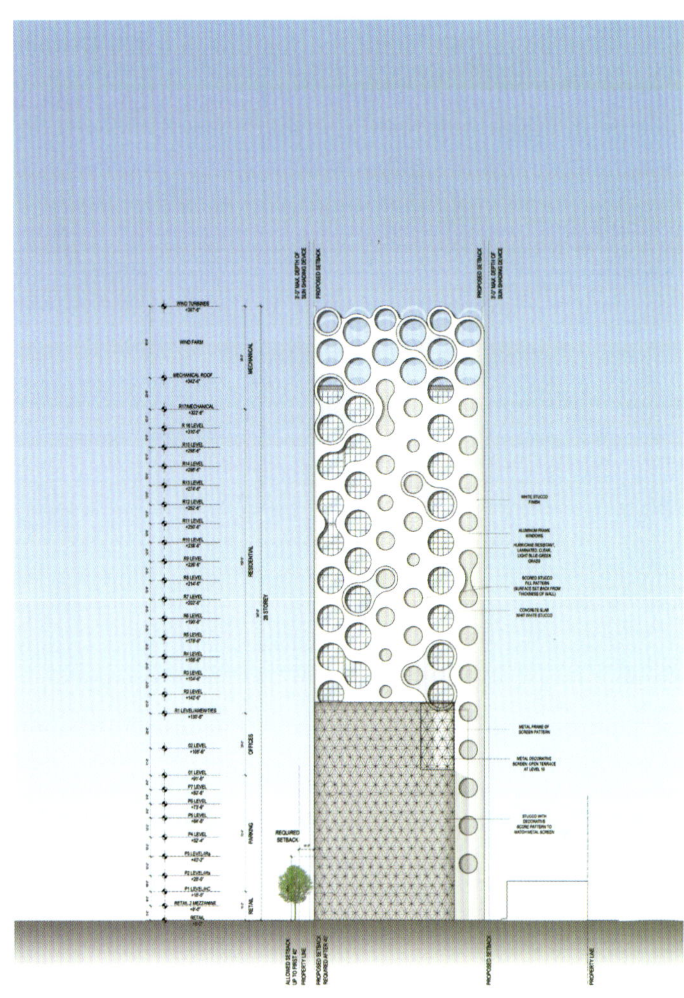

立面图 ELEVATIONS

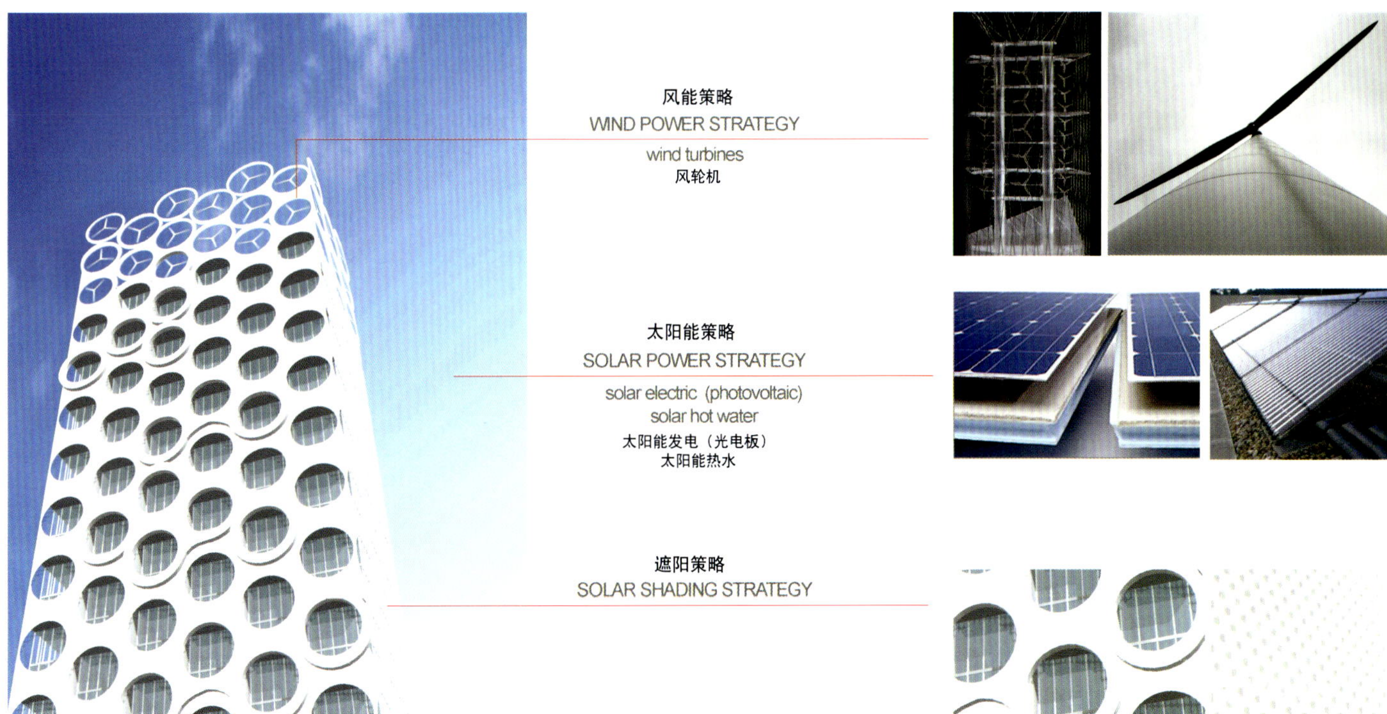

AKERSHUS UNIVERSITY HOSPITAL

Akershus大学医院

地点：挪威奥斯陆阿克斯胡斯
规模：137 000 m²（新建楼118 000 m²）
建筑设计：C. F. Møller Architects
景观设计：Bjørbekk & Lindheim AS；Schønherr Landskab A/S
所获奖项：2009年度优质医疗设施奖最佳国际设计奖；2000年国际竞标一等奖
委托方：Helse Sør-Øst RHF
摄影：Torben Eskerod; Guri Dahl; C. F. Møller Architects

Location: Akershus, Oslo, Norway
Size: 137,000 m² (118,000 m² new building)
Architect: C. F. Møller Architects
Landscape: Bjørbekk & Lindheim AS; Schønherr Landskab A/S
Selected Awards: 2009 Best International Design-Building Better Healthcare Award; 2000 1st Prize in International Competition
Client: Helse Sør-Øst RHF
Photography: Torben Eskerod; Guri Dahl; C. F. Møller Architects

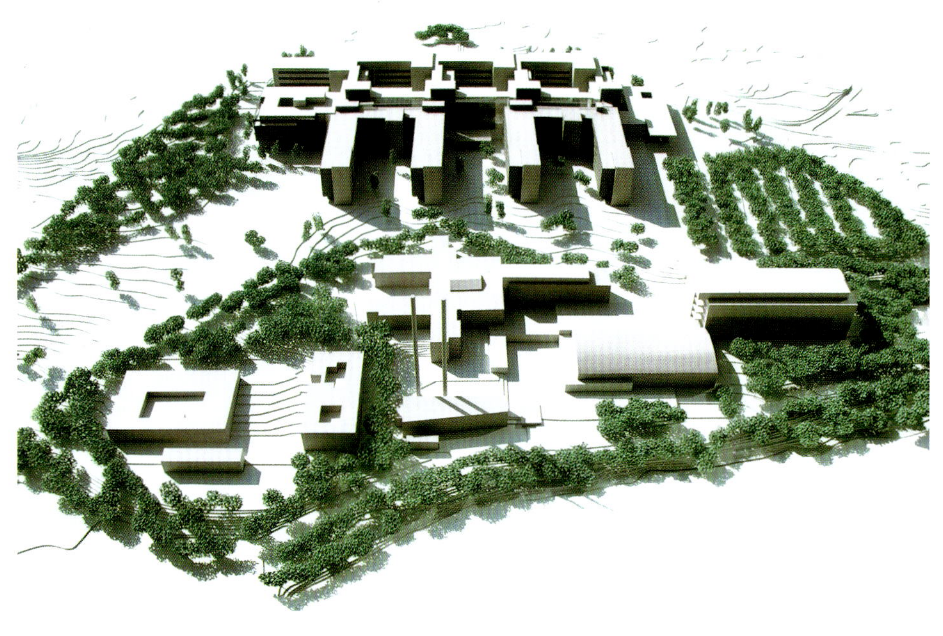

项目概况

位于奥斯陆郊区的阿克斯胡斯大学医院，组织有序的开放空间，为患者及家属营造了友好、轻松的氛围。阿克斯胡斯大学医院的建筑设计强调安全性和通透性，在医院的结构设计中融入日常功能和常用材料。项目采用的材料富于变化，这种变化统一于以面板和玻璃为主要表现形式的建筑主体之中。由此，建筑的各个独立部分相互融为一体，达到了通透、纵深的整体效果。一条玻璃屋顶的中心通道将建筑的各个部分连接起来。这条玻璃街道以大门入口处的休息区为起点，绵延几百米，终点仍设在休息区，同时在小儿科专辟出口。

建筑热环境解决方案

该医院计划建成高水平的可持续项目，减少能源消耗，促进社会可持续发展，实现经济节约。医院在结构设计方面最显著的特点是，整座大楼都需要高质量的自然光。为此，设计师在通道顶部设置了玻璃天花板，中部有与胶合木结合的玻璃部件；病房和诊室装配了宽大的玻璃窗。设计保证了大楼拥有良好的自然通风条件。医院建设过程中就地取材，采用的健康材料能形成良好的室内微气候。

医院85%的热能和40%以上的总体能源消耗主要依靠医院周围的地热发电厂供应。建筑的能源供应就地取材，主要依靠地热能源。医院的能耗总量为每年20 GWh，相当于1300户家庭用电的总和。医院采用了结合岩床储热设备的地热交换系统。由太阳辐射、人体散热、机械设备、冷却及通风设备等产生的过剩热量可以储存在距地面200 m深的350个能量井中。此处配有欧洲同类设施中规模最大的地热电厂之一，它产生的电量中85%用于加热，连同冷却过程在内，可满足整个医院总体用能的40%。此举可使CO_2排放量降至传统医院的一半。

其他可持续设计措施

建筑应用了先进技术，实现了气力输送和机器人等许多自动功能。还应用了其他可持续措施，例如绿色屋顶、LCA、节能设计、可持续规划、雨水收集等，使建筑成为一座健康建筑。

模型图 MODEL DIAGRAM

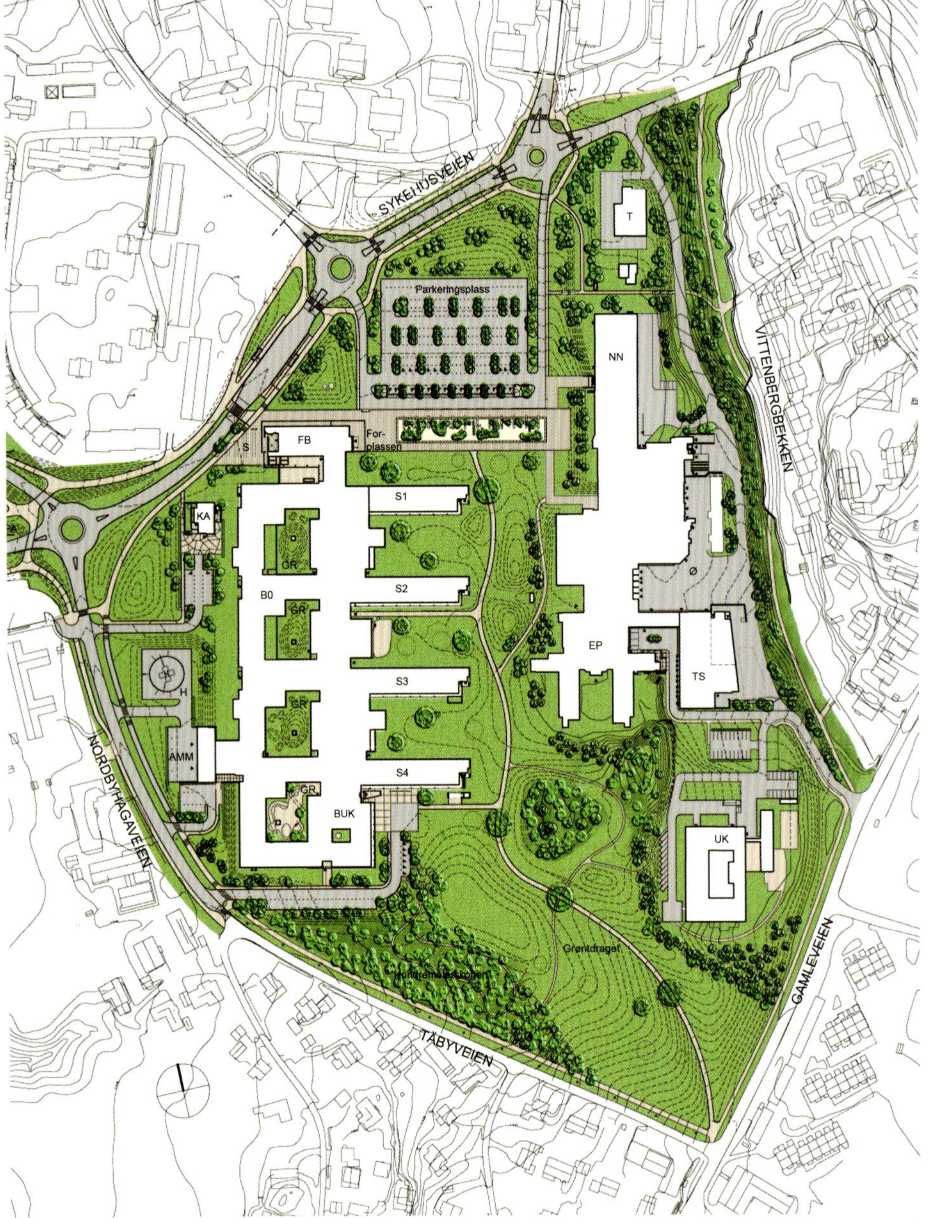

总平面图 SITE PLANS

Program Description

The Akershus University Hospital just outside Oslo is a friendly, informal place with open and well-structured surroundings. These surroundings present a welcoming aspect to patients and their families. Akershus University Hospital has been designed to emphasise security and clarity, where everyday functions and well-known materials are integrated into the hospital's structure. The material expression of the development is rich in variation. Nonetheless this expression is united into a whole by means of a general architectural theme centred on panels and transparency. In this way, a unity is created between the individual parts of the complex, which thereby receive a subtle effect of transparency and depth. A central, glass-roofed main thoroughfare links the various buildings and departments. This 'glass street' begins in the welcoming foyer of the arrivals area, where the main reception desk receives visitors. The street runs several hundred metres in length and concludes in the foyer and the separate arrivals area of the children's department.

Building Thermal Environmental Engineering Solutions

1. Passive measures

The scheme aims at a high degree of sustainability, in terms of energy consumption, social sustainability, and economic liability. The most important characteristic in the physical design of the Hospital has been the central requirement for high quality daylighting throughout the entire complex – right from the experientially rich main thoroughfare, via glass roofs and the impressive glue-lam and glass sections in the middle of the street, to the generously-sized windows of the wards and treatment departments. The design makes sure that the complex keeps good condition of natural ventilation. All materials used in the hospital are healthy materials with a good indoor climate performance.

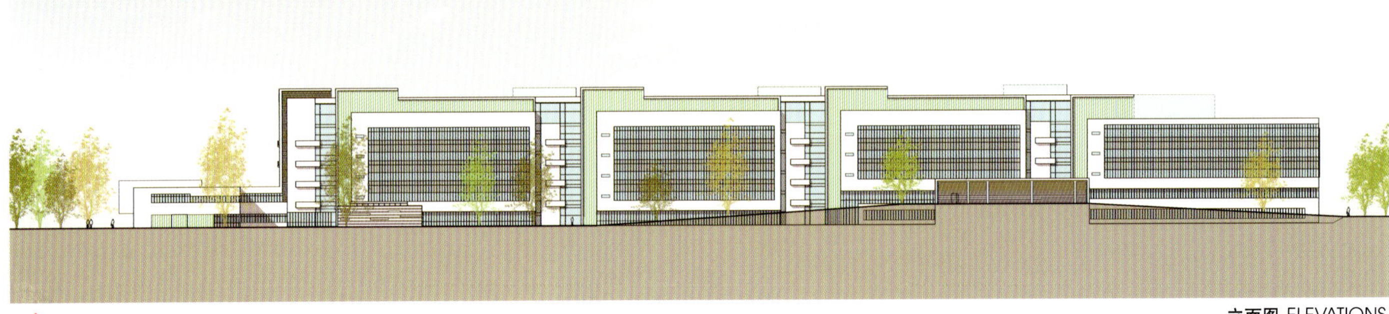

立面图 ELEVATIONS

2. Active measures

85% of the hospital's heating and more than 40% of the total energy consumption are supplied by geothermal plant located on the outskirts of the hospital grounds. The bulk of the building's energy consumption makes use of locally sourced materials, and geo-thermal energy. The total energy consumption of the hospital is approx. 20 GWh/year, similar to the consumption of 1300 single family houses. The hospital uses a ground-heat exchange system, combined with thermal storage capacity in the bedrock, where surplus heat (for instance from solar gain, people, technical equipment, cooling and ventilation plants) can be stored in 350 energy wells drilled to a depth of 200 m. The plant is among the largest installations of its kind in Europe, and it generates 85% of the energy used for heating, and covers over 40% of the total energy consumption in the hospital, including cooling. This reduces CO_2 emissions by over 50% compared to the former hospitals performance.

Other Sustainable Measures

The complex has advanced technology, with many automatic functions such as pneumatic dispatch and robots. Other sustainable measures are adopted such as green roof, LCA, energy efficient design, sustainable planning, rainwater harvesting, making the complex a healthy building.

加热
Heating
冷凝器/ 蒸馏器
Condensator/Vaporizer
Surface
地表
Bedrock
岩床
s energy
echnical
nent
备余热
Borehole thermal
energy storage system
地洞储热系统
地热源
Geothermal
heat source

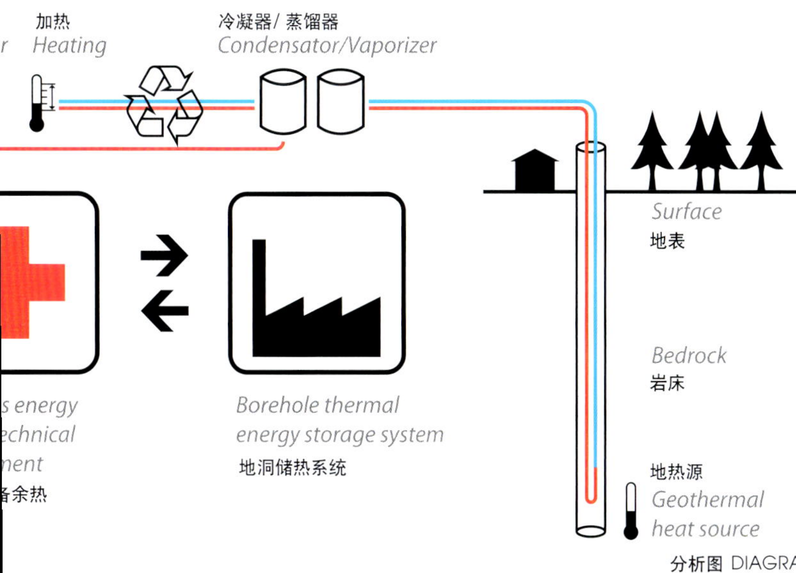

分析图 DIAGRAM

一层平面图 GROUND FLOOR PLAN

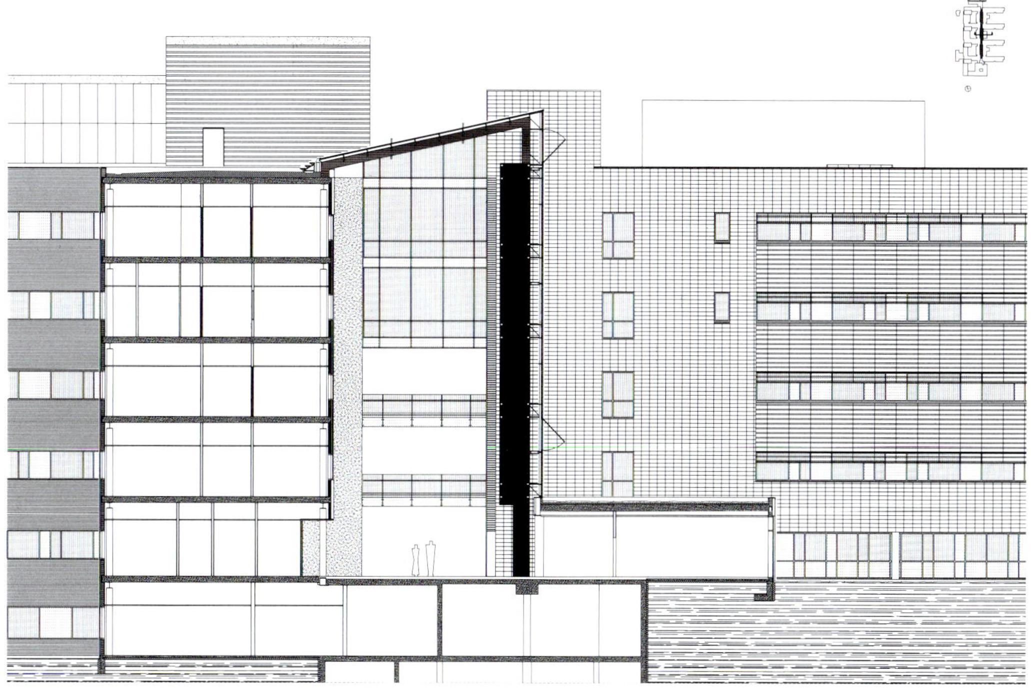

剖面图 SECTION

BEATY BIODIVERSITY CENTRE/ AQUATIC ECOSYSTEMS RESEARCH LABORATORY

Beaty生物多样性中心及水生生态系统研究实验室

地点：加拿大英属哥伦比亚省温哥华
建筑设计：Patkau Architects
景观设计：Phillips Farevaag Smallenberg（Beaty Biodiversity Centre）；
R. Kim Perry & Associates Inc.
（Aquatic Ecosystems Research Laboratory）
所获奖项：LEED金质认证
摄影：James Dow

Location: Vancouver, British Columbia, Canada
Architect: Patkau Architects
Landscape: Phillips Farevaag Smallenberg
(Beaty Biodiversity Centre);
R. Kim Perry & Associates Inc.
(Aquatic Ecosystems Research Laboratory)
Selected Awards: LEED Gold Certified
Photography: James Dow

项目概况

Beaty 生物中心和水生生态系统研究实验室坐落在英属哥伦比亚大学南北面的中央轴线上。它们一同形成了一个相关环境科学研究所的复合体。一个新校区被安排在一个大型的室外庭院空间四周。这个庭院空间被一个新的跨校区的人行道和自行车连接体平分。

Beaty生物多样性中心包含一个自然历史博物馆、一个大型自然历史收藏处，研究实验室和办公室，并设有会议室和其他附属空间。这座建筑规模为11 500 m^2。水生生态系统研究实验室在庭院的北面，天井连接了大楼的四个楼层，它的四周被那些跨领域的研究团队围绕，社交区域位于中庭。这个5150 m^2的大楼联合了这些研究团队所在的空间。教员办公室、学生社团的阁楼空间以及各种会议空间都位于顶部楼层，而大的公共房间位于底层，人们可以在这里参与园区的城市生活。

建筑热环境解决方案

天井对于可持续性设计策略在大楼的应用起到了关键的作用。天井作为一个自然通风管把空气导向大楼，避免了大楼对于传统机械通风系统的需要。夏天，大楼在夜晚自动通风，进而冷却混凝土结构；白天，作为一个可辐射的冷表面，避免了最高三层的空调能耗需要。大楼顶处的玻璃把自然光深入地引向室内，同时又与大楼北面的大量玻璃和光感应控制技术结合，进而使建筑对于电灯的依赖最小化。水生生态系统研究实验室设计获得了LEED金质认证。

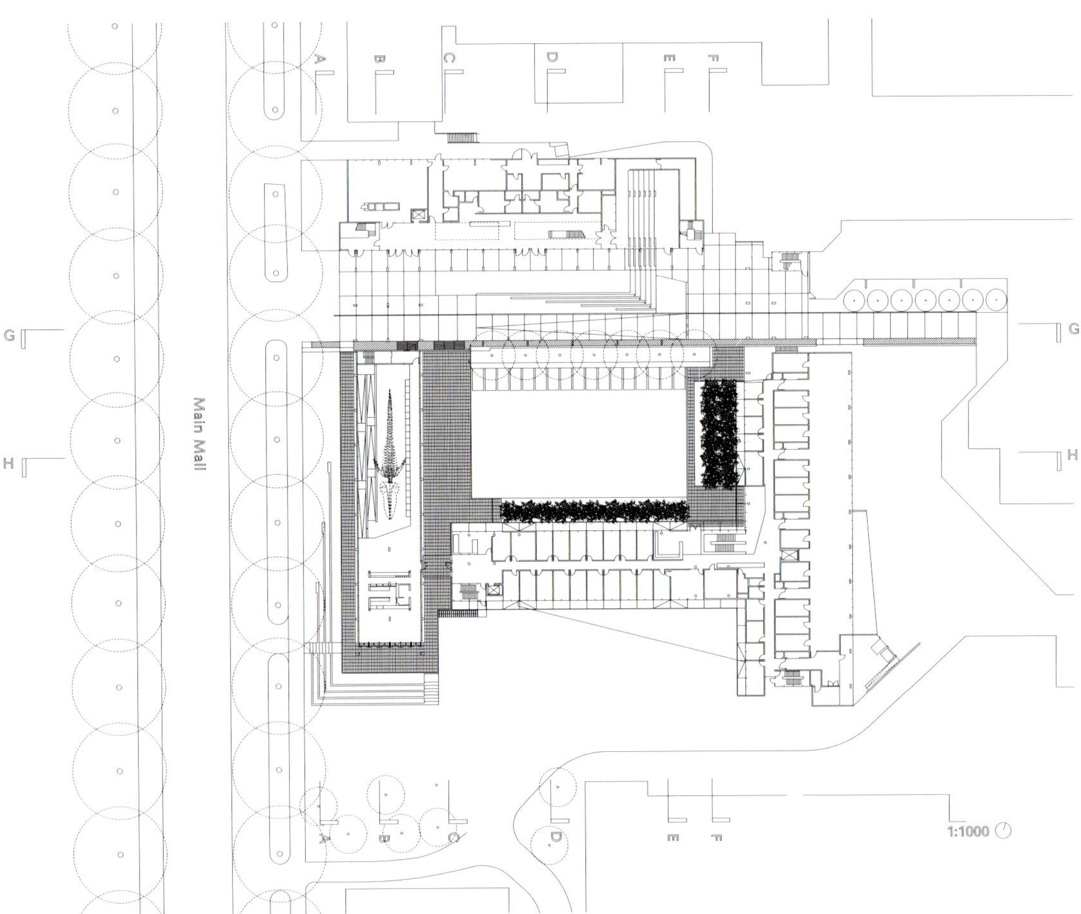

一层平面图 GROUND FLOOR PLAN

Program Description

The Beaty Biodiversity Centre, with the Aquatic Ecosystems Research Laboratory, is located on Main Mall, the central north/south spine of the University of British Columbia. Together they form a complex of related environmental science functions; a new campus precinct organized around a generous exterior courtyard space which is bisected by new cross-campus pedestrian and bicycle connections.

The Beaty Biodiversity Centre comprises a natural history museum, a large natural history collection, research laboratories and offices with related meeting and support spaces. This facility is 11,500 square meters in scale. The Aquatic Ecosystems Research Laboratory is located on the northern side of the courtyard. This 5150 square meter building consolidates interdisciplinary research groups around an atrium that interconnects the four floors of the building. Social spaces are located adjacent to this atrium. Faculty offices, loft spaces / digital laboratories for the student community and a variety of meeting spaces are located on the upper floors, while the large public rooms are located on the ground floor where they participate in the urban life of the campus.

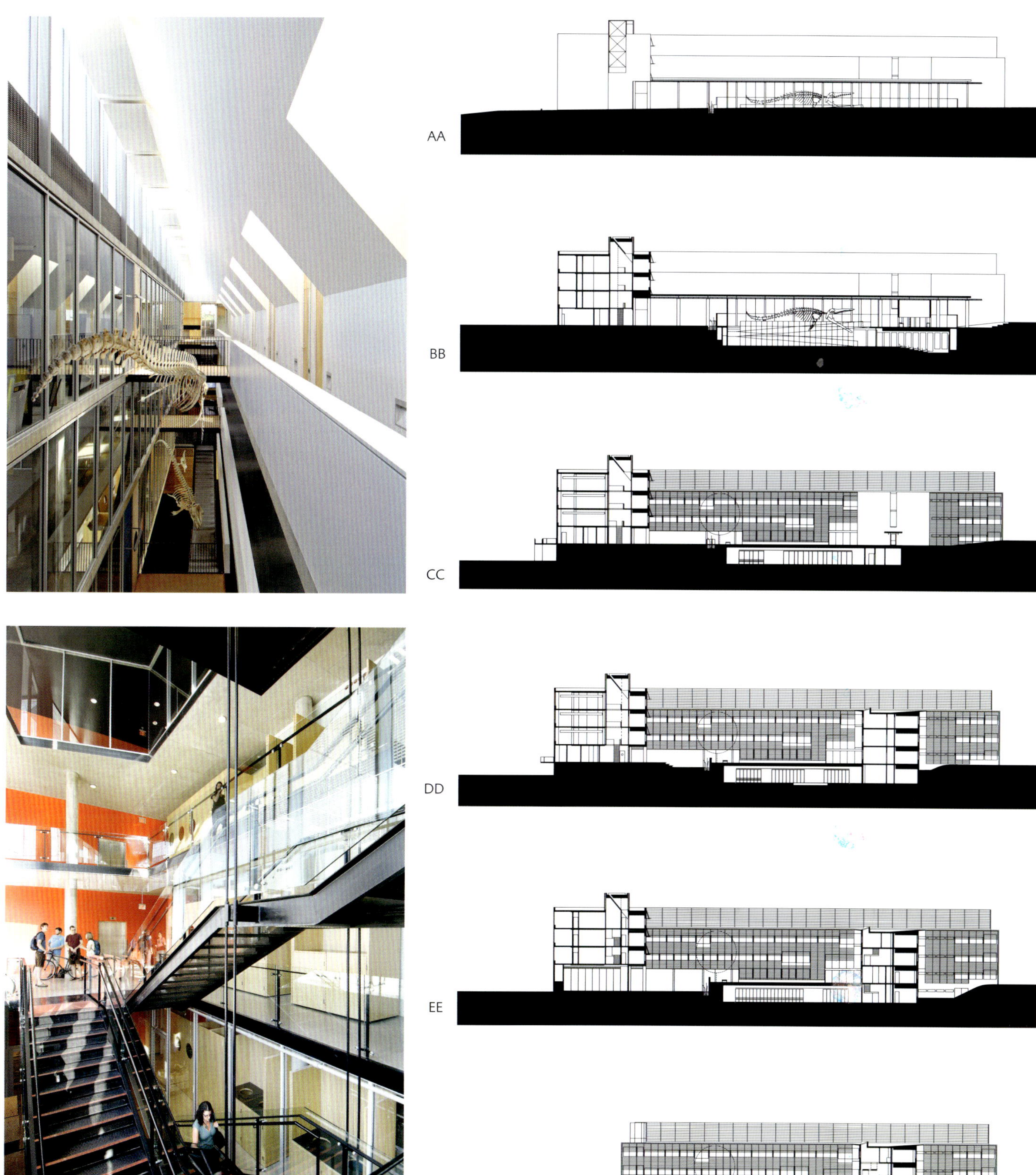

剖面图 SECTIONS

Building Thermal Environmental Engineering Solutions

The atrium plays a key role in the sustainable design strategies employed in the Aquatic Ecosystems Research Laboratory. The atrium also acts as a natural ventilation stack that pulls air into the building, eliminating the need for a conventional mechanical ventilation system. On summer nights the building is naturally ventilated to cool the concrete structure, which acts as a radiant cooling surface during the day, eliminating the need for air conditioning for the three upper floors. Glazed at the top to bring day-light deep into the interior it combines with generous glazing on the north side of the building and photo sensor controls to minimize dependency on artificial lighting. The Aquatic Ecosystems Research Laboratory is certified LEED Gold.

DRAGEN CHILDREN'S HOUSE
DRAGEN 幼儿园

地点：丹麦欧登塞Dragebakken Sanderum
规模：1100 m²
建筑设计：C. F. Møller Architects
景观设计：C. F. Møller Architects
所获奖项：2010年欧登塞市政当局建筑奖
委托方：Odense Municipality

Location: Dragebakken Sanderum, Odense, Denmark
Size: 1100 m²
Architect: C. F. Møller Architects
Landscape: C. F. Møller Architects
Selected Awards: Odense Municipality Architecture Award, 2010
Client: Odense Municipality

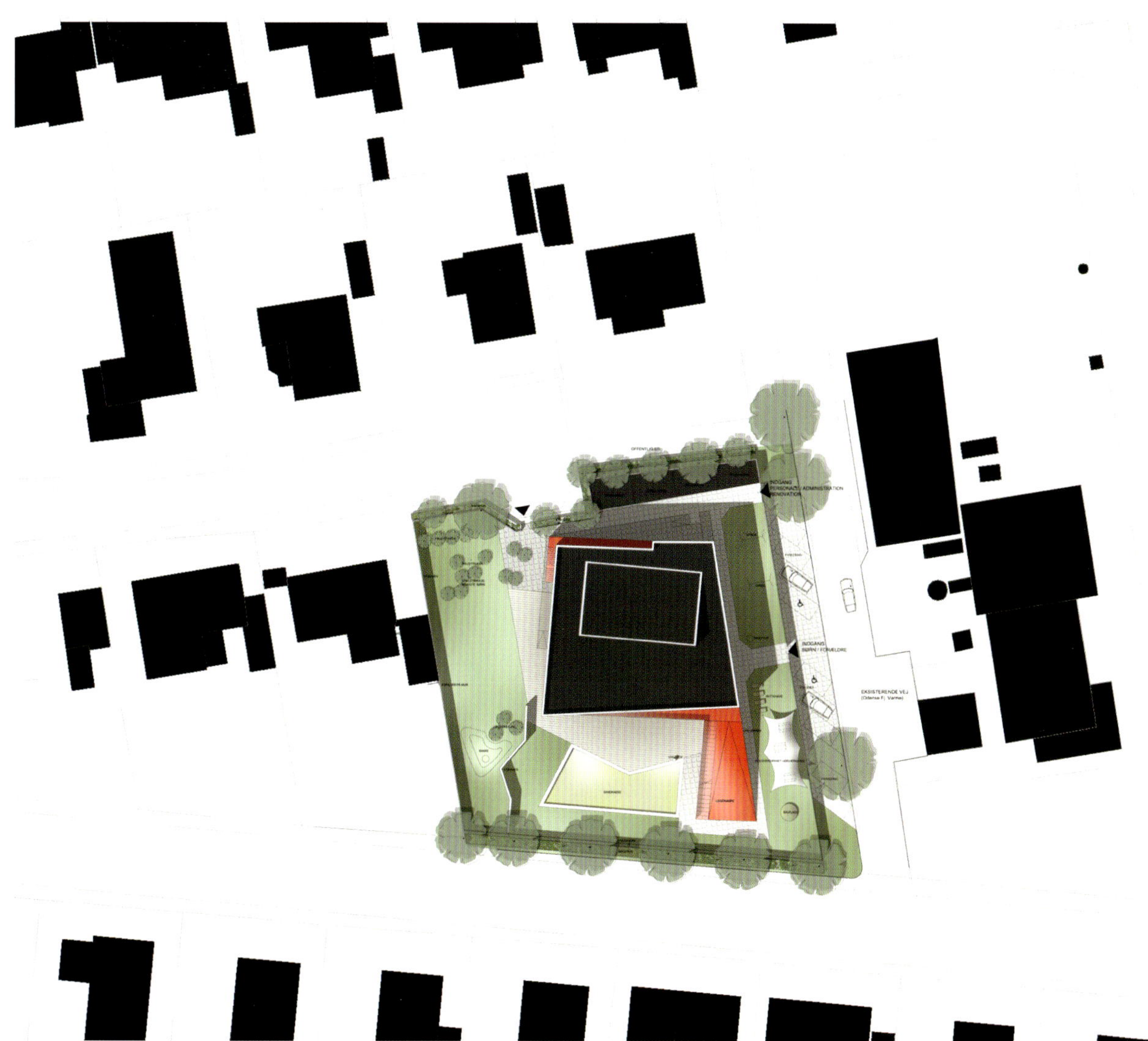

总平面图 SITE PLAN

Dragen幼儿园是一座全新定位的建筑，将可持续理念和教育学思想贯穿于设计始终。建筑采用的组件大部分通过了Nordic Swan环评认证。建筑物被评为“被动式节能房屋”，实现了能源消耗最小化。设计主要围绕满足儿童需求、提高舒适度而展开。

建筑设计的基本理念是设计一个双层的简明清晰的几何形模型，儿童区位于采光最好的南端。同时还设计了楼梯和坡道，以锻炼儿童的协调反应能力和运动技巧。

欧登塞议会主管青少年工作的女议员简·杰珍德说：“设计者为儿童提供的活动空间远远大于传统幼儿园，而且根据教育学的思想对内部装修进行了概念设计。欧登塞市政当局为儿童的成长提出的设想——使他们在愉悦，舒适，新奇、有趣的环境中学习和生活，在这座建筑中得以体现。”

Dragen幼儿园遍布小生境，可供儿童游戏、阅读和休息。游戏空间的总面积为414 m^2，可供88名儿童使用，使用空间远大于建筑标准所规定的268 m^2的最小值。由此可以提供更多的活动空间，减少传播疾病的风险。

此外，幼儿园还有专设的功能空间，为儿童提供各种有利环境，包括一个小剧场、一座画室，一个训练运动技能的房间和室内外教学区。楼内的墙壁上设有小的孔洞，儿童可以穿越于相互间隔的空间。

该建筑不仅让使用者的健康不受损害，而且环保节能。高度隔热的设计相对普通建筑物节能20%。应用健康材料建造的被动节能房屋可以减轻流感的传播，保护了儿童和成人的健康。

建筑由预制木材隔热墙建成，简洁、光滑的外立面可以改善采光，促进被动太阳能加热。此外建筑集成了太阳能热水、太阳能发电和带有热量回收功能的机械通风系统。建筑入口处的触屏为家长提供能源使用的信息，教师也可从中实时收集信息。

在斯堪的纳维亚半岛，Nordic Swan环评认证过去广泛应用于卫生纸、洗手液一类的生活日用品中，现在建筑部件也可获得该项认证。获得Nordic Swan认证的部件应该对环境影响最小，保证室内气候状态。从选料到完工的整个施工进程中，设计者都在考虑如何使用环评认证的建筑部件。

Dragen幼儿园是丹麦最先使用Nordic Swan认证部件进行建设的幼儿园之一，以保证将建筑给环境带来的负担减至最小。该建筑是被动式节能房屋，它的能源消耗和碳排放将大幅度地降低。

通过强化的绝缘材料、高度密闭的结构、井然有序的通风设备和高效率的热循环系统，建筑的能量消耗能够维持在一个很低的水平。室内因人类活动而提高的湿度和温度将由空调系统调节和改善。

房屋建造的全过程也尽可能地应用可持续技术。不选用对周围环境有害的材料，使用预制构件减少装配时间，将建造过程中的能量消耗降到最低。

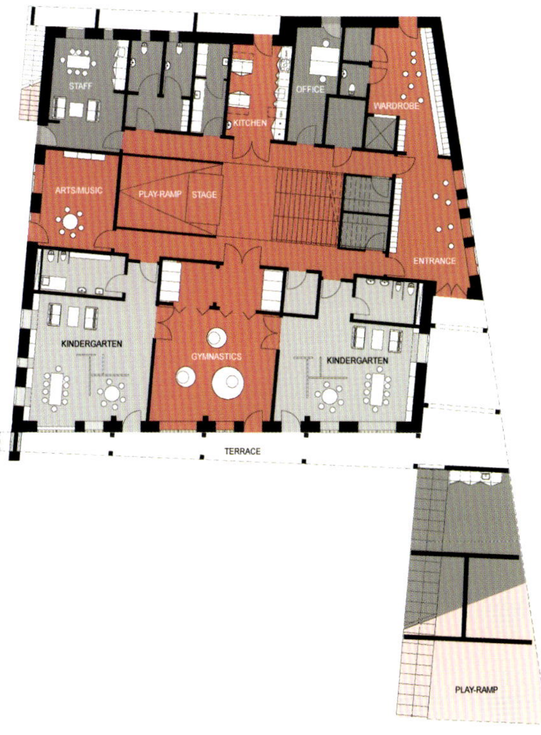

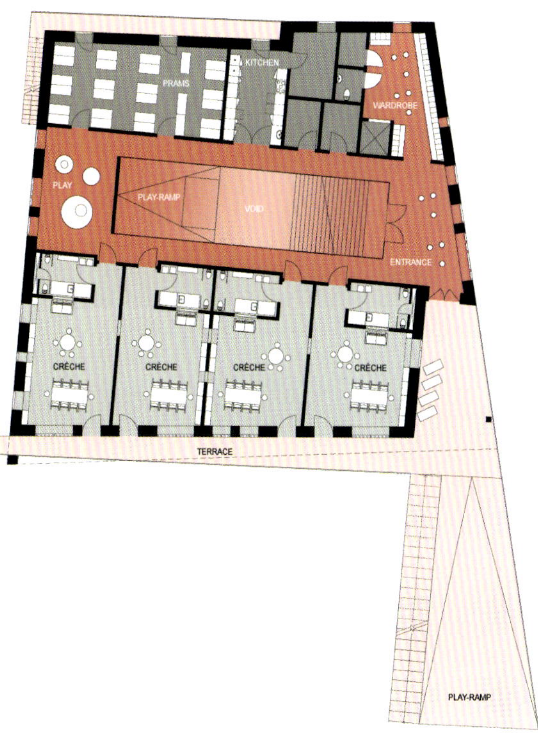

平面图 FLOOR PLANS

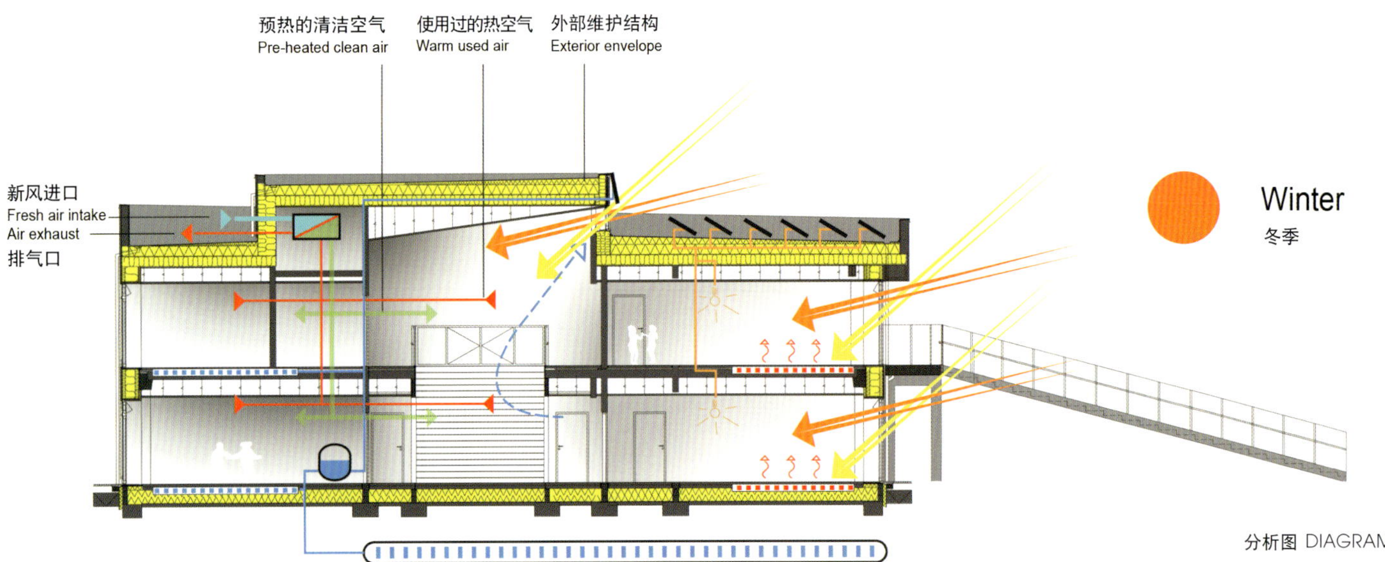

分析图 DIAGRAM

The integrated kindergarten 'Dragen' sets new standards, as a sustainable and pedagogically thought-through design. The components are largely 'Nordic Swan' eco-labelled. The construction is a certified passive-house, using a minimum of energy. And the children's needs and well-being has been the main driver in the design.

The fundamental architectural concept is a simple and clear geometric form on two levels, with the children's areas located in the best-lit southern end. The two levels are linked by staircases and ramps which are designed to stimulate and challenge the children's sensory and motor skills.

"There is far more space available than in traditional kindergartens, and there is a pedagogical idea to the interior design. The entire architecture is supporting the ambitions Odense Municipality has for the children's development - namely that they enjoy attractive and challenging surroundings for learning and growing," says Odense's councilwoman for Children & Young People's Services, Jane Jegind.

As an example, the Children's House Dragen has small niches distributed throughout, where children can play, read or just withdraw. The total area is 414 m^2 of play space for the 88 children (44 in the kindergarten and 44 in the creche), which is far more than the minimum standards of 268 m^2. This will reduce the risk for spreading of illness, and generally make room for more activities.

In addition, there are purpose-built spaces, giving the children special opportunities: There is a small theatre, atelier, motor skills room and pedagogical kitchens indoors and out. Another feature are small 'loopholes' in the walls, allowing kids to play across the room divisions.

剖面图 SECTION

The building respects the environment, energy-savings and not least the health of the children and employees. The highly insulated construction will consume less than 20 percent of the energy used for a standard building.

Passive-houses built of healthy materials have also been proven to reduce the spreading of flues, meaning fewer sick-days for children and adults.

The building is constructed from pre-fabricated wooden insulated wall segments, and generous glazed facades provide daylighting and passive solar heating. In addition the building integrates solar hot water and electricity generation and a mechanical ventilation system with heat recovery. A touch-screen at the entrance informs parents about the current energy-performance, and provides info and updates from the pedagogues.

Scandinavians usually associate the Nordic Swan Eco-label with commodities such as toilet paper and washing-up liquid, but now the label can also be awarded to building components. A component with the Nordic Swan Eco-label is one which has a minimal effect on the environment, and secures a good indoor climate. The use of eco-labelled building components takes into account the environmental factors throughout the entire construction process – right from the raw materials to the finished building.

Dragebakken Kindergarten is one of the first kindergartens in Denmark to be constructed using Nordic Swan Eco-labelled materials, certifying that the building imposes a minimal environmental load. The building is constructed as a passive house, which means that its energy consumption, and thereby its CO_2 footprint, will be drastically reduced.

This low level of energy consumption is achieved through increased insulation, extremely air-tight structures, well-regulated ventilation and highly efficient heat recycling.

In layman's terms, this means that the building is so well-insulated that even the children's activities will cause the interior temperature to rise. Humidity in the rooms will be removed by air conditioning.

The actual construction process is also as sustainable as possible in all phases. Environmentally harmful materials have been rejected, and the energy consumption involved in construction will be minimised by using prefabricated elements with brief assembly times.

MILSTEIN HALL CORNELL UNIVERSITY

康奈尔大学弥尔斯坦大厅

©Philippe

位置：美国纽约伊萨卡
建筑设计：OMA
委托方：美国康奈尔大学，建筑、艺术与规划学院（AAP）

Location: Ithaca, New York, USA
Architect: OMA
Client: Cornell University, College of Architecture, Art and Planning (AAP)

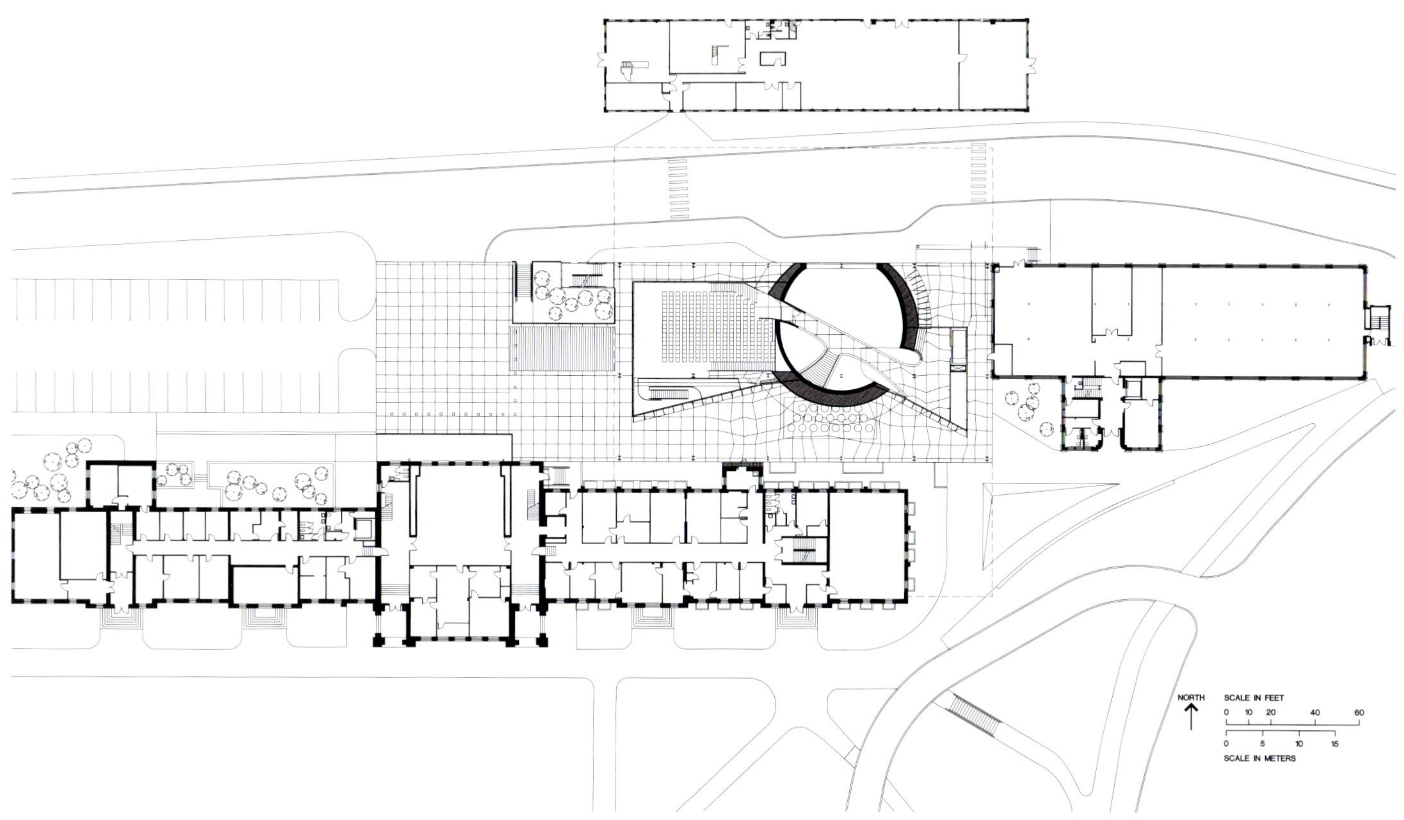

一层平面图 GROUND FLOOR PLAN

©Philippe Ruault

项目概况

弥尔斯坦大厅是康奈尔大学享有盛誉的建筑设计作品、艺术与规划学院（AAP）成立后100多年以来的首座新建筑。弥尔斯坦大厅是对AAP学院楼的扩建，各个楼层室内外空间相互连接构成一个统一的综合体。它为AAP学院提供了4366 m^{2}新增面积，包括工作室、展厅、讨论区和一个253席的礼堂。新增空间使学院设施得以重新规划利用，内部的各功能主体间建立了全新的空间关系。

建筑热环境解决方案

以当地湖水作为冷却水源，一套天花板冷梁系统为工作室提供了舒适的环境，减少了对传统大型HVAC机械系统的依赖，并采用地板式辐射加热。

楼板被提升到APP第二层，与Sibley大厅和Rand大厅相连，并可提供2325 m^{2}的工作室空间，从这里还可以看到周围景观全景。巨大的落地窗加上绿色屋顶上面的41个天窗，这个悬挑建筑向外探出15.24 m，与APP的第三座既有建筑Foundry相连。混合桁架结构让内部空间开阔无阻，激发了互动性并可以在不同时期灵活运用。外露的混合桁架结构造就了大尺度开放空间。天花上的灯光设计采用冷光束，经过精心设计，同时配置定制的高效日光感应器，以保证人造光与日光共同作用达到恒定的光水平。所有的玻璃外幕墙都采用高性能的LOW-E隔热玻璃单元。高效的机械系统和充足的自然采光在这里得到体现。建筑有望获得LEED银质认证，甚至是金质认证。

在工作室东南方，位于建筑上部的悬臂式区域是一个独特的空间，因为位于Arts方庭的行人会最先看到建筑的这一部分；从East Avenue大道眺望，这里也完全呈现于视野中。因此需要特别的解决方案来处理东部和南部的日照方向。一道特质的帘幕设置于此，遮挡来自Arts方庭的视线，保持自然光照，防止过度辐射，防止眩光，使之不对方庭东北入口的行人的视线产生刺激。

2418 m^{2}的屋顶绿化采用景天科植物。通透的外幕墙让工作室内部情形在外面能看得一清二楚。外立面那些像条形码般的垂直线状的石材纹理让建筑的浮动感更加强烈。

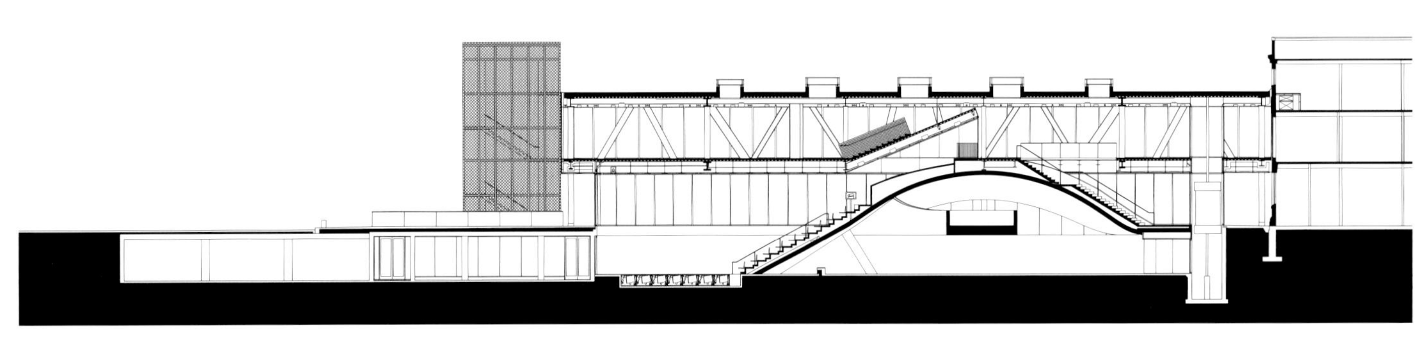

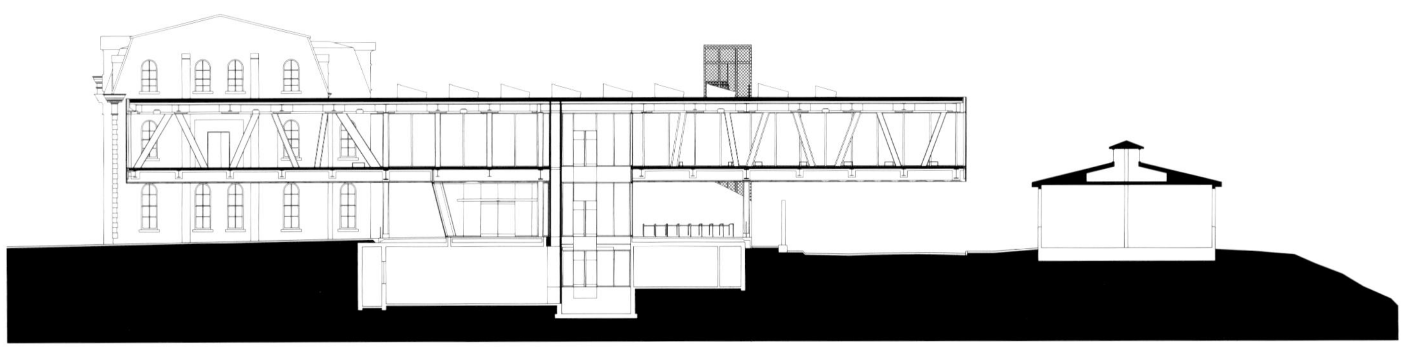

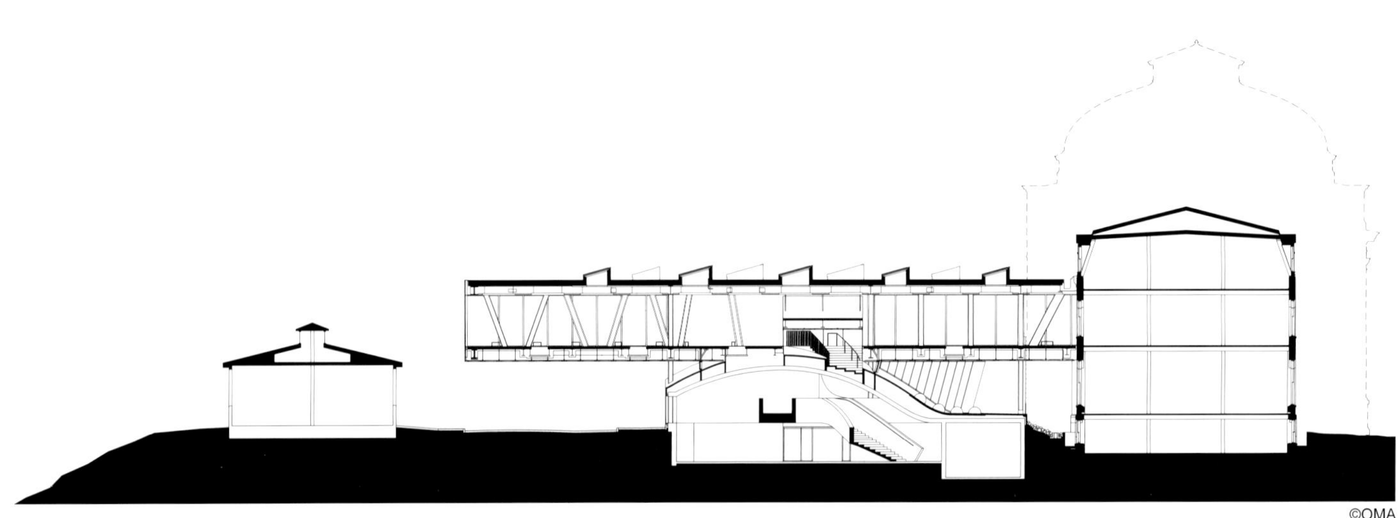

剖面图 SECTIONS

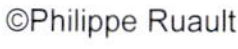

©Philippe Ruault

©Philippe Ruault

©Philippe Ruault

©Philippe Ruault

©Philippe Ruault

Program Description

Milstein Hall is the first new building in over 100 years for the renowned College of Architecture, Art and Planning (AAP) at Cornell University. Milstein Hall is an addition to the AAP buildings creating a unified complex with continuous levels of indoor and outdoor interconnected spaces. Milstein Hall provides 4366 additional square meter for the AAP, adding much-needed space for studios, gallery space, critique space and a 253-seat auditorium The additional space enabled a new master plan of the College's facilities creating extraordinary new spatial relationships between internal programmatic elements.

Building Thermal Environmental Engineering Solutions

The studio comfort environment is maintained by the ceiling's chilled beams that provide cooling by utilizing local lake source chilled water, reducing the need for large traditional HVAC mechanical systems. The heating is distributed through the concrete radiant heated slab.

A large horizontal plate is lifted off the ground and connected to the second levels of the AAP's Sibley Hall and Rand Hall to provide 2325 m^2 of studio space with panoramic views of the surrounding environment Enclosed by floor-to-ceiling glass and a green roof with 41 skylights this "upper plate" cantilevers almost 15.24 m over University Avenue to establish a relationship with the Foundry, a third existing AAP facility The wide-open expanse of the plate – structurally supported by a hybrid truss system - stimulates interaction and allows flexible use over time.

The exposed hybrid trusses were designed to balance structural efficiency at the cantilevers and maintain open circulation within the large open plan A field of custom designed lights and chilled beams were carefully coordinated with the structural and mechanical systems using normally hidden functional elements to define the ceiling plane The lighting is programmed by a highly customizable and efficient Lutron control system connected to daylight sensors to maintain constant light levels that balance the daylight with artificial light. The efficient mechanical systems and abundance of natural daylight are possible through the use of high performance insulated glass units with Low-E coating on all the exterior glass walls. The building is expected to receive a Silver LEED certification with the possibility of achieving Gold

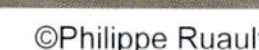

©Philippe Ruault

The south-east cantilevered area of the studios is considered a unique space within the upper plate as it is most visible from the pedestrian walkways to and from the Arts Quad beneath as well as thetransparency seen from East Avenue. Given the east and south exposure a specific solution to moderate the daylight was required. A custom curtain was designed for this prominent corner of the building. The goal was to preserve views out from the studios towards the Arts Cluad, maintain natural daylight without the glare or radiation over-gain and present a striking image at this northeast entry to the Quad.

The 2418 m^2 roof is a sedum covered green roof punctuated by a cluster of northern facing skylights which gradually increase in size towards the darker center of the plate further from the exterior facade.Two different types of sedum create a gradient pattern of dots that transition from articulated small circles near the man-made Arts Quad on the south to a dense, larger pattern of dots towards the natural landscape of the gorge on the north.

©Philippe Ruault

BOSTON UNIVERSITY STUDENT HOUSING

波士顿大学学生公寓

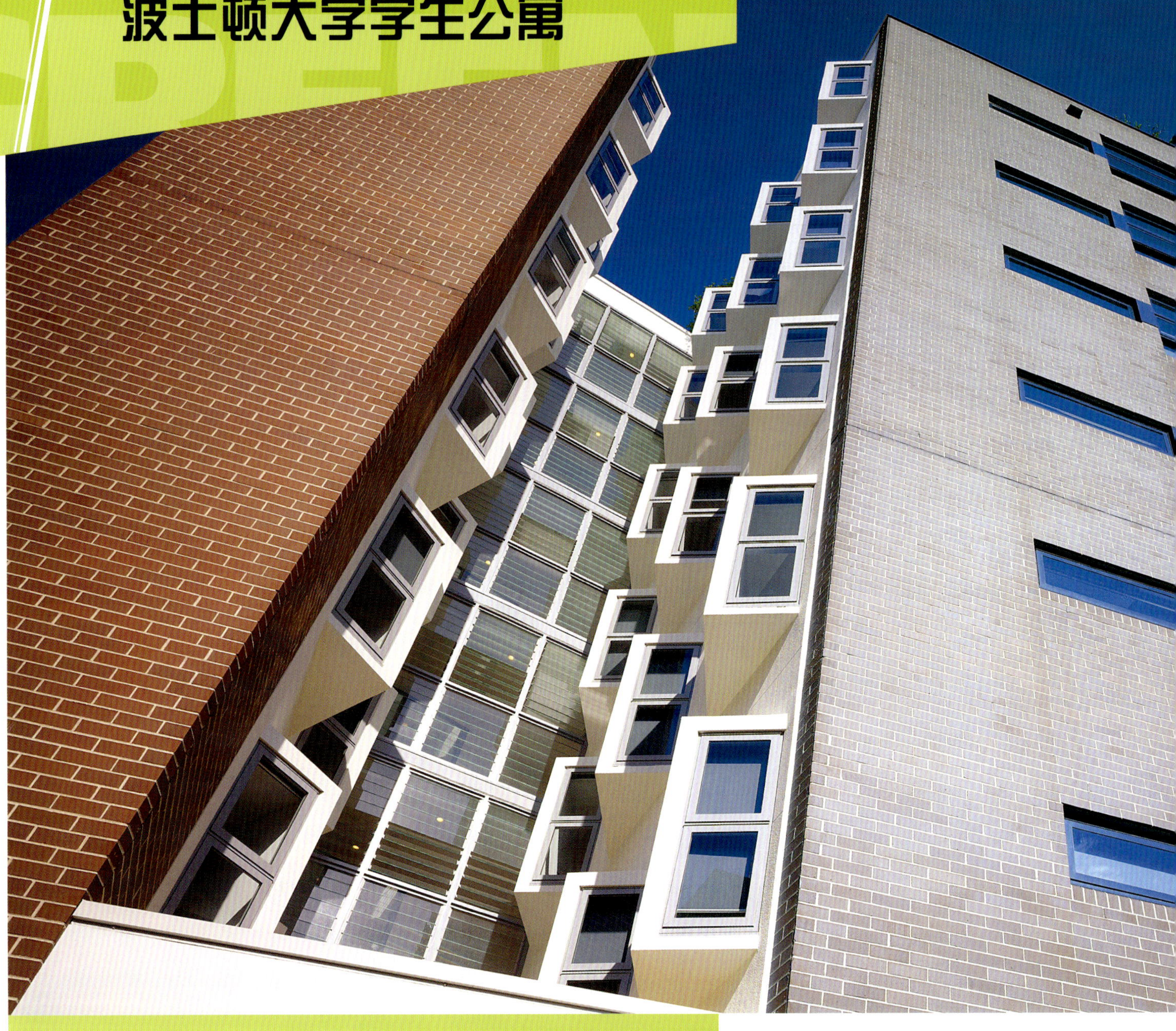

地点：澳大利亚悉尼Regent大街15-25
建筑设计：Tony Owen Partners/Silvester Fuller

Location: 15-25 Regent Street, Sydney
Architect: Tony Owen Partners/Silvester Fuller

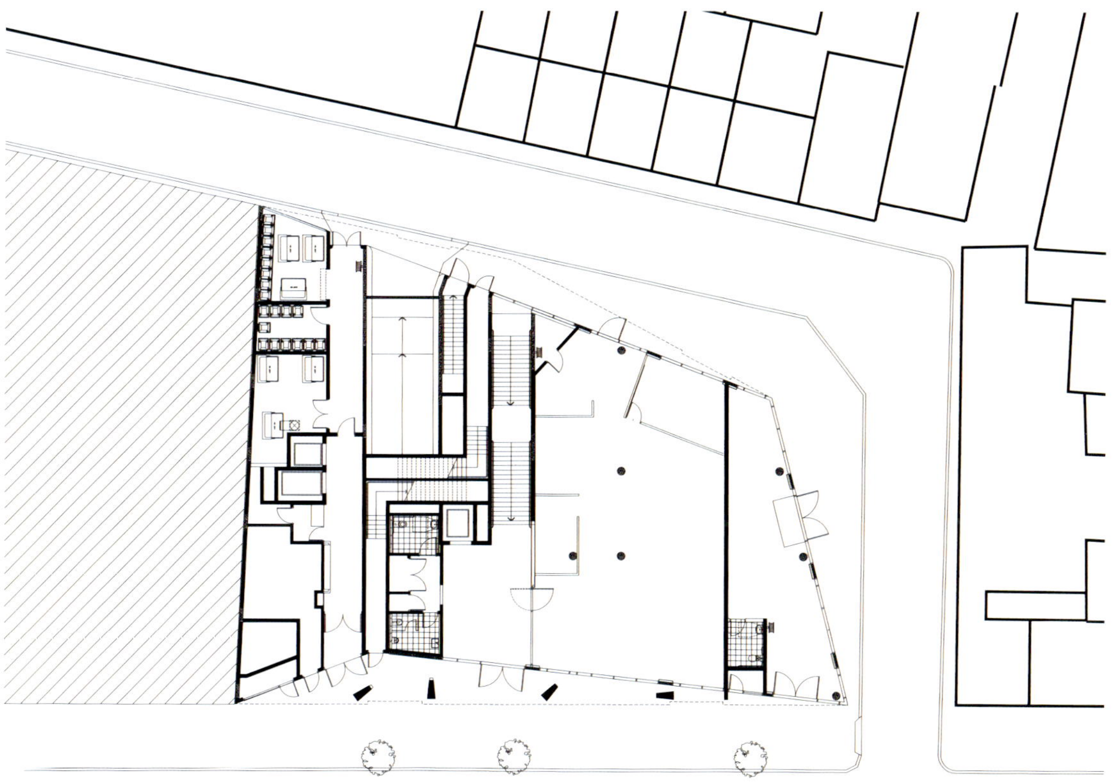

一层平面图 GROUND FLOOR PLAN

Program Description

The new student quarters for Boston University by Tony Owen Partners and Silvester Fuller Architects at 15-25 Regent Street, The eight-level, environmentally-efficient building has been designed to house overseas students visiting form Boston University for a semester in Sydney. It can contains 164 beds as well as a care-taker suite. It also has three lecture halls, a library, an Internet lounge, a rooftop terrace with a timber deck and an adjoining fully-equipped communal kitchen, plus cafe. The unique module houses 4 single bedrooms for every shared lounge and bathrooms.

Building Thermal Environmental Engineering Solutions

Chippendale is a unique design using fissures to provide maximum solar access to bedrooms as well as natural ventilation throughout the building. The design uses large canyon-like slots in the facade which allow sunlight and ventilation to penetrate deep into the building and into each room. The windows in these slots have a rhomboid shape to maximize efficiency, and deliver a bold architectural facade which is illuminated at night through an ever-changing light show. The windows are oriented to trap sunlight whilst ensuring privacy between rooms.

The end walls of the slots are made from glass louvers that are seven storeys high, and the building also contains a seven-storey glass louvered atrium. "Air is drawn through the canyons and passes through the building like gills. It is drawn up through the central void and out the top to ventilate corridor areas, thus allowing the building to breathe naturally. East-facing operable louvers on each level further help to lower ambient temperatures by drawing in fresh breezes. The design allows more light and ventilation into each bedroom, provides good views, and is a sensible model for residential density in the CBD.

剖面图 SECTION

标准层平面图 TYPICAL FLOOR PLAN

VITUS BERING INNOVATION PARK, HORSENS

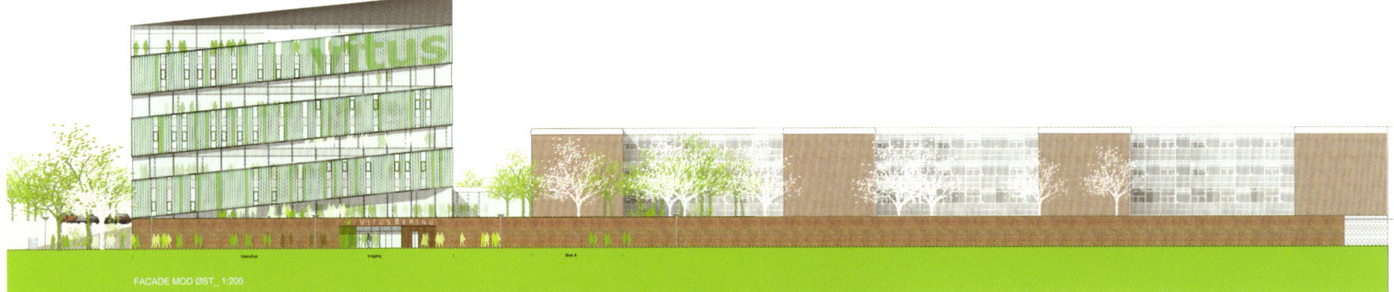

东立面图 EAST ELEVATION

地点：丹麦豪森斯
规模：8000 m²
建筑设计：C. F. Møller Architects
景观设计：C. F. Møller Architects
委托方：University College Vitus Bering Denmark

Location: Horsens, Denmark
Size: 8000 m²
Architect: C. F. Møller Architects
Landscape: C. F. Møller Architects
Client: University College Vitus Bering Denmark

World Green Buildings——Thermal Environmental Engineering Solutions

南立面图 SOUTH ELEVATION

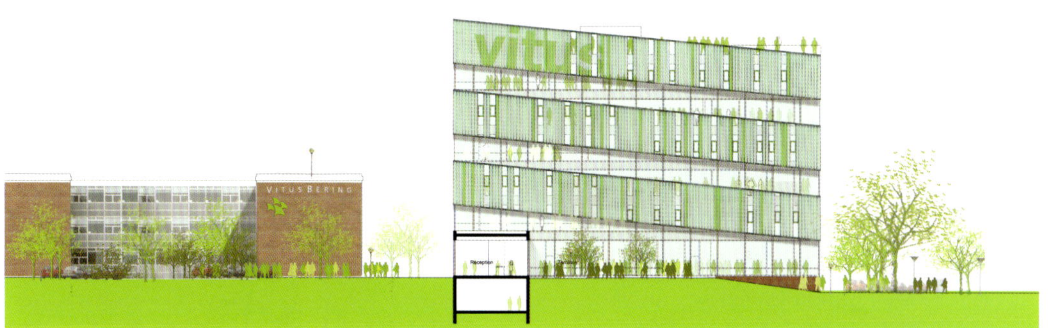

西立面图 WEST ELEVATION

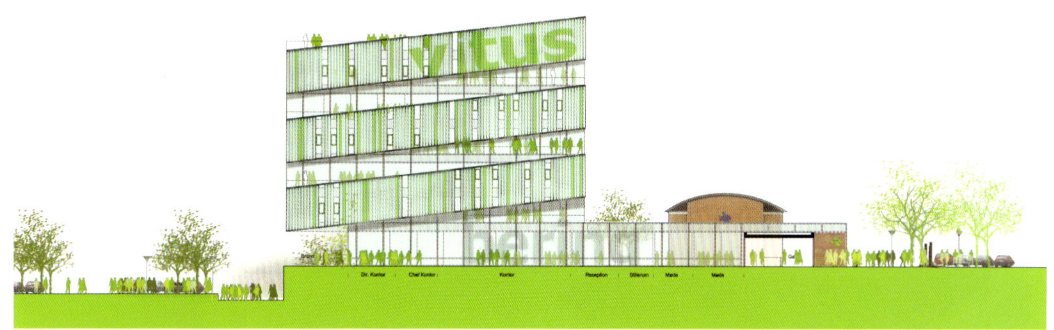

北立面图 NORTH ELEVATION

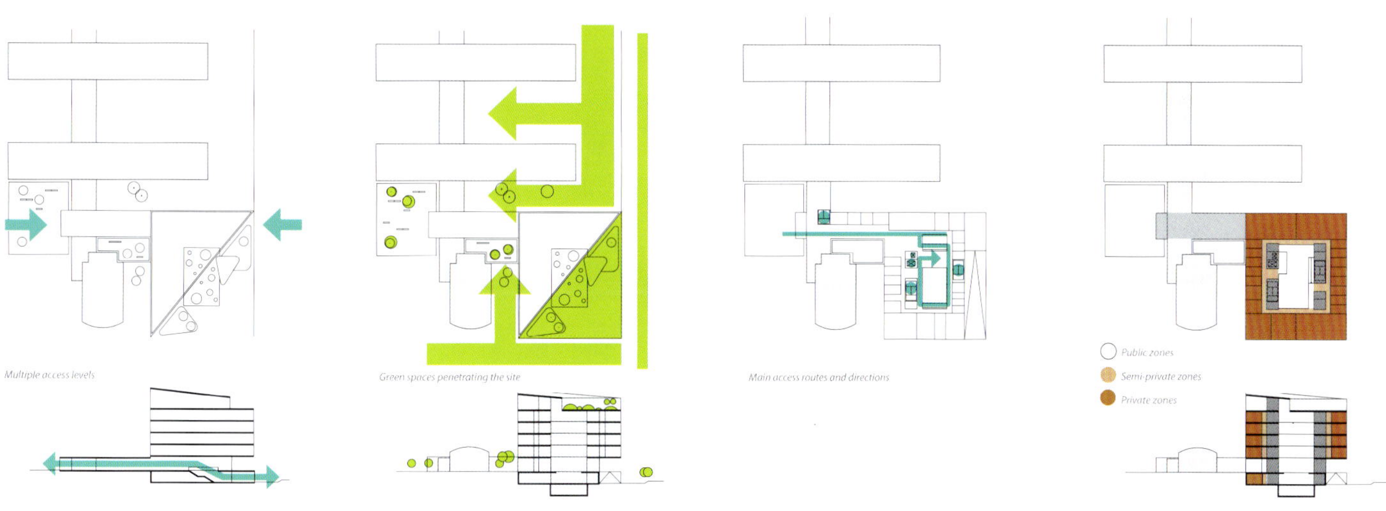

分析图 DIAGRAM

Program Description

Teaching and entrepreneur start-up office facilities side by side – that's the philosophy behind the distinctive extension to the existing 1970's structure of the University College Vitus Bering Denmark in Horsens. The new Innovation building is designed to sit on a brick base, which is a direct continuation of the existing complex architecture, but from there on it is distinctly different and unique. The building's dynamic and innovative character is expressed via its spiral shape. On the facades, the movement is seen in the glazing strips that stretch towards the sky across the six storeys of the building and create the impression of a spiral sequence, while internally it is expressed via the main staircase in green fibre cement, which runs in a spiral form between the storeys in the unifying internal atrium. The inclined forms of the building also have the practical advantage of allowing a necessary fire escape route to be cut through the building.

Building Thermal Environmental Engineering Solutions

The Vitus Bering Innovation Park is one of the first office complexes in Denmark to be classified as low-energy class 1, which means that its energy efficiency is twice that of the minimum required by the Danish building regulations. The low level of energy consumption is achieved through such factors as highly insulating windows and extra insulation on all of the building's external surfaces. Another feature is the building's intelligent air conditioning system, which adjusts itself according to the number of people present in each individual room.

The basic floor plan of the building is a simple and flexible layout, to allow the integration of numerous uses and adaptations. The large and dynamic green stairway element leads to common meeting facilities and a roof terrace with a beautiful view of the Horsens Fjord. The stairs land in a different position on each level, thus activating the entire atrium as the central hub of the building. There are cool air intakes on ground floor, and warm exhaust air rises up in the central open space and flows out via the natural ventilated atrium. The isolated smaller spaces used as offices, meeting rooms etc, are individually ventilated spaces with heat recovery systems. The atrium is covered by a dynamic, diagonally split roof-plane with circular skylights, of which one half forms the common roof terrace.

K
Low-energy clear glazing
Wooden
window sill
Cable channel
Interior drywall
U-glass section
Pre-fab Concrete slab
Aluminium profile
Pre-fab Concrete beam
Emalite glazing
8
Steel section
Aluminium flashing
Glass balustrade
Concrete paver
Roof build-up
Aluminium flashing
Aluminium profile
Pre-fab Concrete slab
U-glass section
Pre-fab Concrete beam
Emalite glazing

8
U-glass section
Emalite glazing
Emalite glazing
K
Interior drywall
Steel section
U-glass section
Aluminium
flashing
Emalite glazing
Steel section
Low-energy clear glazing

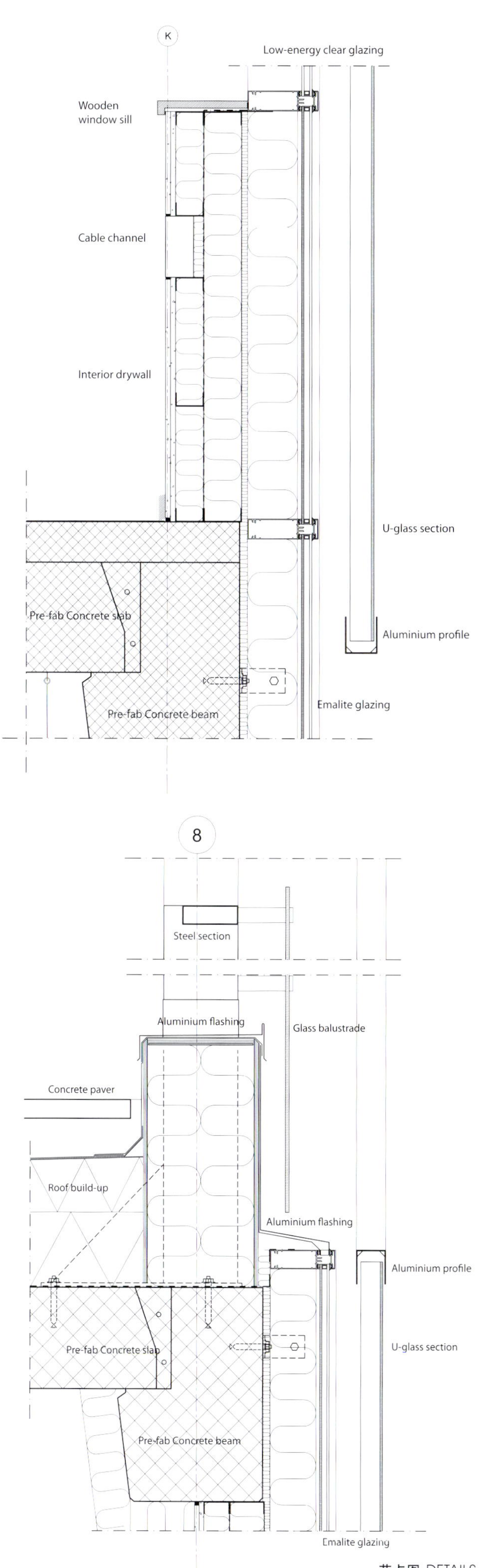

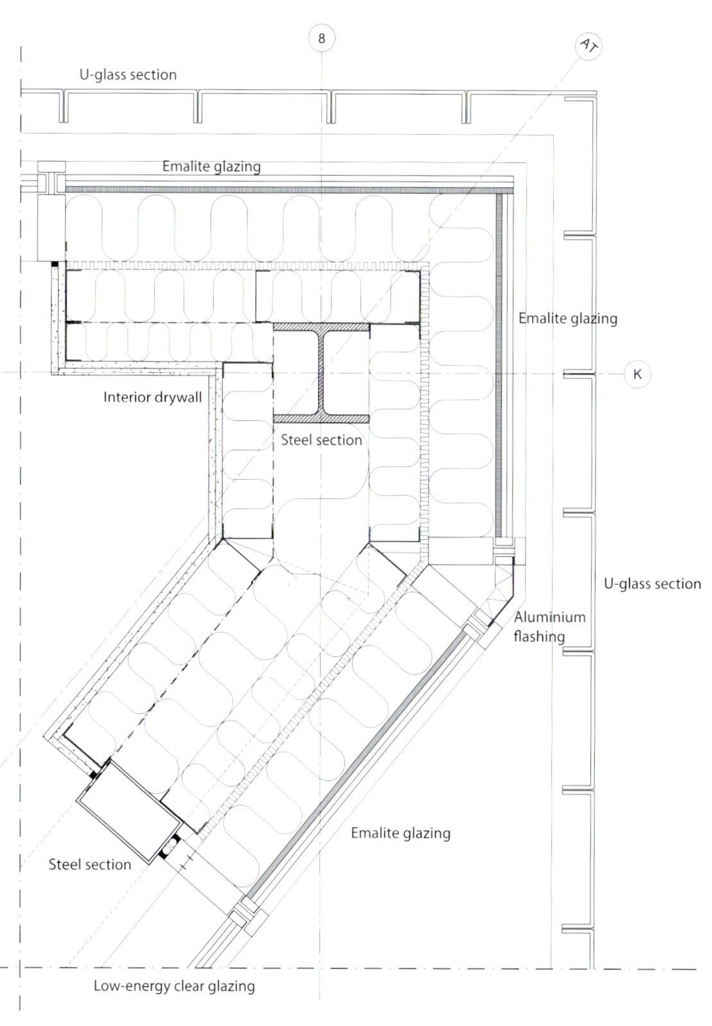

节点图 DETAILS

HONG KONG COMMUNITY COLLEGE (WEST KOWLOON CAMPUS), THE HONG KONG POLYTECHNIC UNIVERSITY

香港理工大学专上学院（西九龙校区）

地点：中国香港
建筑面积：31 600 m²
建筑设计：AD+RG建筑设计及研究所国际有限公司（与创智建筑师有限公司合作）
所获奖项：2010年绿色建筑奖——新建建筑类优异奖；
2009年“《世界建筑新闻》评选六大国际水平项目”教育类奖
委托方：香港理工大学

Location: Hong Kong, China
Building Area: 31,600 m²
Architects: AD+RG Architecture Design and Research Group Ltd. in collaboration with AGC Design Ltd
Selected Awards: "Green Building Award 2010 – New Buildings Category"– Merit Award; "Top-Six World-Class International Projects in World Architecture News"Education Award 2009
Client: The Hong Kong Polytechnic University

Tradition
Transformation
Re-interpretation

floor plate too large hinders the day lighting penetration and air ventilation

better cross ventilation and more efficient floor area

better linkage for communal space

a re-interpretation of high rise campus in the newly-developed West Kowloon of Hong Kong

项目概况

香港专上学院（西九龙校区）是香港理工大学专为工商副学士课程而设的学院。校园环抱维多利亚海港景色，为西九龙区的地标建筑。校园善用空间布局，以高层建筑作垂直发展，于课室、饭堂及图书馆等公用设施附近设置平台花园，为专上教育院校带来全新面貌。配合教学及学生发展的配套设施完备，设有图书馆、多用途礼堂、多媒体教室、马蹄形课室、语言及信息科技学习室，以及五十多个教室等，促进学生的全面发展。

大楼采用了预制式结构组件，这样既可以减少建造期间产生的废料，亦能加强建筑材料监控，加快工程进度，学院的建筑设计尊重自然环境，采用环保物料，符合可持续发展的建筑理念。

建筑热环境解决方案

1.被动措施

（1）空气流通、日光与建筑形态

双塔式设计增加了建筑物的边界范围。三层平台使室内空间摄取足够天然光线，加强对流通风。微风、日光与绿叶和谐地融合在一起，削减了户外户内之别。双塔作为密度低的建筑群，对附近环境的影响亦不大。空间上的留白亦能改善该选址的微气候。

（2）高层绿化

本项目亦表现了高层绿化建筑的双重好处。绿化建筑不但加强了楼宇的室内控温表现及室外空间的舒适程度，而且在都市环境里形成了一个绿化视觉效果。此外，本项目以有效的方法，结合创新意念与环保目标，成功发挥建筑设计，效果良好。作为一个公共建筑，空中花园能为邻近小区提供绿化空间。

2.主动措施

冷空气导流系统。

有空气调节的地方设有连消声器的空气导流管，空气经过走廊后，用过的空气会被送到设有抽气系统的洗手间。由演讲厅抽出的空气会被送到不同层数的制冷机组。这样安排可以尽量利用经过处理的空气，又可以减少建筑物的碳足迹。

其他设计措施

1.可持续性设计

项目在内部走廊设置活动窗，引入阳光减少耗电，并使用太阳能发热系统协助节能。同时加强天然能源及耗能控制，课室和个别办公室设有自动动作传感器控制照明和风机。T5光管及电子火牛含有较少的水银、玻璃及包装物料，更为环保。

2.设计期间对构造物料、结构及用途的考虑

本设计采用了灵活的“模块式空间布置”，加强未来的转变及可塑性、灵活性。预制的钢筋混凝土建筑可以符合环保及建筑管理上的期望及要求。机电设备妥善地配合结构设计，机电的主回路沿走廊铺设，然后分支到各动能室，这种安排可灵活配合将来改动的需要。所有设计均从实用角度深入研究，例如媒介和材料的选择、绿化的植物品种及配套设施、实际维修和运作的问题。

3.水循环

设有一个非化学的水处理装置，负责处理由风机（Fan Coil Units）与空气处理器凝固下来的水分，而在地下低层的储水槽（bleed off tank）会收集冷却塔滴下来的水分。这些水分会被再用，或传送至冲水槽作冲厕用。

4.健康与安全：可持续免疫建筑

新的建筑理念包括建构一个安全、健康、舒适及节能的环境，规划完整的、病理学上有根据及有系统的健康管理建筑。

5.采用预制件建筑方法

减少了工地的污染工序、施工废料的产生。同时，减少了施工成本。

9/F Sky Deck Garden
Layout Plan
1. Teaching West Wing
2. Sky-deck Garden
3. Teaching North Wing

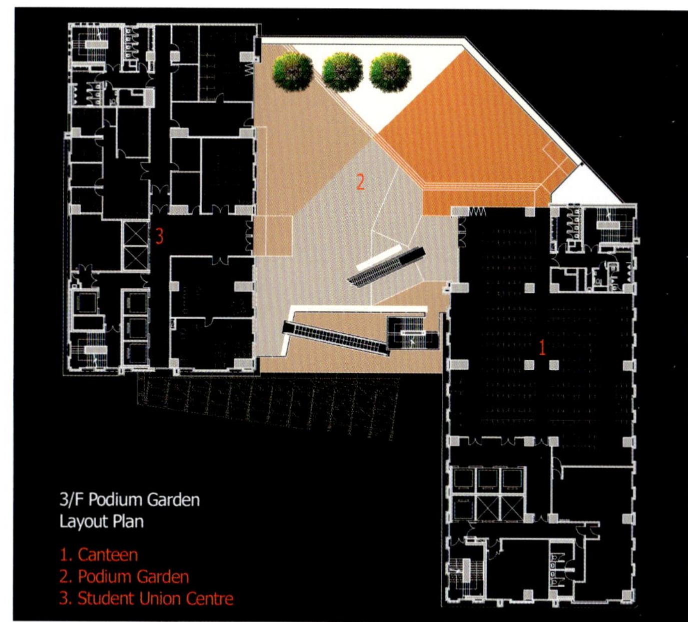

3/F Podium Garden
Layout Plan
1. Canteen
2. Podium Garden
3. Student Union Centre

平面图 FLOOR PLANS

Longitudinal Section
1. Multi-purpose Hall
2. 3/F Podium Garden
3. 9/F Sky Deck Garden
4. 14/F Sky Deck Garden

剖面图 SECTION

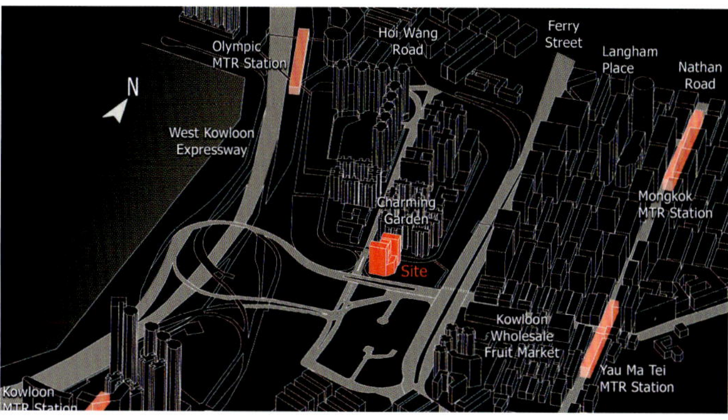

Program Description

The Hong Kong Community College (West Kowloon Campus) provides vertical campus life for pre-tertiary students. It is set with an orientation that maximize the sea view of the Victoria Harbor. It becomes the landmark of West Kowloon. The spatial organizations match with the other campus of The Hong Kong Community College – the Hong Kong Polytechnic University. With the concept of "Vertical Campus", sky decks forming sky gardens are surrounded by various function rooms like classrooms, canteen and library. Students enjoy great accessibility in this interactive learning environment.

Pre-cast beam and semi-precast slab systems are adopted for the project which helps to minimize waste during construction and facilitate the construction progress. The building design embraces natural environment and use of environmentally friendly materials.

Building Thermal Environmental Engineering Solutions

1. Passive Measures

(1)Air Ventilation, Day-lighting and Building Form

The design of a twin tower increases the periphery of the building block. The 3-level podium allows unobstructed views of nearby developments and enhances the natural lighting and ventilation of internal space. The wind, sunlight and greenery integrate harmoniously with the inside with no boundary between indoors and outdoors. The linked towers also minimize their impact to the surrounding as being a less dense massing. Communal voids are designed to improve the site's micro climate.

(2)High Rise Green Building

The campus presents the dual advantages of a high rise green building. A Green Building can enhance indoor temperature control and the level of comfort in outdoor spaces. It also provides a green visual impact in an urban environment. The project combines creative ideas and environmental objectives with effective methodology. As a public architecture, the sky garden can provide green space for the nearby communities.

2. Active Measures

Air Transfer. Transfer air duct with silencer is provided for air-conditioned areas for transferring air to the corridors, from which the 'used' air is transferred to the toilets where exhaust system is installed. Exhaust air from Lecture Theatres is transferred to Chiller Plant at different levels. This can fully utilize the treated air before exhaust and reduce the carbon footprint for the building operation.

Other Design Features

1. Sustainability

The provisions of open able windows are maximized to the internal corridors in order to take advantage of natural light and reduce electricity consumption. Solar heat water system was adopted for better energy conservation. Meanwhile occupancy motion sensors for lighting and FCU control in classroom and individual staff offices are provided. T5 Tubes and Electronic Ballasts are used, which are more environmental friendly for having relatively smaller amount of mercury used, less glass is used in the slimmer tube, less package material.

2. Consideration of fabric, structures and services at design stage

A flexibility 'modular system of spatial combination' capable of future transformation has been developed to facilitate the need of adaptability / flexibility. Pre-cast R.C. construction can be adopted for better environmental and construction management consideration. Building service is well integrated with the structural system to cater for future alterations. The main service trunks run along the corridor, with branch pipes branch off into function room. All treatments and planning of the building involve in-depth research on the practical use viewpoints, medium and materials selection, plant species for greening and its supporting devices, actual maintenance and operation issues.

3. Water Recycle

A non-chemical water treatment plant for recycling condensation water from fan coil units and air handling units is provided and a bleed off tank located at LG/F collects bleed off water from cooling towers. The treated condensation water and the bleed off water will be re-used or transferred to flushing water tank for flushing purpose.

4. Health and Safety: Sustainable Immunized Building (SIB)

New principles are introduced for strategic building modeling and its system on the delivery of a safe, healthy and comfortable environment having the effective use of energy. The principles formulate various holistic, pathological and system indices for total approach of health management in a building for people, space and system.

5. Pre-cast Construction are adopted

The pollution and nuisance caused to the surrounding environment is reduced.

MDIS RESIDENCES

新加坡管理发展学院学生公寓

建筑设计：Ong&Ong Pte Ltd.
室内设计：Ong&Ong Pte Ltd.
景观设计：Ong&Ong Pte Ltd.
委托方：新加坡管理发展学院

Architect: Ong&Ong Pte Ltd.
Interior: Ong&Ong Pte Ltd.
Landscape: Ong&Ong Pte Ltd .
Client: Management Development Institute of Singapore

项目概况

新加坡管理发展学院（MDIS）要为学生创造一个互动的学习环境，以适应高等教育的需求和急速变化的世界。MDIS住宅楼位于Stirling大街，是MDIS园区里开发的新的管理和公寓楼群。发展区域里的空间是为了把室内的学习空间带到室外，然后让新的大楼融入到现有的园区中。

项目的主要构成：796个学生宿舍，1642个床位，四种不同的房间布局；自然通风的天井和食堂；为学生锻炼建设的五星级公寓大厅；有500个座位的演讲大厅；景观交流空间，屋顶花园。

建筑热环境解决方案

大楼坐北朝南，极大地利用自然通风。四层高的天井大厅两侧设置了开放交流空间，以实现完全的对流通风。需要空气调节设备的空间比如办公室、教室、演讲厅都从大楼外轮廓凹进，以便使大楼在直射阳光下的暴露最小化，同时减少热量吸收。

景观露台作为立面的延伸在为周围环境降温的同时也成为学生和工人的绿色休息区。宿舍房间在两个分开的翼楼之下，在太阳照射下彼此遮阳。所有的酒店房间和办公室都用了低辐射双层玻璃窗户，这样就允许了阳光的大量传输，进而也避免了过多的热吸收，同时将大楼的ETTV值降至39.7 W/m^2，不住人的空间比如楼梯间，电梯间被有策略地放在了东面和西面，同时它们作为缓冲空间也减少了房间里的热吸收。这些外墙都被铝板覆盖。铝板是100%可回收的，而且对太阳辐射得到的能量可以高反射、低辐射，进而可以作为大楼的热辐射遮阳系统。环保的干壁分区，低挥发性涂料在内外墙的广泛应用及无溶剂的聚氨基甲酸乙酯和环氧树脂的走廊地板都提高了室内的空气质量。

有效的空气调节设施。大楼安装了高效的空气调节设施。796个宿舍房间采用由面状的风机盘管组成的冷水墙装置，代替了管道式风机盘管。这样保证了空气调节设备的低能耗。

可再生能源。每天学生需要大概64 000 L的热水供应，这些都由服务于宿舍厕所的总计182个淋浴点提供。一个包含真空太阳能收集管和热泵的混合系统将用于生产需要的热水。屋顶架子作为建筑屋顶设计特色的一部分，被一个130 m^2的真空太阳能收集管覆盖，每天生产411 kWh热量的热水。一个储藏生成热量的中央热水箱被放在了屋顶上。

第二套供给热水系统来自放在屋顶的四组热泵，通过吸收环境中的热空气来对水加热。推荐的混合系统充分利用了新加坡丰富的太阳能资源和高温气候。热泵作为真空太阳能收集管的第二套供给系统，保证了平时，甚至在雾天和夜晚，产生热水时消耗更低的能量。每年预计节约大约612 000 kWh的电能。

其他可持续设计措施

1. 有效的采光系统

所有的公共厕所的机械通风扇和采光设施都是交错布置的，同时它们被运动传感器控制，这些传感器采用微波技术有策略地布置在公厕里。停车和办公区域安装了T8的灯光装置，这些装置采用的是高频率电子镇流器。

2. 雨水收集器

另外一个绿色建筑特征是为景观灌溉而使用的雨水收集器。雨水在帐篷状的第五层的景观露天平台和屋顶处被收集，然后储藏在水箱里。被收集的雨水将会用于灌溉第一到第五层的景观植物。

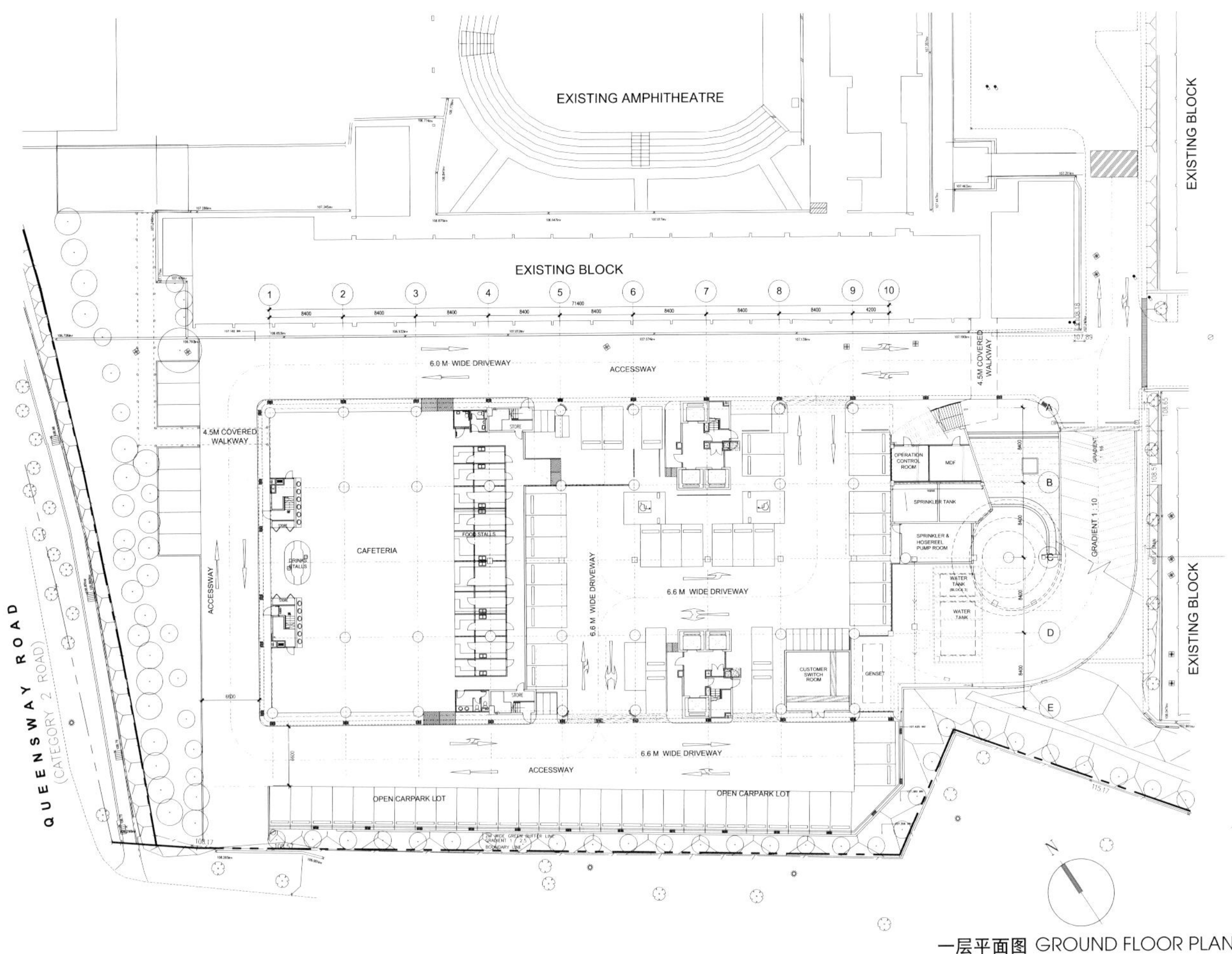

一层平面图 GROUND FLOOR PLAN

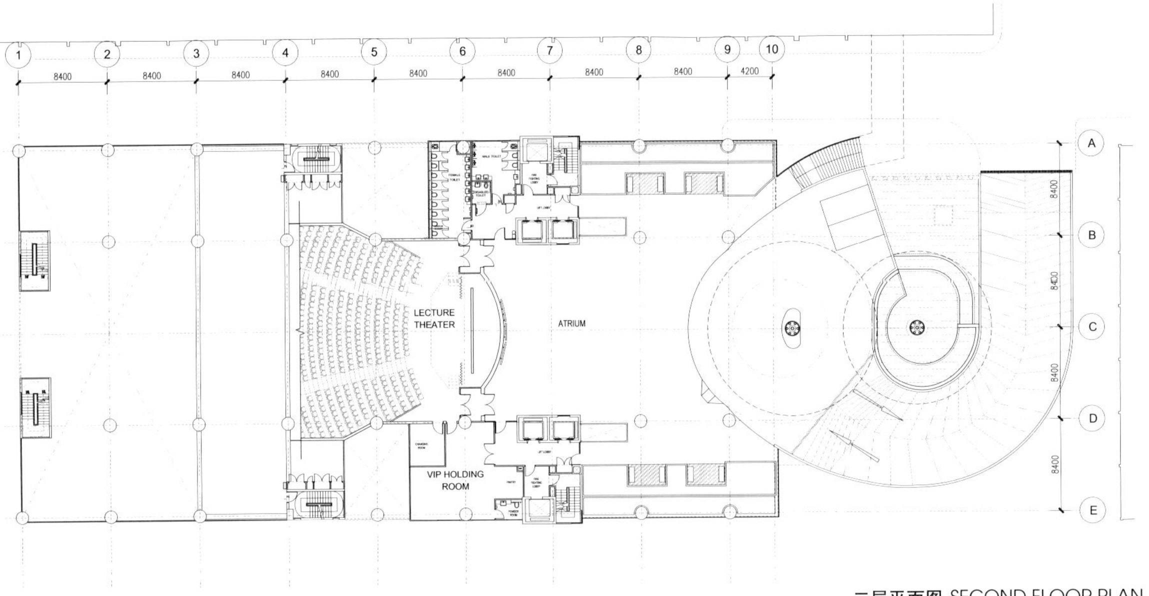

二层平面图 SECOND FLOOR PLAN

Program Description

The brief from Management Development Institute of Singapore (MDIS) was to create an interactive learning environment to suit the trend of tertiary education and fast-changing world which would facilitate and enhance the attitude of continual learning beyond classrooms; as well as maintenance of community-orientated and team-based working spirit in the younger generation. MDIS Residences @ Stirling, the new administration and hostel block developed within MDIS campus is designed for flexibility to suit the changing trade and technology of education. The architecture should be sustainable so as to achieve an environmental-friendly and energy-efficient building and certainly, the green way is the way forward. Spaces within the development were designed to bring indoor learning outdoor and to integrate the new building with existing campus.

Key components of the development: accommodation for 796 students, 1642 beds; 4 different types of room configurations; Natural ventilated atrium lobby and canteen; 5-stars hotel lobby for student training; auditorium hall for 500 seats; Landscaped communal spaces and roof top sky gardens.

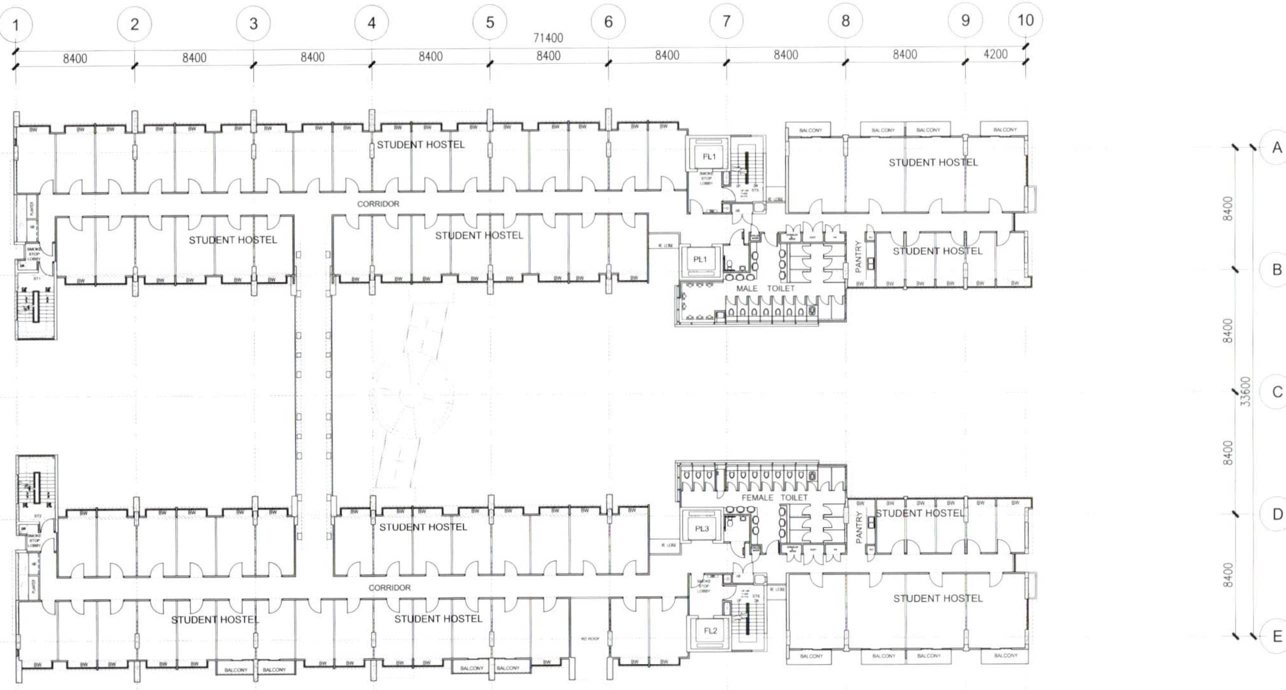

标准层平面图 TYPICAL FLOOR PLAN

Building Thermal Environmental Engineering Solutions

The development is designed to create a sustainable learning and living environment for students. The building is orientated North-South and to capitalize on natural ventilation. The 4-storey high atrium lobby is designed with penetration of open communal spaces on both sides for thorough cross ventilation. Air-conditioned spaces such as office, classrooms, auditorium are recessed from the building outline in order to minimize expose to direct sunlight and reduce heat gain.

The elongated facade lined with landscape terraces cools the ambient atmosphere and also serves as green relief for students and workers. The hostel rooms are housed under 2 separated wings shading each other from the sun path. Low emissivity double glazed windows are used at all hostel rooms and offices to allow high transmission of light without excessive heat gain and lowering the building's ETTV to 39.7 W/m². Non-habitable spaces such as staircases, M&E shafts were positioned strategically at the East and West facing facade, also acts as buffer space to reduce heat gain into the rooms. These external walls are clad with aluminum panels. The aluminum panels are 100% recyclable and has a high reflectivity of radiant energy with low emissivity; further acts as radiant thermal shield for the building. Eco-friendly dry wall partitions, extensive uses of low VOC paint for all internal and external walls; solvent free polyurethane epoxy flooring for corridors improve the indoor air-quality of the development. Another green architectural feature is the use of rain harvester for landscape irrigation purpose. Rain water is harvested at the canopy features located drop off, at 5th storey landscape deck and top roof garden and stored at the water tank. All the landscaped areas are zoned with drip irrigation system. Right Water collected from roof top, 5th storey landscape deck and rainwater harvester is store in water tank

located at 5th and 1st storey. Rainwater collected will be used to irrigate landscape area from level 1 to 5. Bottom Hybrid solar vacuum tube collectors and heat pumps system.

Efficient air conditioning equipment. Highly efficient air conditioning equipment is installed for the building. The 796 hostel rooms are served by chilled water wall-mounted fan coil units, instead of ducted fan coil units, which ensured low energy consumption of the air distribution equipment.

Renewable energy. A total of 182 shower points serve the hostel toilets with an estimated 64,000litres of hot water required daily for the students. A hybrid system consisting of vacuum tubes solar collector and heat pumps are proposed to be used to generate the hot water required. The roof trellis, which is part of the architectural feature of the building's roof design, is covered with 130 m² of vacuum tubes solar collectors, generating 411 kwh/day of hot water. A centralized hot water tank to store the hot water generated is located on the roof.

A secondary source of generating hot water supply comes from 4 sets of heat pumps placed on the roof to draw the hot ambient air to heat up the water. The proposed hybrid system takes advantage of and fully utilizes Singapore's abundant solar energy and high ambient temperature. The heat pumps are installed as a secondary source to the vacuum tubes solar collectors to ensure much lower energy consumption for generating hot water even on a cloudy day and during the night. An estimated savings of 612,000 kwh/year of electrical energy is expected.

Other Sustainable Design Measures

1. Efficient Lighting System

All communal toilets' mechanical ventilation fans and lighting are interlocked and are controlled by motion sensors using microwave technology strategically placed in the toilets. The carpark and office areas are installed with T8 light fittings using high frequency electronic ballasts.

2. Rain Harvester

Another green architectural feature is the use of rain harvester for landscape irrigation purpose. Rain water is harvested at the canopy features located drop off, at 5th storey landscape deck and top roof garden and stored at the water tank. All the landscaped areas are zoned with drip irrigation system. Right Water collected from roof top, 5th storey landscape deck and rainwater harvester is store in water tank located at 5th and 1st storey. Rainwater collected will be used to irrigate landscape area from level 1 to 5.

TECHNICAL FACULTY SDU ODENSE

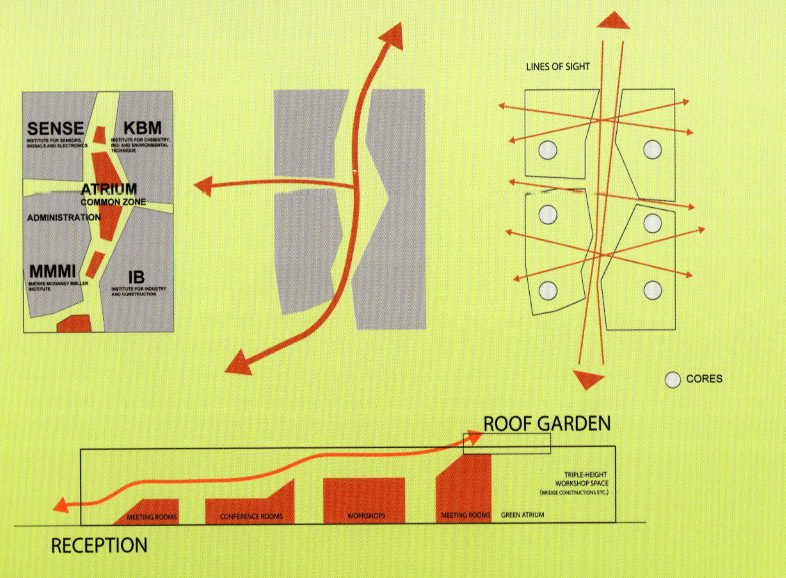

地点：丹麦欧登塞南丹麦大学
规模：21 000 m²
建筑设计：C. F. Møller Architects
景观设计：Schønherr Landskab A/S
委托方：University and Building Agency

Location: Odense, Denmark
Size: 21,000 m²
Architect: C. F. Møller Architects
Landscape: Schønherr Landskab A/S
Client: University and Building Agency

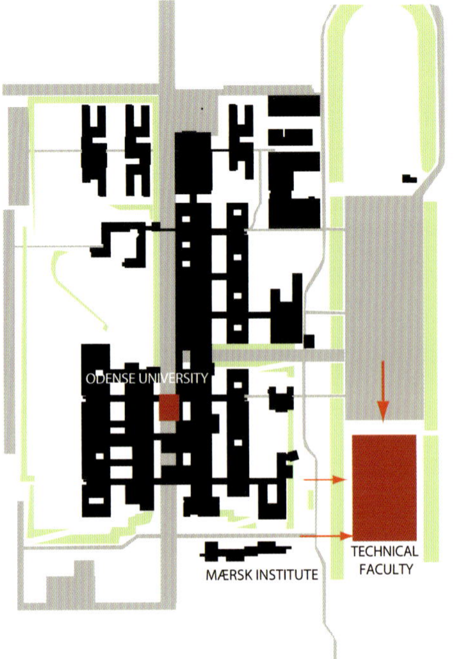

区位图 LOCATION

项目概况

南丹麦大学工学院位于丹麦城市欧登塞，为4个学院提供研究和教学环境。 这座建筑有一个巨大的外壳，内部由5座教学楼组成，教学楼的核心部位通过位于不同楼层的天桥互相联通。这座外形与家具类似的建筑包括常用功能区域和若干会议室，也是通往屋顶花园休息区的入口。多种关系的聚合带来更多的人流、更多的交流和知识分享。

建筑热环境解决方案

大楼设计为一座玻璃建筑，建筑外部根据暴露程度和朝向的不同，由多种材料构成一组屏帷。东侧和西侧为纤维性混凝土制成的精美屏帷，屏帷墙之下是太阳能幕墙和自然通风系统。南侧外立面全部为形制统一的太阳能电池和太阳能遮板。新颖的屏帷墙设计反映了创新精神和不同学院的性质特点。

建筑内部的布局设计灵活多样，可移动墙体与位置固定的承重墙组合，可根据需要分隔内部空间。大型实验室位于一楼，便于使用者出入。

SDU工学院教学楼符合BR95(丹麦建筑规范)体系中的一级低能耗建筑标准，在建造及使用的全过程中，达到最少的能源消耗，创造良好的室内微气候，使用对环境影响较小的建筑材料。

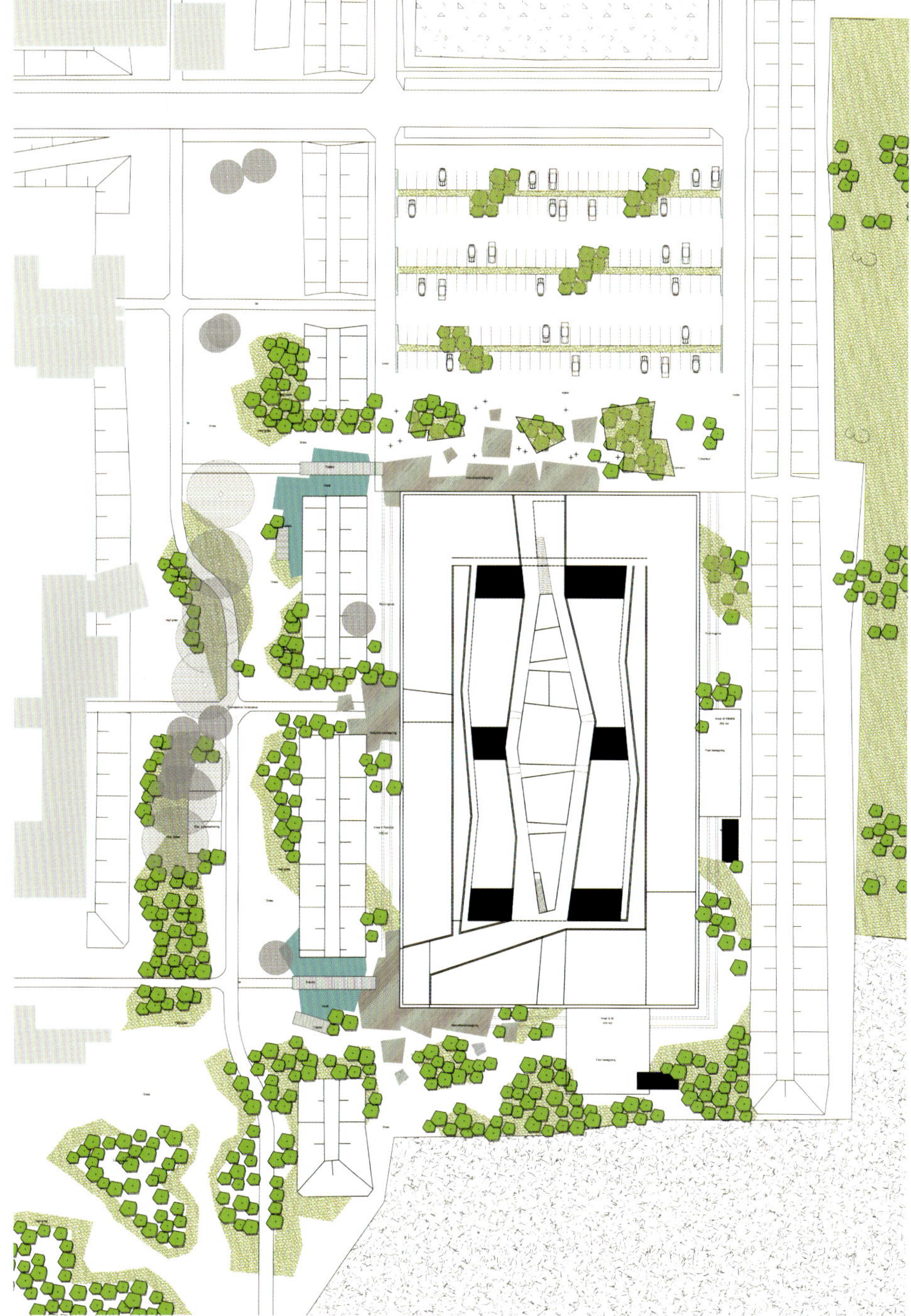

总平面图 SITE PLAN

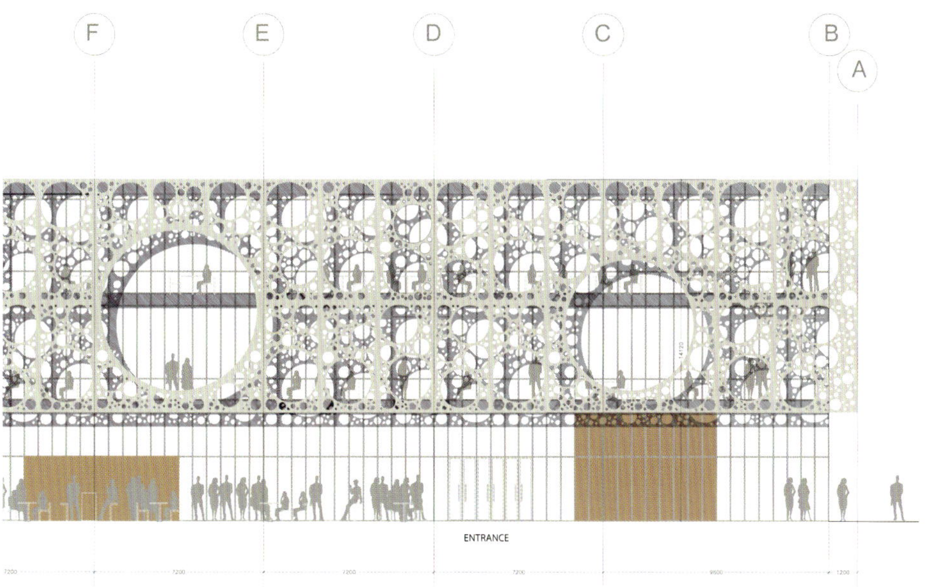

东立面图 EAST ELEVATION

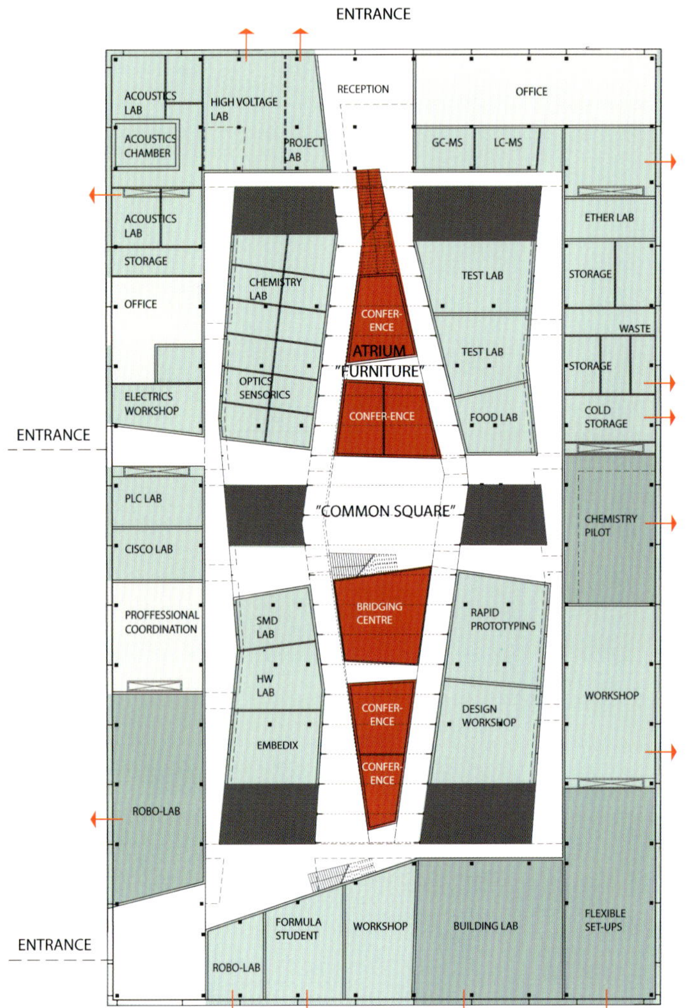

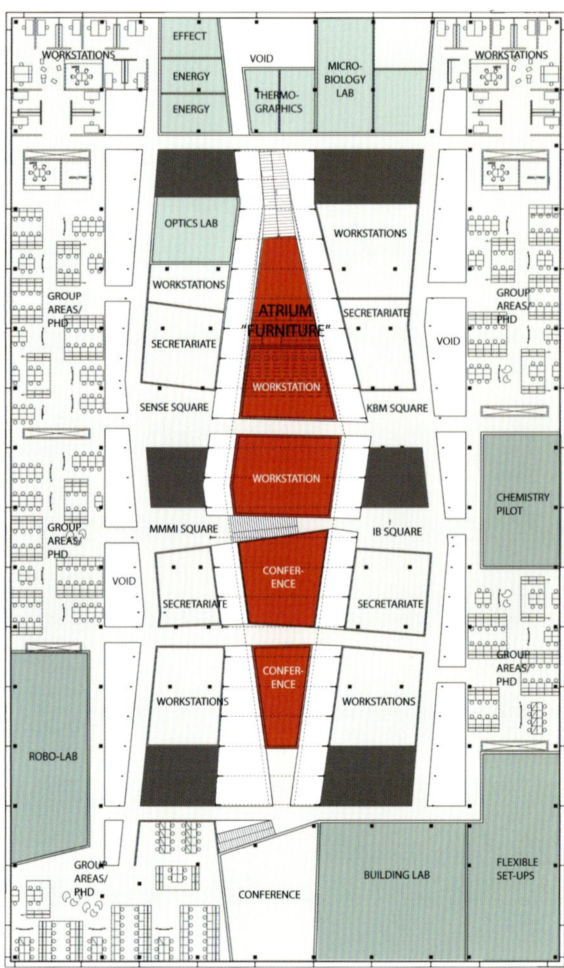

Project Description

The Technical Faculty is part of the University of Southern Denmark in Odense, and constitutes a shared research and education environment for four different institutes. The building is designed as one big envelope consisting of 5 houses connected by bridges at multiple levels crossing the heart of the house, a "piece of furniture" containing common functions and meeting-rooms, and giving access to a roof garden/cafe/lounge area. The many connections allow for more fluid boundaries, and more community and knowledge sharing.

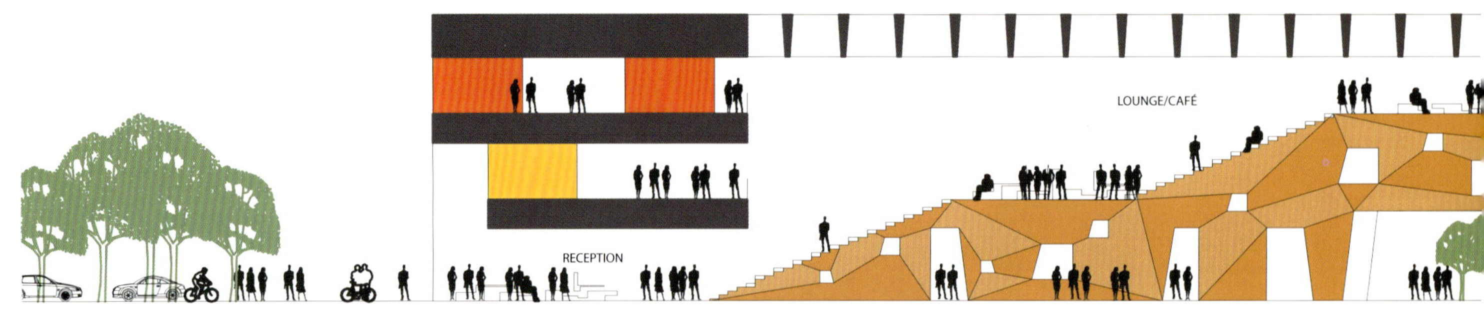

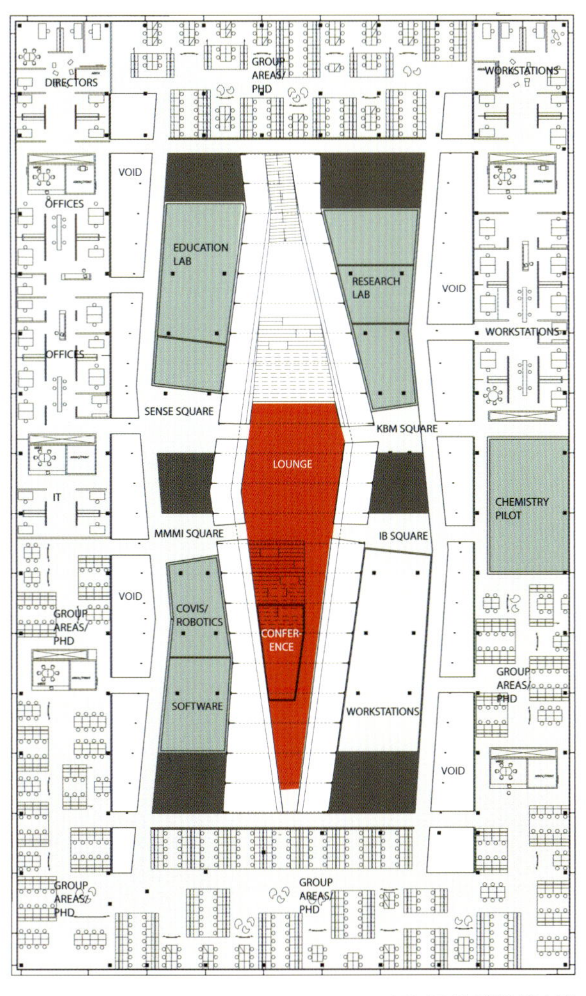

平面图 FLOOR PLANS

中庭立面图 COURTYARD ELEVATION

CIRCULAR SKYLIGHTS
SKYLIGHT
SKYLIGHT
GROUP AREAS
STUDIES
LOUNGE
GROUP AREAS
GROUP AREAS
GROUP AREAS
WORKSHOPS
CONFERENCE
GROUP AREAS
GROUP AREAS
LABORATORY
LABORATORY
CONFERENCE
LABORATORY
LABORATORY
DUCTS

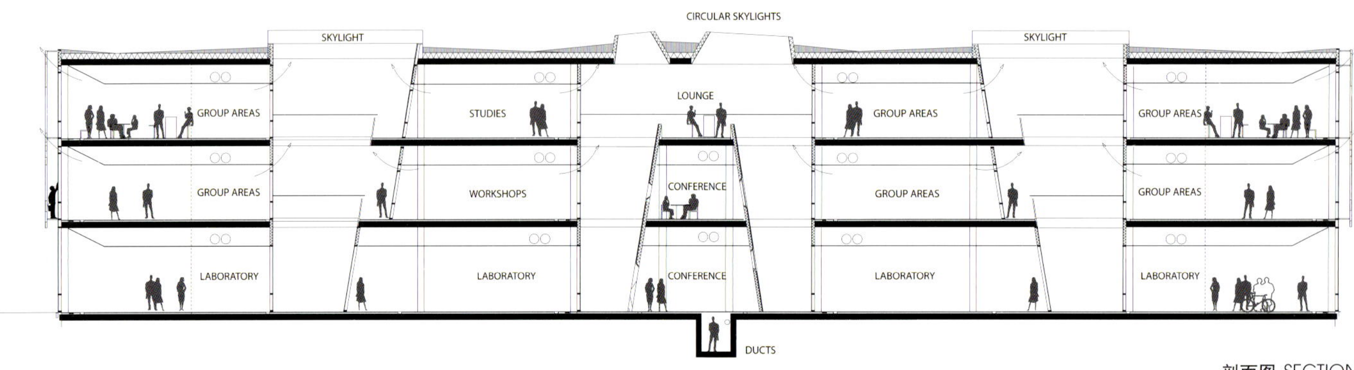

剖面图 SECTION

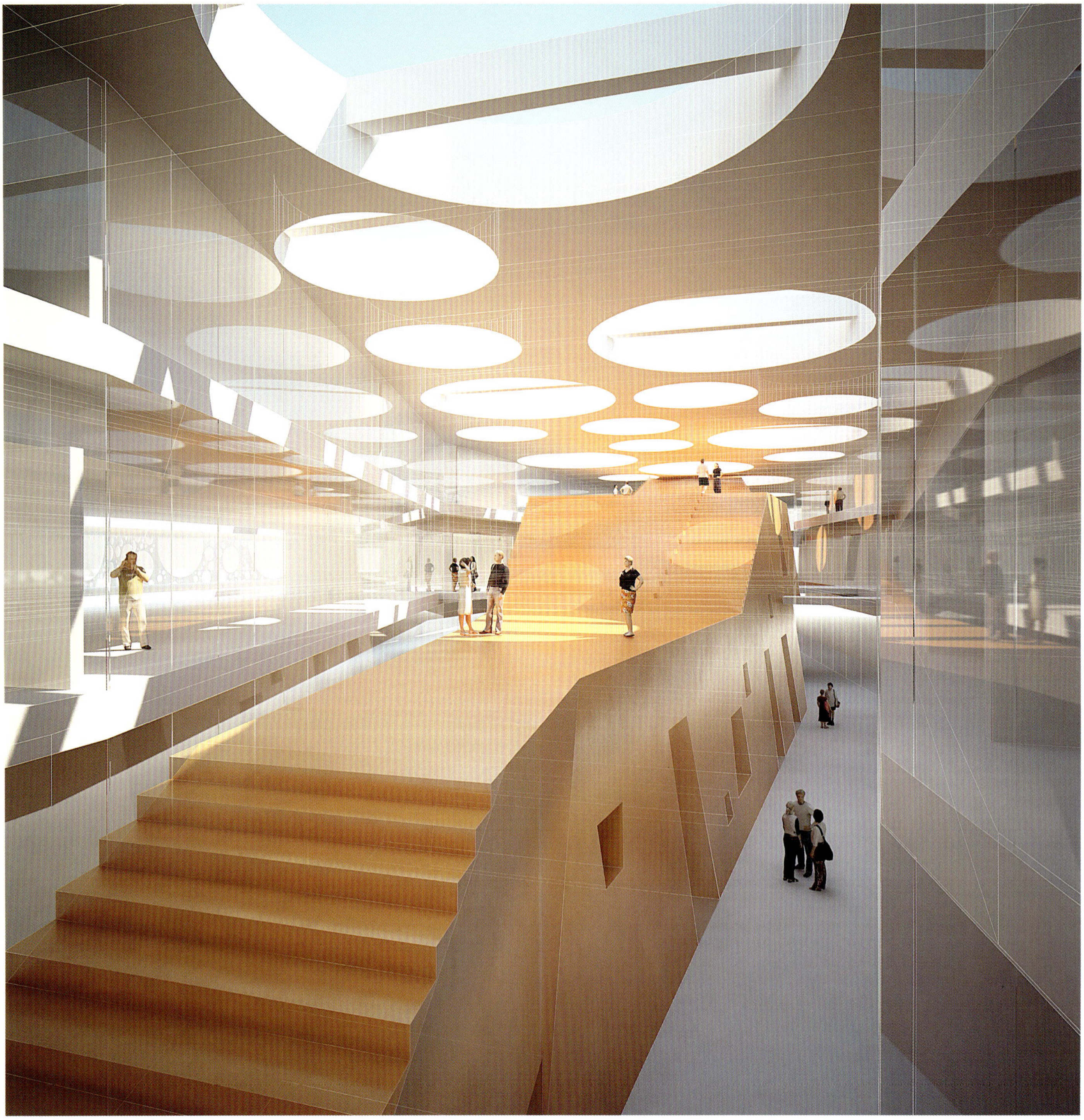

Building Thermal Environmental Engineering Solutions

The building is designed as a glass house with an external screen of varying materials, depending on exposure and orientation. An elegant screen of fibre-reinforced concrete is designed to the west and east, with an underlying solar screen and natural ventilation, whereas the south facade is fully glazed with solar cells and solar shading in a similar pattern. The unusual screen reflects the innovation and creativity that characterises the various institutes.

The interior layout creates great flexibility, by a combination of solid cores and sliding wall system for adaptible sub-divisions depending on group sizes. The larger labs are located on the ground floor, for easy access to the terrain and opportunity for outdoor activities.

The Technical Faculty at the SDU is to meet the requirements for low energy class 1 according to BR95 (Danish building codes). This means minimal energy consumption, good indoor climate and use of materials with a low environmental impact in a life cycle perspective.

UNIVERSITY OF CALIFORNIA, SANTA BARBARA STUDENT RESOURCES BUILDING

加利福尼亚大学圣芭芭拉学生服务楼

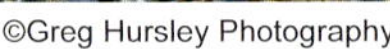
©Greg Hursley Photography

地点：美国加利福尼亚州圣巴巴拉
规模：6457 m²
建筑设计：Sasaki Associates, Inc.
所获奖项：LEED银质认证
摄影：Robert Benson; Greg Hursley

Location: Santa Barbara, California, USA
Size: 6457 m²
Architect: Sasaki Associates, Inc.
Selected Awards: LEED Silver certified
Photography: Robert Benson; Greg Hursley

@SASAKI

项目概况

这座学生服务楼获得LEED银质认证。自2007年建成以来，成为校园社团的聚集地，并为兴趣迥异的使用者提供了活动空间，成为校园的一道独特的风景线。

这座建筑是在设计团队和学生团体的紧密合作下完成的。学生们投票建议增加经费投入，兴建一座服务于学生组织的大楼。学生要求建一座色彩丰富的现代空间，其中设置多种功能部门和服务中心，包括妇女中心、学校学习和援助服务处、一座幼儿园和美国印第安人中心。

建筑热环境解决方案

建筑为翼状分布的两栋三层楼围合一个纵向的广场空间。这个开放区域在建筑自然通风系统中同样起到了关键作用——从各个楼层吸走废气，从通风窗排出。同时，作为各项活动的中心，广场在视觉上和自然环境上将各个楼层的功能资源相互连接，促进了使用者的互动交流（平日学生和教职员工通过露天的金属桥穿越小广场，树木掩映下形成视觉屏障，有选择地创造了一些私密空间）。广场通往多功能厅——一座整体性的椭圆形建筑，与南边两翼形制规则的楼宇形成了对比。多功能厅向一个"自由言论"讲坛开放，后来演变为庆典和公共活动的礼堂。

建筑设计师在完成项目的过程中坚定不移地将可持续理念贯彻始终，并把获得LEED认证作为实现目标。他们尽了最大努力为学生活动提供健康的环境。建筑主体的设计综合多项可持续技术，包括自然采光、气候控制技术和自然通风。建筑内部选用的建材包括含有再生物质成分的岗石地板、橡胶地面，含有废旧尼龙成分的地毯，低VOC涂料，经美国联邦科学委员会认证的木材，再生吸声砖（只在需要处使用），和由羊毛及再生尼龙制成的室内陈设等。根据24-2001能耗对照条款计算，本建筑的年均能耗比同等规模建筑标准的能耗减少了15.1%。

©Robert Benson Photography

©Robert Benson Photography

©Robert Benson Photogr

©Robert Benson Photography

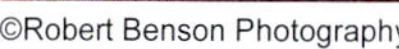
©Robert Benson Photography

©Robert Benson Photography

Project Description

Since its opening in 2007, the LEED Silver certified Student Resource Building has become a community hub and source of campus pride that reflects the diverse interests of its users. The building was the result of close collaboration between the design team and the student body, which voted to increase its fees to fund a building dedicated to student organizations. The students requested a colorful, contemporary space to house a variety of departments and resource centers, including the Women's Center, Campus Learning and Assistance Services, a Child Care Center and the American Indian Center.

The building is comprised of two three-story wings flanking a vertical Forum space. Designed to be the center of activity, the Forum visually and physically connects resources on all floors to encourage interaction. Students and staff routinely gather on the open metal bridges traversing the Forum while wood screens create selective privacy zones. This open area also plays a key role in the building's natural ventilation system, drawing air through the floors and out the clerestory.

The Forum opens to the Multi-Purpose Room - an organic, oval form inflecting towards the entry to contrast with the refined geometry of the north and south wings. The Multi-Purpose Room in turn opens to a "free-speech" plaza which transforms into a loggia-like staging area for celebrations and public events.

Inspired by nature and cultural arts, the color palette echoes the building's vibrant programming. Decanting through clerestory windows, daylight reflects off of a cadmium yellow ceiling, while the building's transparency engages passersby with an open view of the nearly dozen colors on its interior accent walls. The bright, lightweight furniture was selected for its ease of movement, offering students the ability to control the spaces they inhabit.

Sustainability and LEED certification were resolute goals for the project and every effort was made to promote a healthy environment. The building mass is oriented in an ideal configuration for day-lighting, climate control, and natural ventilation. Interior materials include engineered stone flooring with high recycled content, rubber flooring, carpet tiles with post-industrial nylon content, low VOC paints, FSC certified woods, highly recycled acoustical tiles (used only where necessary), and furniture upholstery composed of wool and recycled nylons. According to the Title 24-2001 energy cost comparison, the Proposed Building Annual Energy Use is an impressive 15.1% less than the Standard Energy Budget for Building.

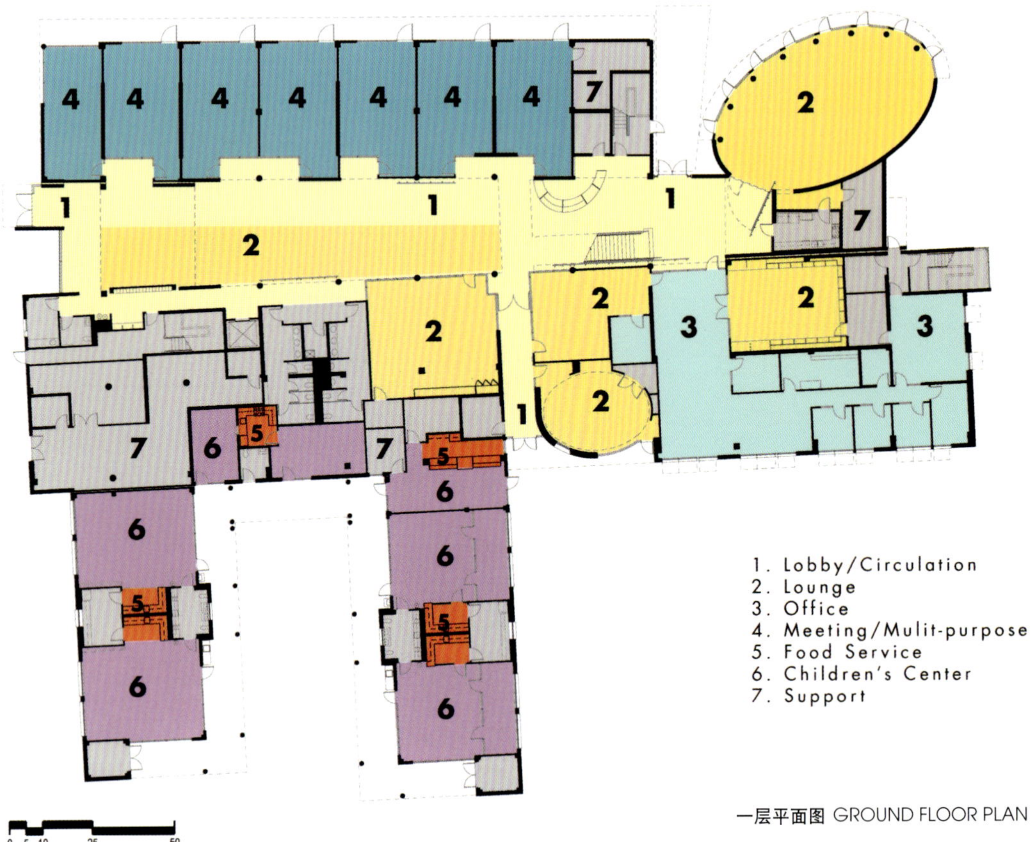

一层平面图 GROUND FLOOR PLAN

Placed at a major campus intersection, the building is a component of the arrival gateway that connects the adjacent residential community to the core campus. Every day, more than 10,000 students, staff, and faculty approach the campus from this direction and use it as their "living room" to meet, study, and socialize.

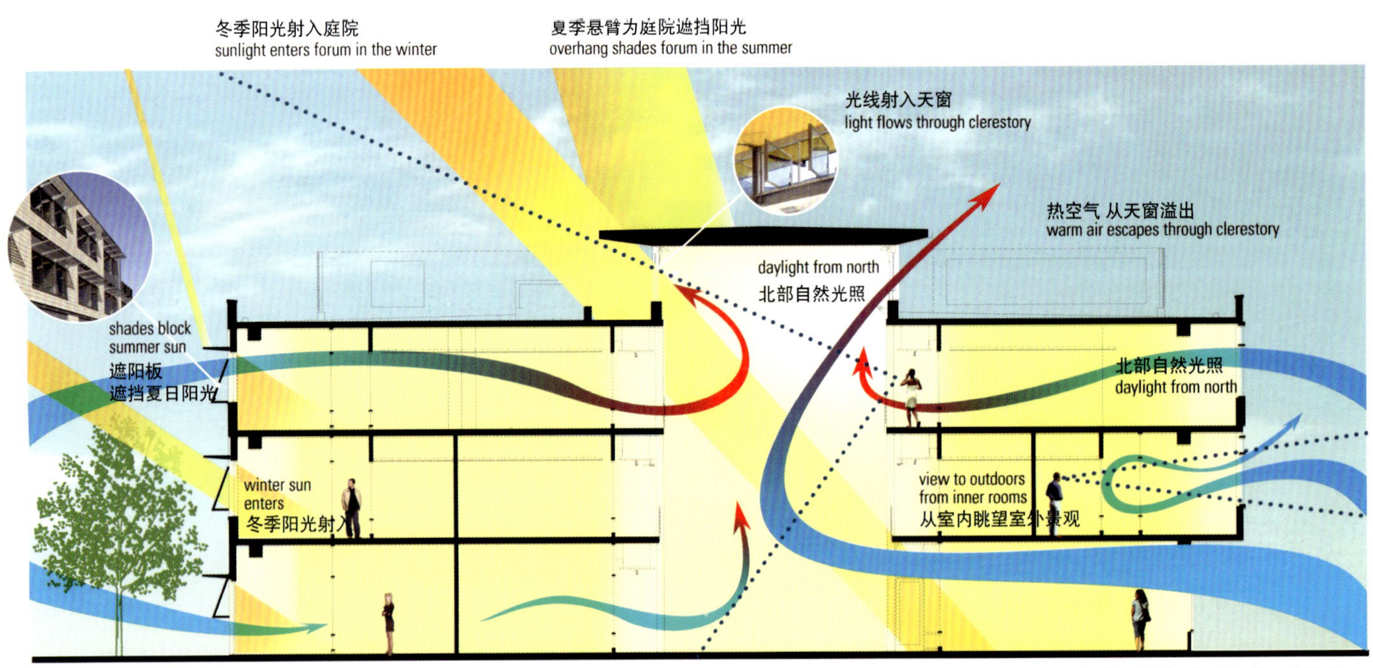

分析图 DIAGRAM

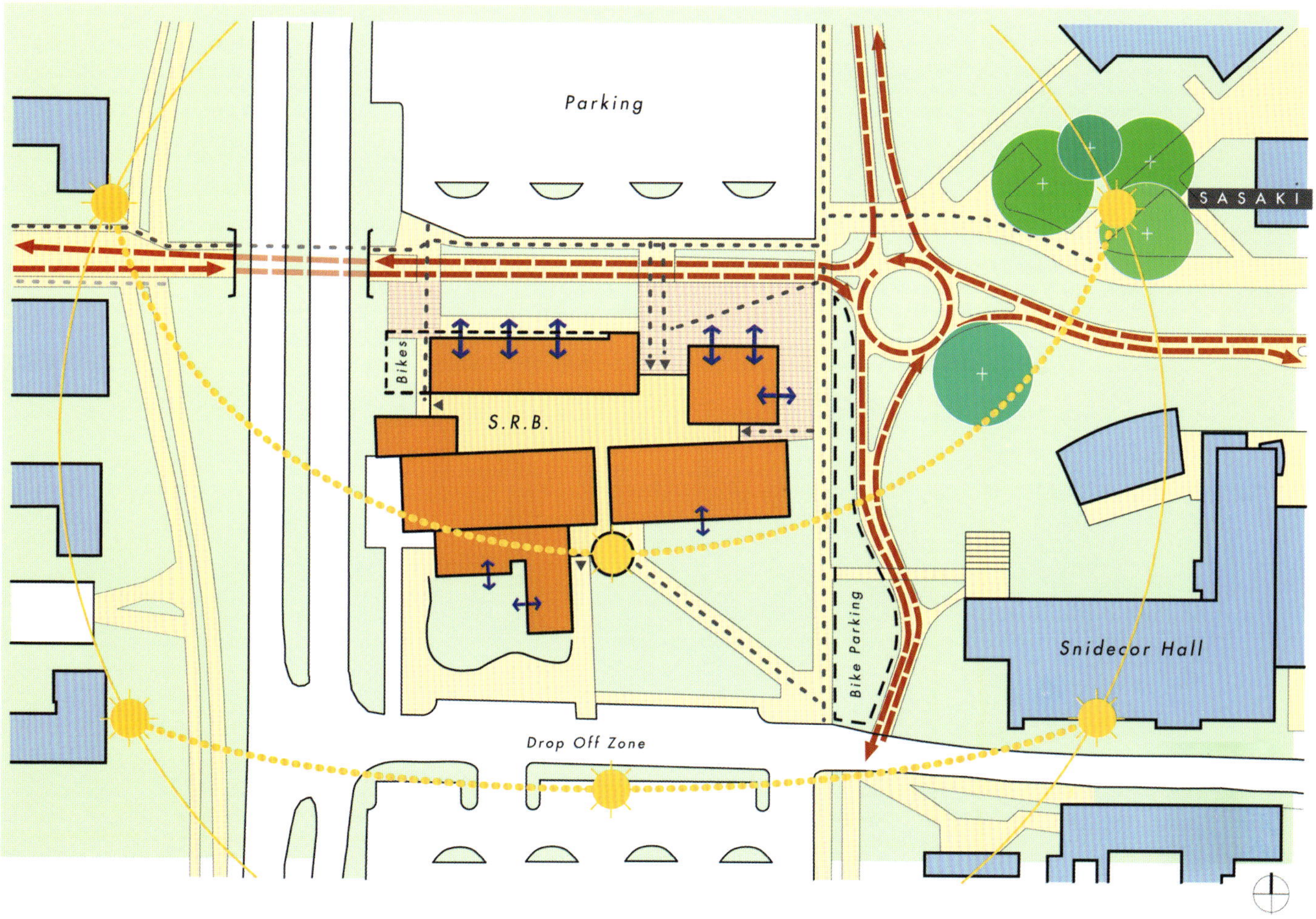

Parking
SASAKI
Bikes
S.R.B.
Bike Parking
Snidecor Hall
Drop Off Zone

分析图 DIAGRAM

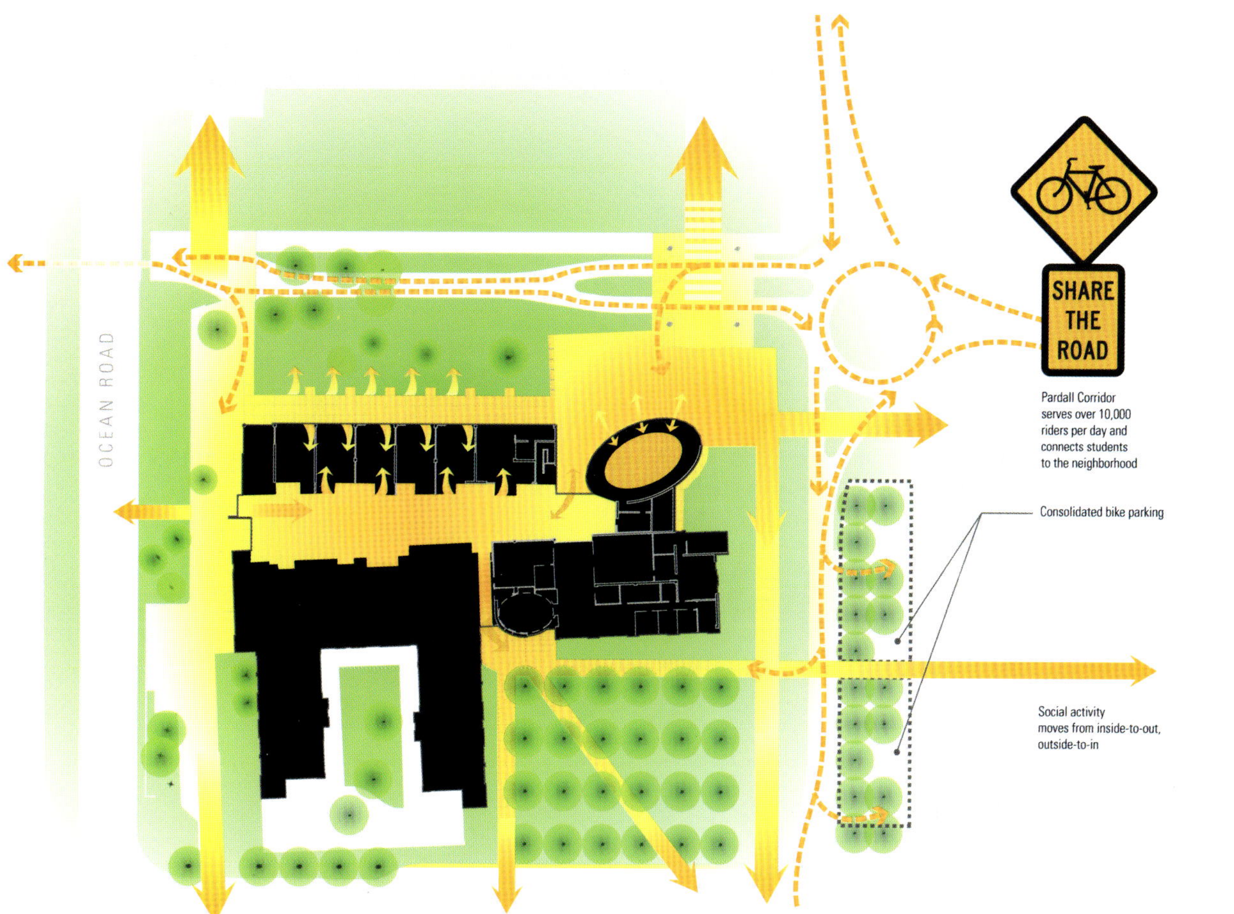

交通组织分析图 TRAFFIC ORGANIZATION DIAGRAM

UNIVERSITY OF CALIFORNIA, DAVIS, GRADUATE SCHOOL OF MANAGEMENT

管理学研究所及莫瑞斯·格拉夫会议中心

©Bruce Da

地点：美国加利福尼亚州戴维斯
规模：7618 m²
建筑设计：Sasaki Associates, Inc.
所获奖项：2011年10月获LEED铂金认证；2011年10月美国设计建造协会（DBIA）卓越建筑奖——教育设施类；2010年美国建筑师协会（AIA）旧金山设计奖节能及可持续设计优异奖
委托方：University of California, Davis
摄影：Bruce Damonte

Location: Davis, CA, USA
Size: 7618 m²
Architect: Sasaki Associates, Inc.
Selected Awards: U.S. Green Building Council, LEED Platinum Certification, October 2011 National Design-Build Excellence Award - Educational Facilities, Design-Build Institute of America (DBIA), October 2011 Award of Merit for Energy + Sustainability, AIA San Francisco Design Awards, 2010
Client: University of California, Davis
Photography: Bruce Damonte

©Bruce Damonte

项目概况

管理学研究所及莫瑞斯·格拉夫会议中心（GSM）位于校园南部人流密集的区域，对面是罗伯特及玛格利特·蒙德表演艺术中心。GSM旁边的方庭将作为校园的新校门，GSM与邻近建筑相互呼应，有助于促进人们对新校门的认同感。GSM项目7618 m^{2}的规划面积包括三层办公楼和两层会议中心，内设饭店、办公空间、会议室和舞厅。

由于建设工期较紧，建筑设计项目组和校方有效合作提高效率。在研究创新方案的过程中，不同学科背景的专家进行了多次交流探讨，研究出了独特的系统性解决方案。通过实行综合方案，项目组可以以最低的成本引入革新系统，并且在全部24项指标中，最终达到节能27.1%的效果，在建设过程中节能32.8%。另外，根据能源之星网站的统计，该项目的能源利用率较一般办公楼提高74%，大大超出了为应对2030年环境挑战而提出的现阶段目标。主要基于上述节能措施，最近该建筑获得了LEED铂金认证，是加利福尼亚州第一所获此荣誉的商学院。

建筑热环境解决方案

建筑的屋顶将安装102 kW太阳能光伏电源阵列。这个144 000 kW/h的供电系统将解决大楼20%的电力需求。建筑采用了辐射核心冷却/加热系统、地源热泵系统，并设置大窗用于自然采光，通过这些措施控制室内微气候。与地缘热泵系统配合使用，发热楼板系统可以按照两种模式运行。因为空间可以由辐射能加热或冷却，所以用户可以任意开关窗子进行通风，而对供热系统几乎没有影响。

浅颜色的铺面材料和白色屋顶有助于建筑的降温。大楼外表面以玻璃、石材和灰泥覆面，简约美观。建筑师在建筑表皮外设计了通风屏障系统，同时可以起到挡雨的作用。屏障系统配有可开关的窗子，开口接合和空腔促进了空气流通。

其他可持续设计措施

大楼能够收集落到其表面上的雨水，大约25%的雨水可以被回收。还设置了人体感应灯，减少能耗。建筑物周围栽种了耐寒植物，高效的给排水管网可节水30%～40%。其他可持续措施还包括循环利用75%的建设废料，以及在建造中使用30%的再生材料。

通过采用BIM技术，项目组从一开始就可以精确地定量测算施工成本。由于在建设全过程中可以得到实时的费用信息反馈，使技术人员可以评估各种建筑系统的优劣，找到有利于实现可持续目标的整体性方法。

©Bruce Damonte

©Bruce Damonte

©Bruce Damonte

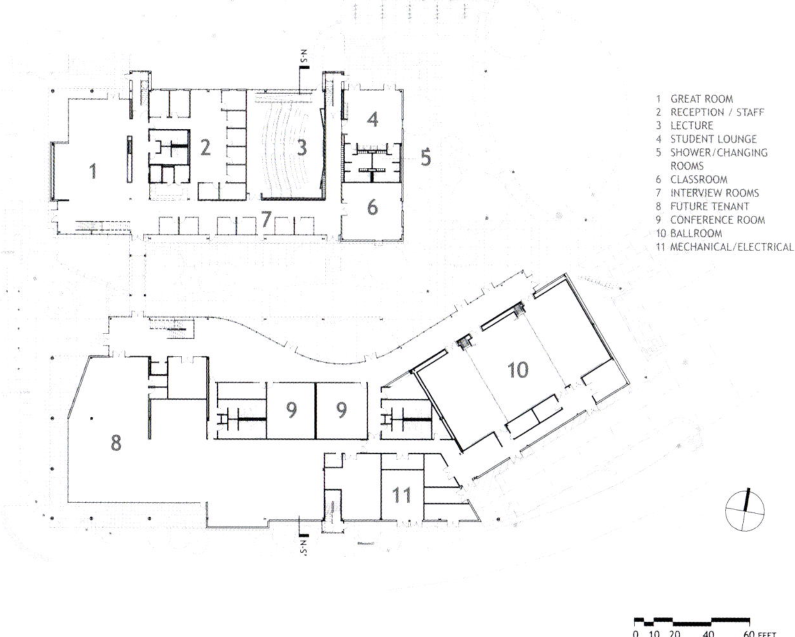

一层平面图 GROUND FLOOR PLAN

©Bruce Damonte

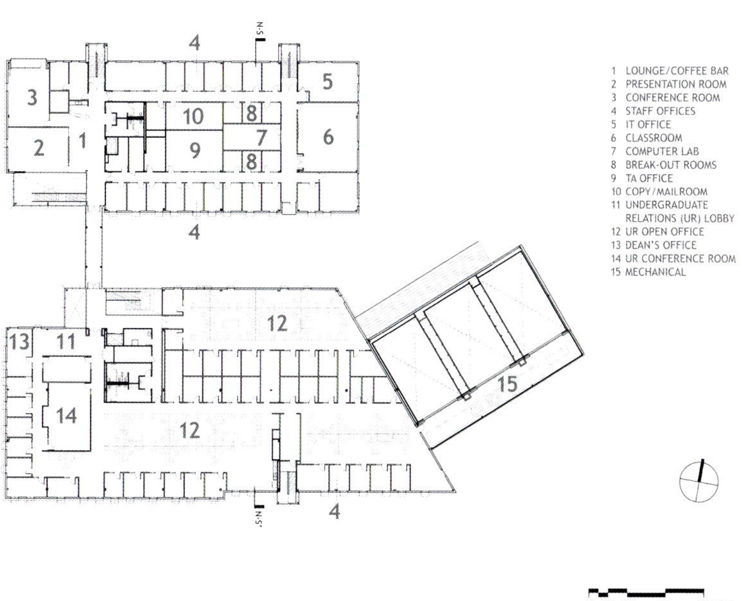

二层平面图 SECOND FLOOR PLAN

©Bruce Damonte

Project Description

The Graduate School of Management and Conference Center/Maurice J. Gallagher Jr. Hall (GSM) is located at the highly public, south edge of the campus, across from the Robert and Margrit Mondavi Center for the Performing Arts. Complementing its neighbor, the GSM plays a significant role in defining the entry quad as the new front door to the campus. The 7618 m² project includes a three-story office structure for the GSM and a two-story Conference Center that includes a restaurant, office space, meeting rooms, and a one-story ballroom.

An accelerated schedule compelled alignment and collaboration between the design-build team and the University. In search of innovative solutions, a number of interdisciplinary brainstorming sessions explored unique systematic approaches. Integration allowed the team to deliver innovative systems for the lowest first cost, and contributed to the project's final calculated energy savings of 27.1% over Title 24 requirements or 32.8% excluding process energy. In addition, according to the Energy Star website, the project's energy usage reflects a 74% improvement over the typical office building, well ahead of the current goals for the 2030 Challenge. These energy savings are a major contributor to the building's recently bestowed LEED Platinum certification, the first business school in California to receive this designation.

©Bruce Damonte

Building Thermal Environmental Engineering Solutions

The roofs of the buildings are planned to house 102 kW photovoltaic solar power array. The 144,000 kW/h system is expected to generate about 20% of the building's electricity requirements. The building is equipped with radiant core cooling and heating system, ground source heat pump array, and tall windows for natural light, which contribute to regulate interior microclimate. Coupled with the ground source heat pump array, the radiant floor slab system allows mixed mode operation. Since spaces are cooled or heated through radiated energy, occupants can open windows to control ventilation air with minimal impact to the system.

©Bruce Damonte

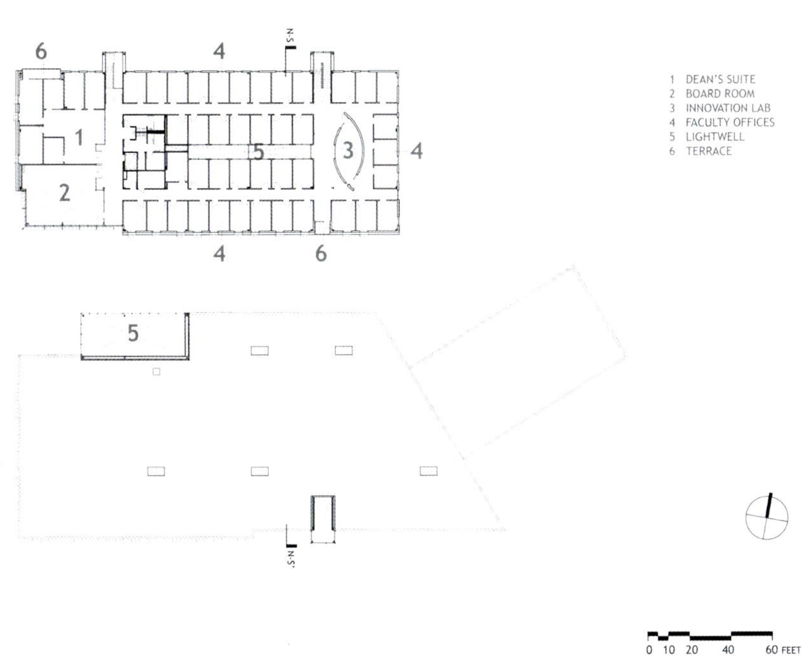

三层平面图 THIRD FLOOR PLAN

Cool temperature is maintained by light-coloured pavings and white roof. The building is clad in glass, stone and stucco for aesthetics. The facade is designed as a ventilated barrier system which serves as rain water covering. The barrier wall with operable windows has open joints and cavity for free flow of air.

Other Sustainable Design Measures

The storm water runoff from the building is reduced by 25% and is recycled. Motion-sensitive lighting can reduce the energy consumption. Drought-tolerant native plants are used for landscaping and efficient plumbing fixtures reduce water wastage by about 30% to 40%. Other sustainable features included recycling of 75% of the construction wastes and usage of 30% of recycled materials for construction.

The use of BIM technology enabled the team to pursue an accurate approach to quantifying construction costs from the start. Having real-time cost feedback during the process allowed the team to evaluate the relative benefits of various building systems, yielding a more holistic approach to sustainability.

发热楼板：与地源热泵系统配合使用，发热楼板系统可以按照两种模式运行。因为空间可以由辐射能加热或冷却，所以用户可以任意开关窗子进行通风，而对供热系统几乎没有影响。

RADIANT FLOOR SLABS. Coupled with the ground source heat pump array, the radiant floor slab system allows mixed mode operation. Since spaces are cooled or heated through radiated energy, occupants can open windows to control ventilation air with minimal impact to the system.

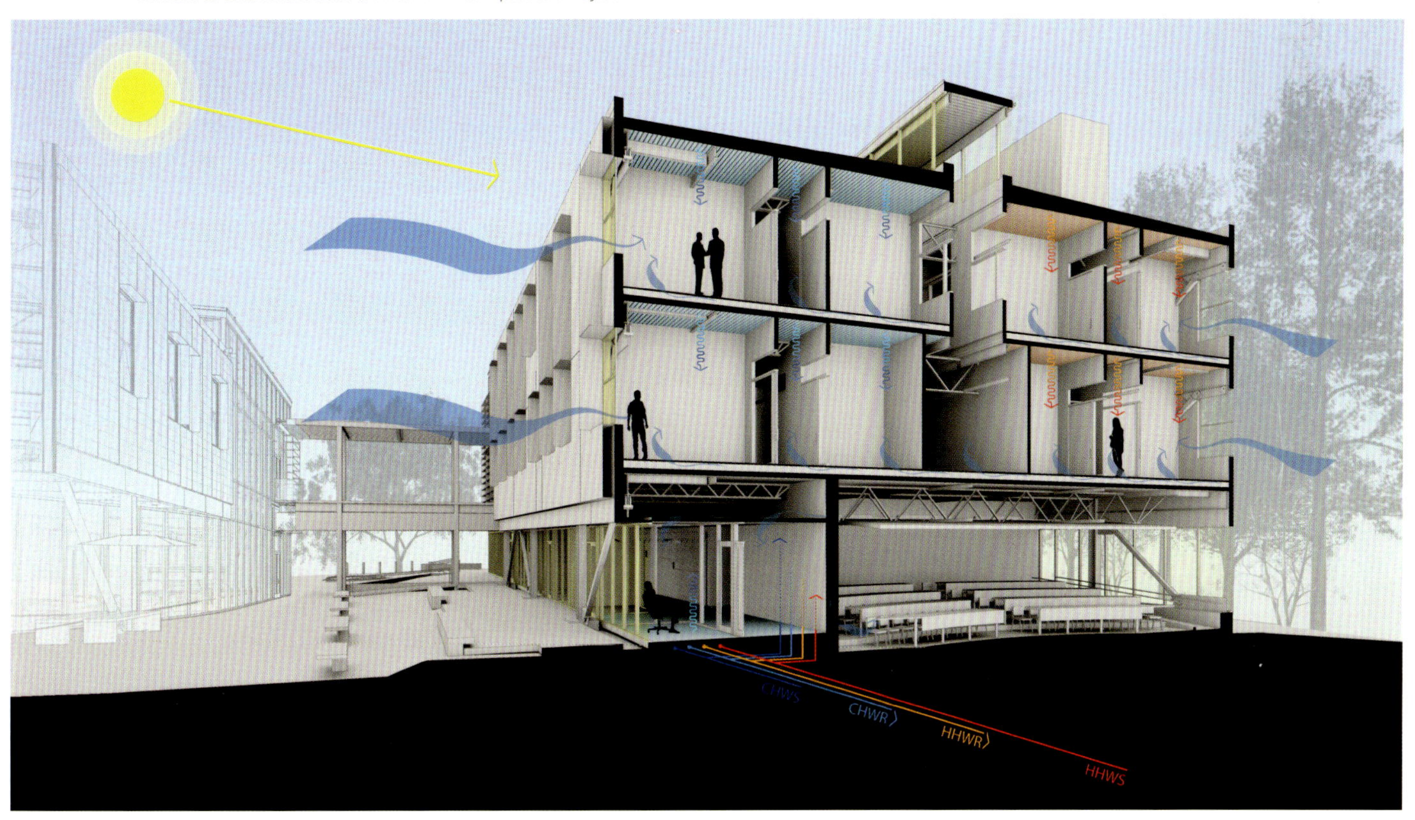

CHWS 冷冻供水
CHWR 冷冻回水
HHWR 加热回水
HHWS 加热供水

NO.5600 WILSHIRE STREET
WILSHIRE大道5600号

地点：美国加利福尼亚州圣巴巴拉
规模：6457 m²
规划及建筑设计：van Tilburg, Banvard & SoderBergh, AIA
所获奖项：LEED银质认证
摄影：Robert Benson；Greg Hursley

Location: Santa Barbara, California, USA
Size: 6457 m²
Planning and Architect: van Tilburg, Banvard & SoderBergh, AIA
Selected Awards: LEED Silver Certified
Photography: Robert Benson; Greg Hursley

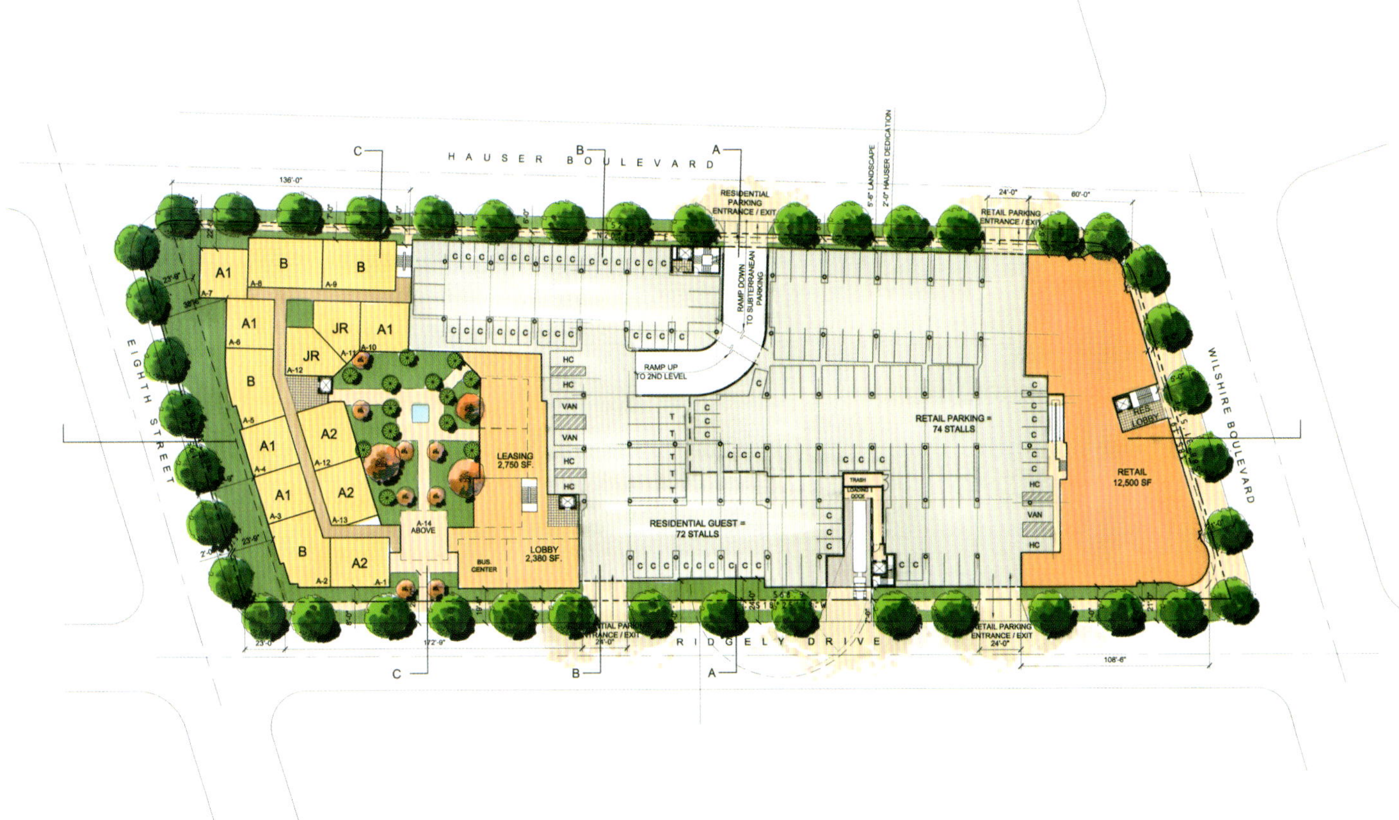

总平面图 SITE PLAN

项目概况

"Wilshire大道5600号"位于公共交通方便的洛杉矶市主城区，为避免居民出行使用私家车辆提供了条件。住区中专设可满足15%居民需求的自行车停车设施，进一步减少了居民出行对汽车的依赖。为了鼓励居民使用低排污节能型汽车，停车库中指定了5%的机动车停车位为该类车型专用。主要的节能措施包括选用"能源之星"级的节能厨房和洗衣家电。

建筑热环境解决方案

采用各户独立控制的照明系统和热能系统；设置高效暖通空调系统；选用双层玻璃、低传热型门窗以提高外墙体的保温隔热性能，降低采暖和制冷的能耗。为缓解城市热岛效应，设计中选用了具有高日照反射系数（SRI）的绿地、铺地和屋面材料。为了提高封闭性、使75%的空间日照采光良好及改善室内空气质量，设计中选用了低挥发性的涂料、地毯、黏合剂和密封剂。

其他可持续设计措施

因洛杉矶地区的排洪管线最终汇入太平洋，严格控制排水的污染是当地的一项重要政策。因此，住区中排出的所有雨水是就地净化处理后才排入市政管网的。在节水措施方面，设计中选用了节水厨卫器具和园林灌溉系统，以及耐旱景观植被，使得生活和灌溉用水的用量分别减少了30%和50%。该项目在施工过程中在工程污染控制、工程垃圾的分类和处理等方面也做到了环保，并成为LEED认证过程中的重要得分项。建材和部品选用了距离当地805 km范围内生产的产品，以降低因长途运输造成的能耗。同时20%的产品源自于回用材料。

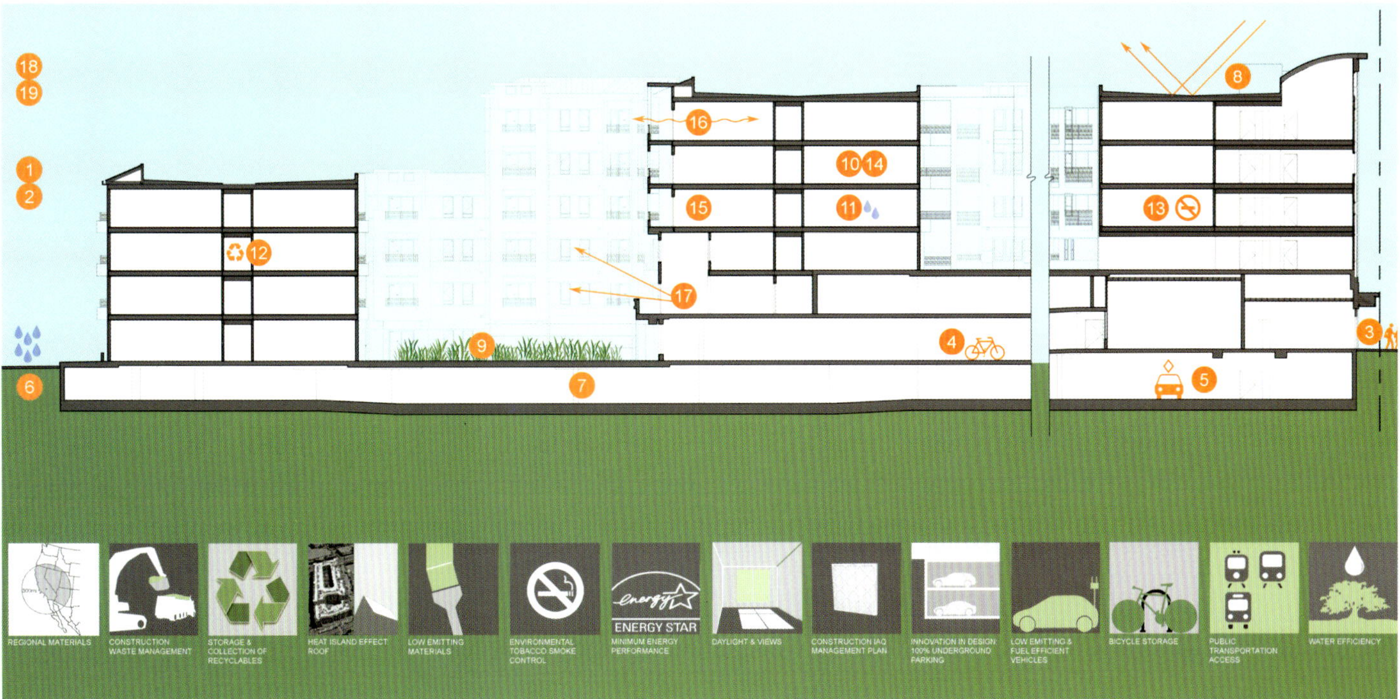

①Remediation and Re-use of a Brownfield Site;
②Use of an Urban Infill Site;
③Within Walking Distance of Many Basic Services;
④Secured Bicycle Storage for Residents;
⑤Preferred Parking for Fuel Efficient Vehicles & Carpools;
⑥100% of Storm Water Collected and Filtered on Site;
⑦100% of Parking Located Underground;
⑧White Single-PLY Polymer Roofing Material;
⑨Native Plants and Drip Irrigation for Water Efficiency;
⑩Materials with High Recycled Content and Extracted & Processed Regionally;
⑪Water Efficient Fixtures;
⑫Separate Trash and Recycling Chutes;
⑬Non-Smoking Building;
⑭Carpets, Sealants and Paints with Low VOC Content;
⑮Energy Performance Exceeds Minimum Standards by 21%;
⑯Individually Controlled Lighting and Thermal Systems;
⑰Daylight and Views for a Minimum 75% of Spaces;
⑱Classes to Educate Residents on Green Practices;
⑲Construction Process Included Pollution Prevention Recycling of Waste, and Maximizing Indoor Air Quality.

①整治并重新利用棕色地带；
②利用城市填平区域；
③步行距离内基础设施齐备；
④为居民设置安全的自行车存放处；
⑤机动车车库；
⑥就地收集净化社区排出的所有雨水 ；
⑦停车全部采用地下车库；
⑧白色单层聚合物屋面材料；
⑨栽种当地植物，节水滴灌系统；
⑩建材使用当地生产加工的高度再生原料；
⑪节水装置；
⑫垃圾分类回收；
⑬无烟建筑；
⑭低挥发性涂料、地毯、黏合剂和密封剂；
⑮能耗性能仅仅超过最低标准值20%；
⑯各户独立控制的照明系统和热能系统；
⑰75%的空间日照采光良好；
⑱为社区居民提供绿色环保知识讲座；
⑲施工全过程注重环保，减少污染、循环利用废弃物，优化空气质量。

Project Description

"No.5600 Wilshire street" locates in the main city area of Los Angeles in order to provide convenience for private car communication. The living area specially arranged bicycle parking infrasturcture for 15% of the residents so that dependence on cars could be further reduced. In order to encourage the residents to use low emission and energy-saving car, the garage allocates 5% of vehicle parking spaces for this type of cars. Major energy-saving strategy includes "energy star" energy-saving kitchen and laundry appliance.

Building Thermal Environmental Engineering Solutions

Each house unit uses independent lighting and thermal system, setting up high-efficiency thermal air system; using double glazing, Low heat transfer type door and window in order to reduce heating and cooling consumption. Design chose high sun reflectivity coefficient (SRL) green land, paving and roof material for buffering Urban Heat Island Effect and also chose low volatile paint, carpet, adhesives and sealants for improving closure, nice lighting of 75% space and interior air quality improvement.

Other Sustainable Design Strategy

Since Drainage pipeline in Los Angeles will finally converge to Pacific Ocean, to control sewage pollution strictly becomes a significant local policy. Therefore, all the rainwater eliminated from residence area has already been purified on the spot then flowed into municipal pipe network. In water-saving strategy aspect, design choose water-saving kitchen appliances, landscape irrigation system and drought resistance of landscape plants, which reduces the living and irrigation water amount by 30% and 50%. This project achieve environmental protection in constructing, project pollution control and project waste classification and these strategies become scores-getting parts for LEED certificate. Building Material has chosen local products with 805 km to cut down the energy consumption of long-distance transportation and 20% of the energy derives from recyclable materials.

GREEN
WORLD GREEN BUILDINGS 世界绿色建筑——热环境解决方案
THERMAL ENVIRONMENTAL ENGINEERING SOLUTIONS

办公 Office 158

PARKVIEW GREEN
侨福芳草地

地点：中国北京
建筑设计：综汇建筑设计有限公司，北京市建筑设计研究院
建筑物理：奥雅纳
结构工程：奥雅纳
所获奖项：2011年BCI FutureArc™ 绿色领导奖（商业建筑）；LEED-CS认证
委托方：北京侨福置业有限公司

Location: Beijing, China
Architects: Integrated Design Associates, Beijing Institute of Architectural, Design & Research
Building Physics: Arup
Structural Engineering: Arup
Selected Awards: BCI FuturArcTM Green Leadership Award, Commerical Architecture, 2011; LEED-CS Registered
Client: Beijing Chyau Fwu Development Company

©Arup

©Integrated Design Associates Architects & Designers

©Zhou Ruogu Architecture Photography

©Zhou Ruogu Architecture Photography

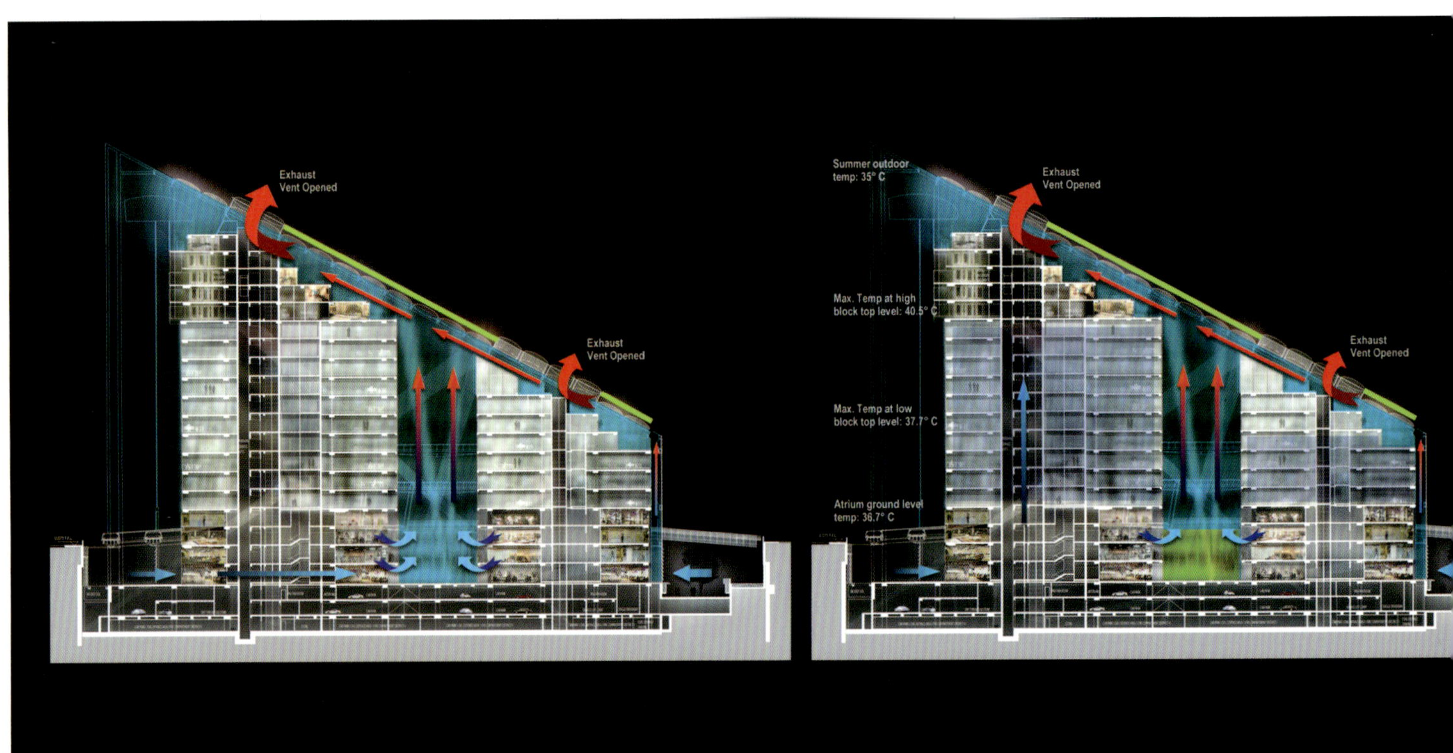

©Integrated Design Associates Architects & Designers

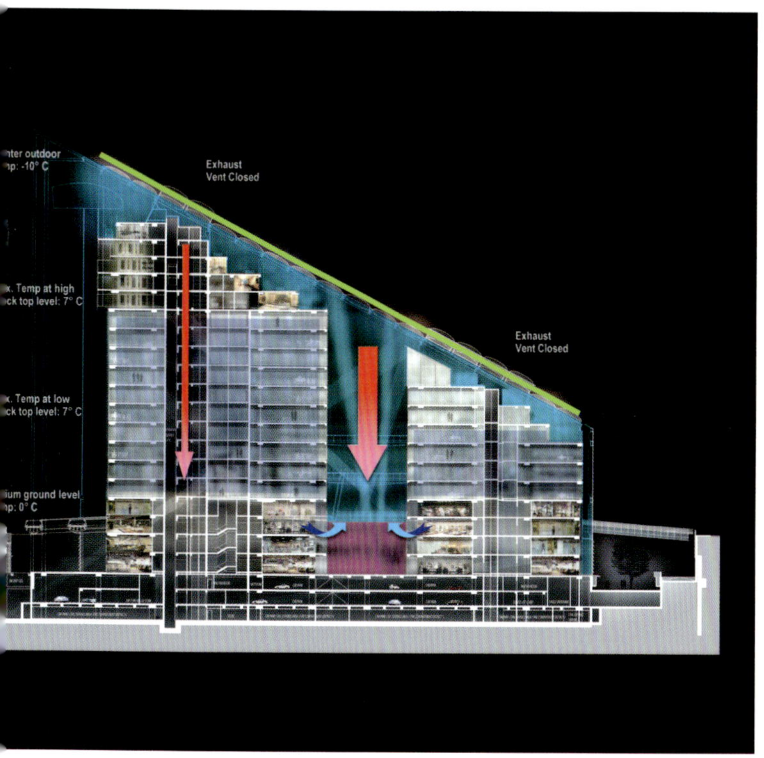

©Integrated Design Associates Architects & Designers

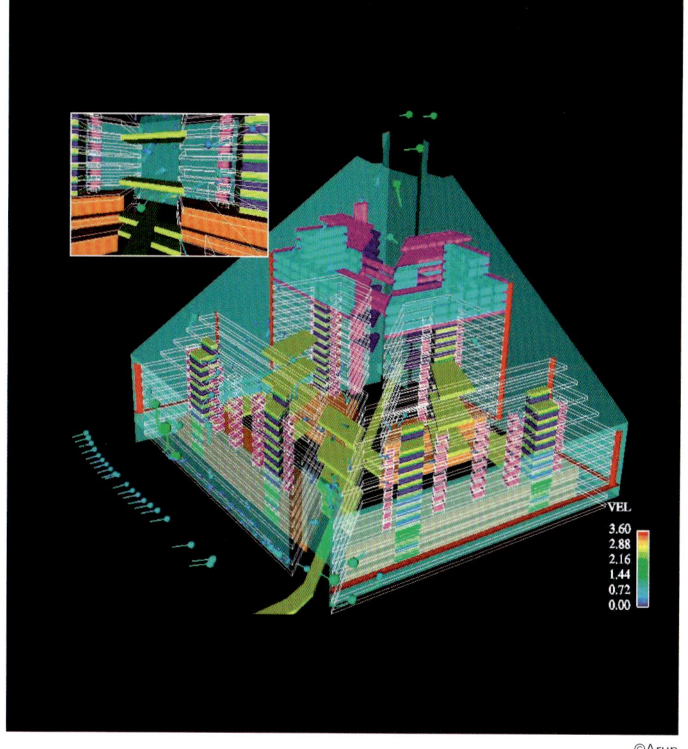

©Arup

ONE ISLAND EAST
港岛东中心

地点：中国香港
建筑设计：王欧阳（香港）有限公司
结构工程：奥雅纳
所获奖项：2010年香港优质建筑大奖优异奖(非住宅类项目)；2009年香港工程师学会/英国结构工程师学会优秀结构工程大奖（香港项目类）；2009年英国结构工程师学会中国建筑银奖；香港BEAM建筑环境评级：铂金级
委托方：太古地产

Location: Hong Kong, China
Architect: Wong & Ouyang (HK) Ltd.
Selected Awards: Quality Building Award Merit Award (Non-residential Project Category), 2010; HKIE/IStructE Structural Excellence Grand Award (Hong Kong Projects Category), 2009; IStructE China Silver Award, 2009; BEAM (Building Environmental Assessment Method) Rating: PLATINUM
Client: Swire Properties Limited

©Arup

©Arup

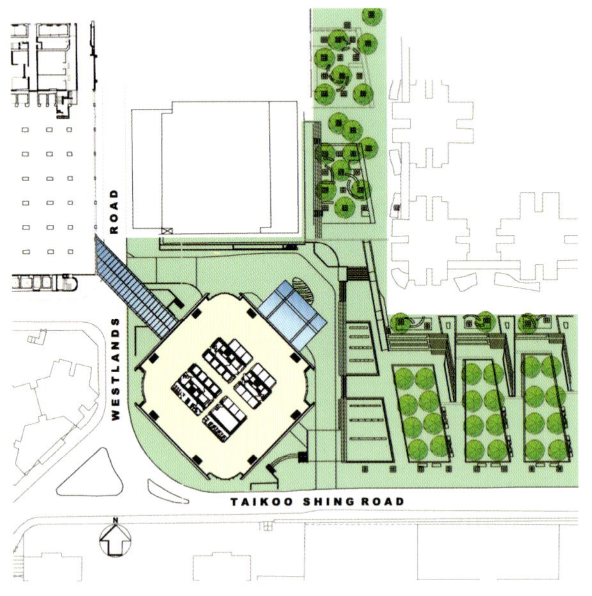

总平面图 SITE PLAN

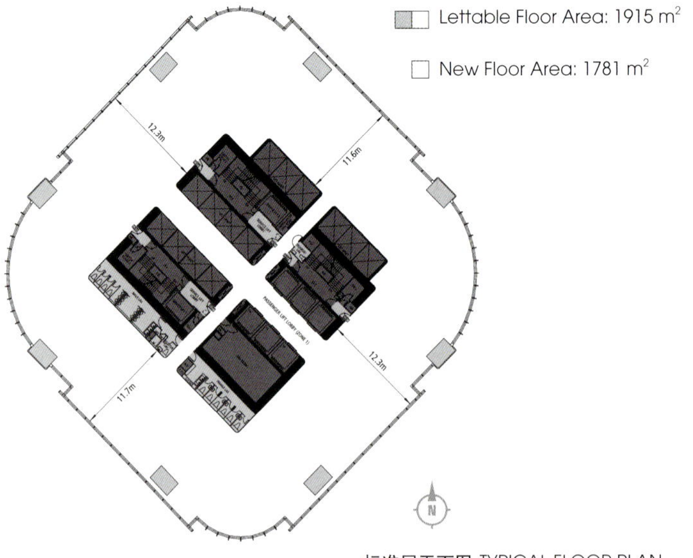

标准层平面图 TYPICAL FLOOR PLAN

港岛东中心是一座高308 m的A级办公大厦。该项目率先使用了100号高强混凝土。与60号混凝土的性能相比较，使用100号高强混凝土，使大厦使用面积增加3700 m^2，减少排放3500 t CO_2，减少1200 t废物。BIM被用于检查关键的结构细节。据估计，应用BIM进行自动冲突识别，可在桁架建构方面节省约20 d时间，减少建筑废物250 m^3。

顶部的楼层外形倾斜、侧移，以“开放”面向北部和南部的部分，从而提供更好的视野。为了尽量减少不透明拱肩的深度并提高幕墙的透明度，施工人员采用了消防工程方法，成功地将拱肩深度从900 mm降低到650 mm。在基座铺装、墙板和空气调节装置等部件、设备的配置过程中采用了标准化设计，使设备和材料的效能得到最大发挥，同时将浪费减到最小。对室内环境采取高标准，采用健康高效的设施，并将环境影响降到最低。在办公空间和照明设施的设计上，设计了开敞的办公空间，最大限度地使用自然采光，提高工作效率。

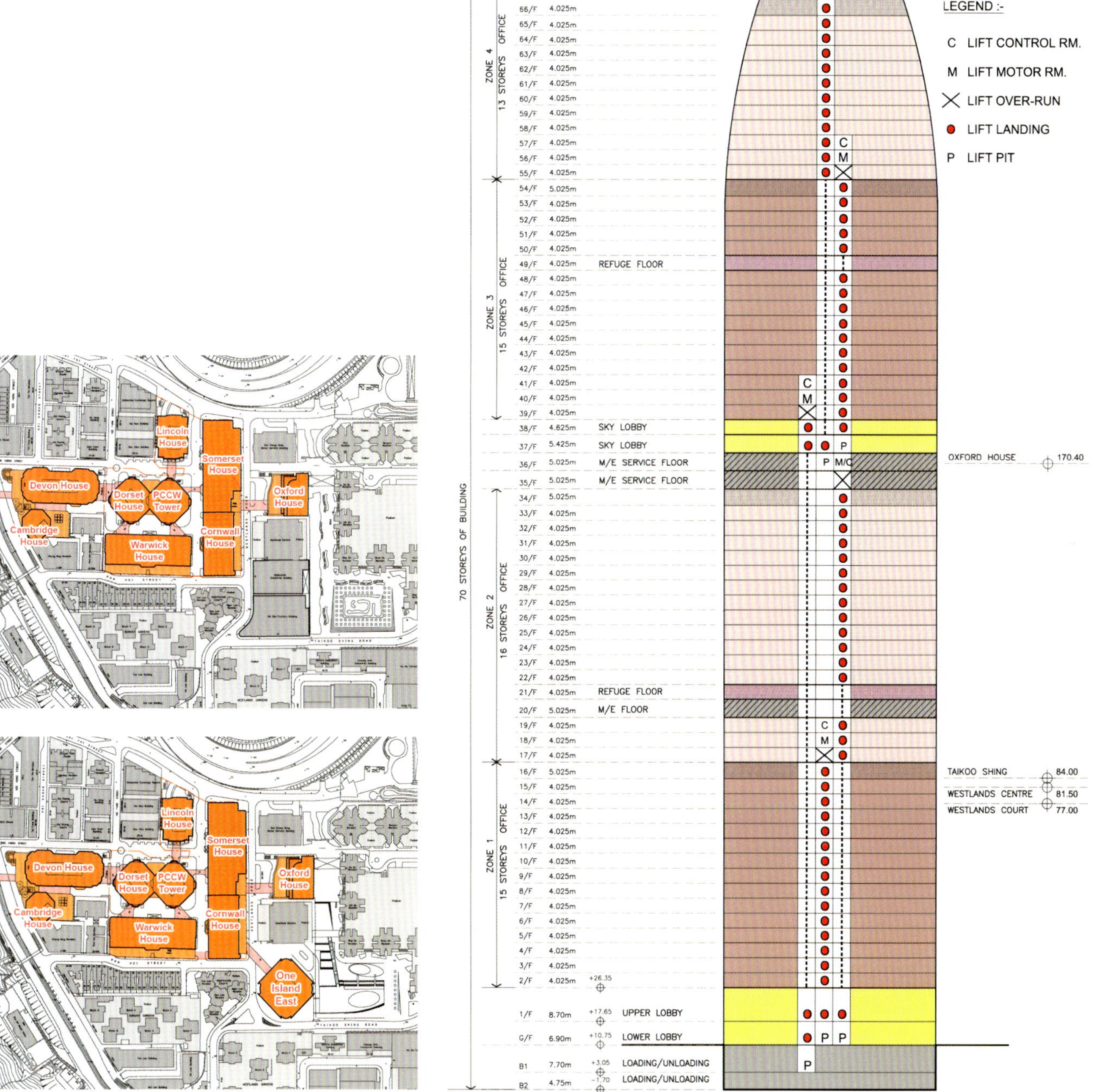

剖面图 SECTION

©Arup

©Arup

©John Nye

©Frank Fischbeck

©John Nye

©John Nye

©John Nye

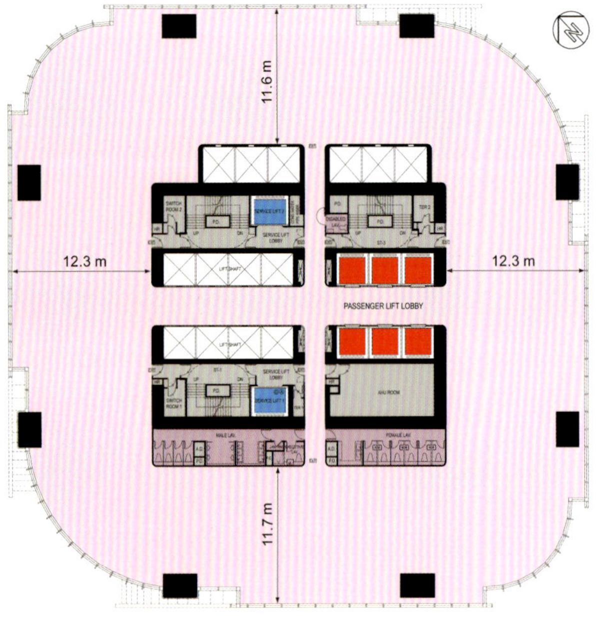

低区标准层平面图 LOW ZONE TYPICAL FLOOR PLAN

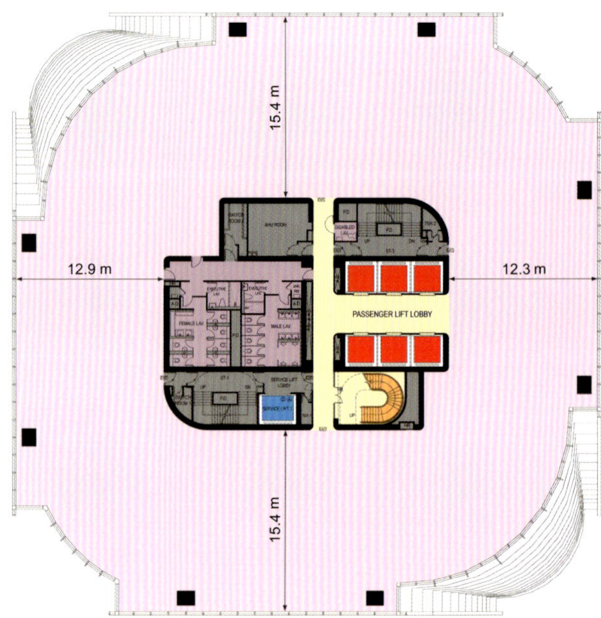

高区标准层平面图 HIGH ZONE TYPICAL FLOOR PLAN

©John Nye

©John Nye

One Island East is a 308 m tall grade-A office tower. The project was pioneering in terms of its extensive application and use of grade 100 high-strength concrete. Using grade 100 high-strength concrete, we increased useable floor area by 3700 m^2 compared to grade 60 concrete; reduced carbon dioxide emissions by tonnes and waste generated by 1200 tonnes. BIM has been used to check critical structural details. It was estimated that applying BIM to conduct automatic clash identification saved approximately 20 days in the construction of the outrigger, and 250 m^3 of construction waste.

The columns at the uppermost floors have been raked sideway to 'open up' the corners facing north and south to provide better views. To minimise the depth of the opaque spandrel and improve the transparency of the curtain wall, we employed a fire engineering approach to successfully reduce the spandrel depth from 900 mm to 650 mm. Modular and standardised designs, such as movable external pedestal paving, wall panels, light troughs and air handling units, were adopted for One Island East to maximise production and material efficiency as well as to minimise waste generation. The standard of interior environment sets a high benchmark for fit-outs that are healthy, productive places to work, maximise operational efficiency, and have a minimised environmental footprint. The open plan offices and lighting have been designed to maximise natural lighting and enhance work productivity.

CITIC PLAZA, SHANGHAI
中信广场

地点：中国上海
建筑面积：109 828 m²
建筑设计：日建设计

Location: Shanghai, China
Building Area: 109,828 m²
Architect: Nikken Sekkei

World Green Buildings—— Thermal Environmental Engineering Solutions

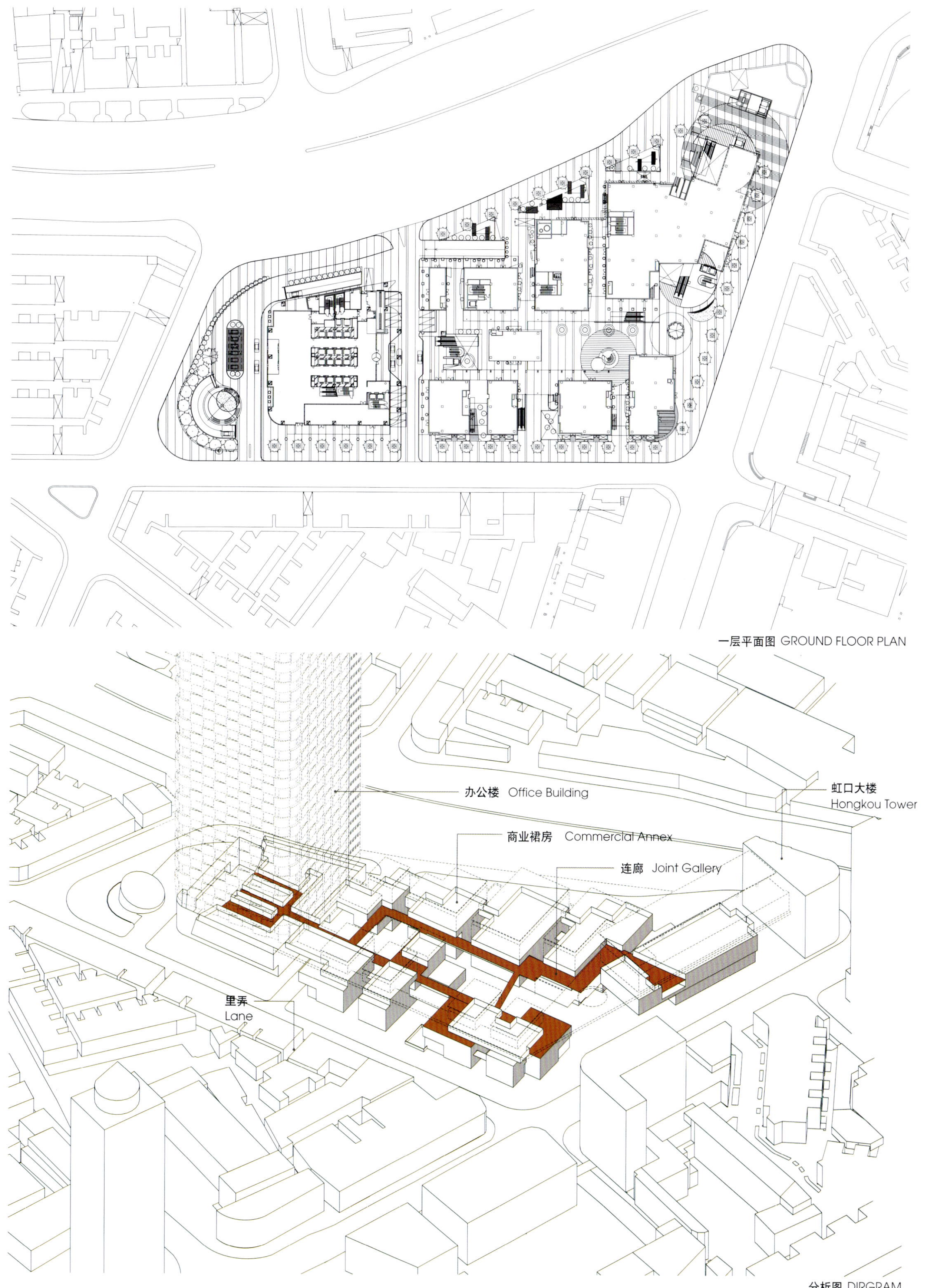

一层平面图 GROUND FLOOR PLAN

分析图 DIRGRAM

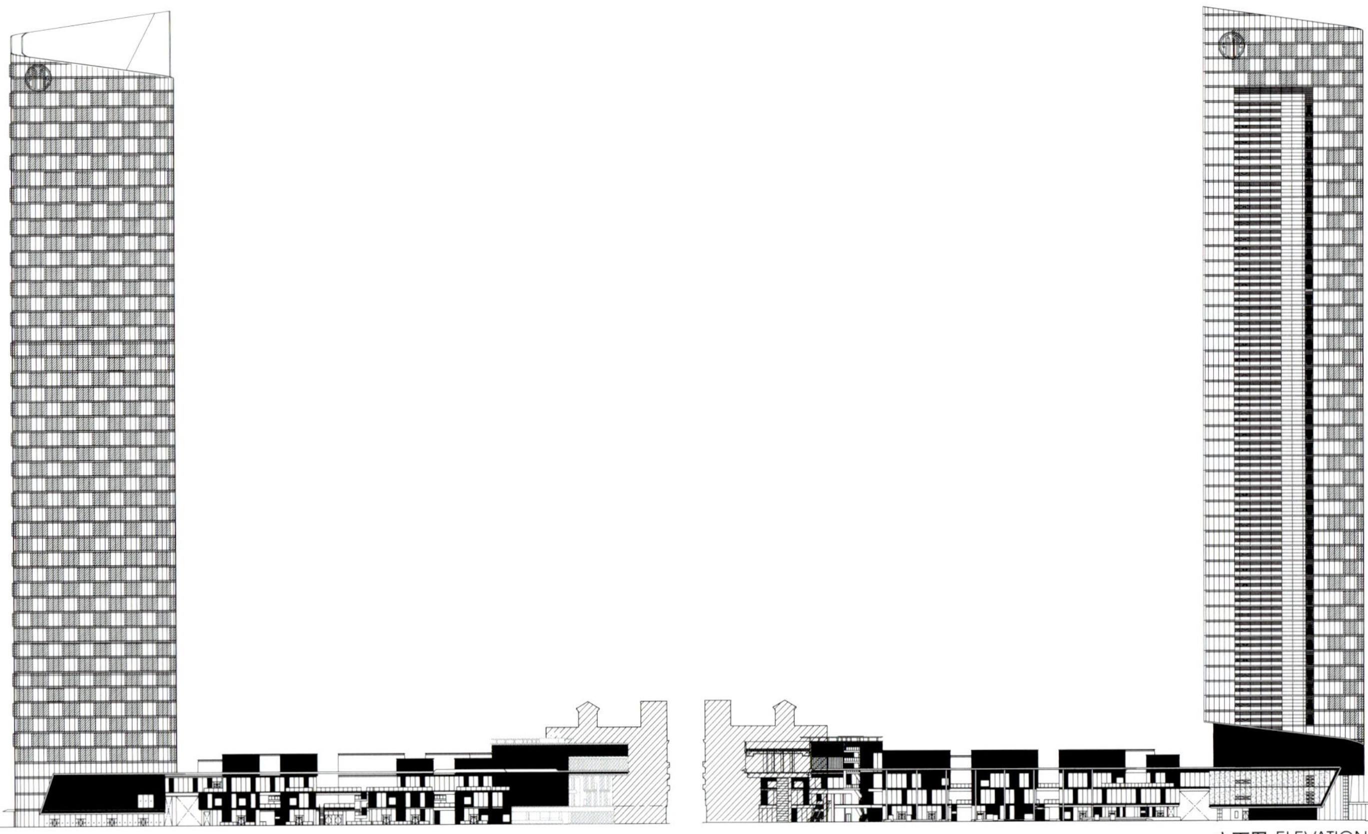

立面图 ELEVATIONS

项目概况

上海外滩中信广场位于上海市四川北路，占地面积约为5.8 hm^2。该地块位于四川北路的南端，开发潜力巨大。随着周边地铁和高架路不断完善，该建筑将会成为上海市中心的高端办公和商业楼。在总体规划和设计中，该建筑物要实现：四川北路传统商业街的二次开发；与周边的城市规划和设计相融合；历史建筑的保护和再利用；建成具有商业和办公功能的多功能建筑物；最大限度地利用城市景观。

建筑热环境解决方案

设计引入创新技术，超高层办公的自然通风口设计就是其中最典型的例子。建筑师将4.5 m×4.5 m见方的幕墙单元在平面上以千鸟格状有规则地倾斜错位，面向上风方向顺序展开,利用在鳞片状的幕墙单元的侧面设置的格栅实现高效的自然通风。鱼鳞状的幕墙既有利于减小风压的影响，又能表现建筑表皮的进深感和光影变幻。茶红色的格栅能够防止雨水侵入，同时在特定的视角上呼应了低层部分的外墙设计元素，强调了设计的整体性。

NOKIA CHINA CAMPUS
诺基亚中国园区

©Ben McM

地点：中国北京
建筑面积：74 000 m²
建筑设计：奥雅纳
所获奖项：2008年香港绿色建筑奖优异奖；
中国首座获LEED-NC金质认证楼宇
委托方：诺基亚（中国）投资有限公司

Location: Beijing,China
Building Area: 74,000 m²
Architect: Arup
Selected Awards: Hong Kong Green Building Award (Merit), 2008;
LEED-NC Gold-certified
Client: Nokia (China) Ltd.

©Ben McMillan

©Ben McMillan

©Jerry Lee

©Jerry Lee

根据委托方对于可持续设计的要求，奥雅纳公司的建筑师和工程师设计了一座低碳大楼。项目采用了双层幕墙、遮阳系统、优化自然通风和采光，以及节水系统。

园区设计结合了三十多种不同的节能技术。双层幕墙在室内和室外气候之间形成了一道屏障。自然通风窗格通过温控空洞吸入或排出空气，对室内微气候进行调节，较美国采暖、制冷与空调工程师学会及北美照明工程协会制定的90.1-2004标准节能20%。

园区是业主的一项示范性绿色资产，对于增进中国区的业绩大有益处。奥雅纳为诺基亚北京园区提供了整体设计服务，使其成为中国第一座获得LEED-NC金质认证的楼宇。诺基亚园区是一个综合研发基地，集合了办公设施、研发实验室、信息会议区、礼堂、餐厅、停车场和自行车棚。

通过计算流体力学、热工及能量模型、最优化结构和建筑持续工具，合理使用空调设备调节室内微气候，实现总体节能14%；通过水资源循环利用等节水技术，较其他同类设施节水37%。

用户舒适度在设计中也是主要考虑因素之一。97%室内空间——包括体育馆、公共食堂和幼儿园在内——都有观景窗。开敞的中庭开有天窗，引入自然光照。

©Jerry Lee

©Jerry Lee

BOSCH CHINA HEADQUARTERS

地点：中国上海
建筑面积：78 000 m²
所获奖项：上海市节能办授予2010年度“上海节能模范单位”；
2011年住房和城乡建设部绿色建筑评价标志（二星级）
委托方：博世中国

Location: Shanghai, China
Building Area: 78,000 m²
Selected Awards: 2010 Energy-Conservation Exemplar Unit of Shanghai; 2011 Evaluation Standard for Green Building (Two Star) Credited by MOHURD
Client: BOSCH China

World Green Buildings——Thermal Environmental Engineering Solutions

项目概况

博世中国总部大楼位于上海市长宁区（浦西），建成后成为博世集团在中国的业务枢纽。整个建筑群由三幢高层建筑通过联廊连接，主体办公大楼为九层，高度39.6 m，其二层、三层设置餐厅及会议区，西南角为另外两座九层办公楼。主要建筑结构形式为钢筋混凝土框架抗震墙结构。博世集团在建筑智能化技术领域是世界领先的技术及服务供应商之一。大楼应用了大量节能环保技术，其中部分技术由博世集团子公司提供，将绿色建筑的概念化为现实（据估算，博世总部大楼每年能减少约2000 t CO_2排放并能节约7000 MWh的电力。依据温室气体盘查议定书：1 MWh电能=787 kg CO_2，1 MWh天然气=220 kg CO_2进行换算，并通过设备正常条件下功率计算得出）。

建筑热环境解决方案

1.生活热水系统

主要由太阳能集热系统、蓄热系统、二次加热、循环系统等子系统组成。太阳能集热系统通过热交换器，将收集到的太阳能传递给蓄热系统备用，市政供水（冷水）首先跟蓄热系统进行 热交换，如因天气、季节等原因导致交换后的水温达不到使用的要求，该热水会同专用热水锅炉进行热交换（二次加热），使水温始终维持在设计使用的温度（60 ℃），然后进入热水循环系统供应到各用水点。热水循环系统采用全日制机械循环，热水循环泵的启闭由设在热水回水管上面的电接点温度计自动控制，从而在确保热水24 h不间断供应的情况下实现了电能的节省。另外对于部分集中供热不方便的区域，灵活地采用了分散式电热水器作为补充。

2.地源热泵系统

项目设计包括制热（锅炉），制冷（冷水机组）和地源热泵系统以满足采暖和制热的需求。

系统基本信息：
地热桩基数目：300；
钻孔深度：120 m；
加热/制冷功率：2200 kW / 1900 kW。

冬季状态下，热泵机组承担热负荷的同时，将未使用的冷负荷输入地源系统交换。夏季状态下，热泵机组承担冷负荷的同时，将未使用的热负荷输入地源系统交换。

3.热回收系统

热回收系统的主体为热转轮全热交换器。安装在组合式空调机组内。该设备为蓄热、吸湿性蜂窝状转轮。冬季，室内回风过滤后通过转轮处理，转芯温度升高，水分含量增加，当转轮与室外新风侧接触时，会释放出热量和水分。夏季与之相反，转轮会降低新风的温湿度。从而有效地节约了处理空气所需要的能量。

4.外遮阳系统

博世大楼玻璃幕墙外立面，增加了智能百叶系统，通过对照度的感应和太阳照射角度的计算自动控制百叶的开度，可以遮挡太阳直射，节约空调费用；引入自然光线，节约照明费用；消除眩光对眼睛的直射。因此，在防止阳光直接射入室内的同时，尽可能张大百叶角度，能够保持良好的视觉效果，让自然光反射到室内天花板，增加室内的亮度。

5.光伏发电系统

博世大楼两个屋顶上，由博世（中国）建设了太阳能光伏并网电站，有晶硅电池组件474块，单晶硅电池组件308块，总装机容量为123 kWp，全年产生的绿色电能约122 000 kWh。

Program Description

Bosch China Headquarters, the central official hub of Bosch China business, locates in Changning District of Shanghai. The complex contains 3 tall buildings connected by corridors. The main office building has 9 floors with the height of 39.6 m. Restaurant and meeting areas are set in the 2nd and 3rd floors. The rest two office buildings are in the south-west of the site. The buildings are constructed in RC frame-shear wall structure. Bosch Group is one of the world's leading technology and service providers in the field of building intellectual technologies. Quite a few energy saving technologies are applied in the buildings, many of which are supplied by subsidiary companies of Bosch Group, taking the concept of green building into reality. (Approximately based on the calculation, some 2000 metric tons of CO_2 and 7000 megawatt hours of energy per year could be saved in Bosch China Headquarters.Conversion according to Green House Gas Protocol (GHG): Electric Energy 1 MWh=787 kg CO_2, Natural Gas 1 MWh=220 kg CO_2)

Building Thermal Environmental Engineering Solutions

1. Domestic Hot Water System

Mainly composed of solar energy collector system, heat storage system, reheating system, the circulatory system and other subsystems. The solar energy collector system transfers the collected solar energy to heat storage system by heat exchangers. Municipal water supply (cold water) exchanges heat with heat storage system first. If the temperature of water could not reach the standard after the exchange course due to weather and climate impacts, the hot water will exchange heat with an exclusive boiler (reheating), until the temperature of water maintains at 60 °C. Then the water enters into the hot water supplying facilities. The hot water circulatory system is adopted with mechanical circulation all-day, and the hot water circulatory pump is controlled by the thermometer setting on the return pipe of hot water, ensuring energy conservation with uninterrupted supply of hot water 24/7. Besides, in some area where central heating could not be easily accessed, distributed electric water heaters as an alternative are installed as perfect supplement to the whole domestic hot water system.

2. Geothermic Thermal System

The design includes heating (boiler), refrigeration (water chiller) and geothermic heat pump system to appease the demand of heating.

Basic information:
Geothermic pile foundations: 300;
Depth of drill: 120 m;
Power of Heating/Cooling: 2200/ 1900 kW.
In winter, the heat pump units bear the heating load, meanwhile imports the untapped cooling load into geothermal system for exchange. In summer, the heat pump units bear the cooling load, meanwhile imports the untapped heating load into geothermal system for exchange.

3. Heat Recovery System

The principal part of heat recovery system is ECSE thermal heat exchanger, installed in the assembled air-conditioning units. The equipment is a kind of heat-harbouring and moisture- absorbing honeycomb type runner. In winter, interior return air is disposed by the runner after filtration. The core of the runner is calefactive continuously, and moisture increases. It will release heat and moisture when the runner meets the exterior fresh air. Contrarily, in summer the runner will decrease the temperature of fresh air, thus save energy for colling down the temperature.

4. External Sunshading System

Intelligent Venetian Blind system is equipped on the glazed wall of the headquarters. The opening degree of grilles is automatically controlled based on the induction to luminosity and calculation of sun elevation angle. The sunshading system keeps the sun from shining in without any shelter which saves energy for air-condition system, but meanwhile it allows certain natural daylight in, to support interior illumination which also saves certain energy in lighting Thus in general, even though the sunshading system stops the sun from shining directly in, it is still able to maintain relatively good lighting effect by opening the blades as much as possible to allow daylight in and to reflect to the ceiling.

5. Photovoltaic System

On the roofs of Sgh201and Sgh202, Bosch built a pv grid-connected plant in Bosch China Headquarters with 474 pieces of thin film modules and 308 pieces of mono crystalline modules. The gross installed capacity is 123 kWp, and the green electricity power generated in a year is about 122 thousand kWh.

ADVICE HOUSE, LYSHOLT PARK
LYSHOLT园区咨询大楼

地点：丹麦瓦埃勒
规模：5000 m²
建筑设计：C. F. Møller Architects
委托方：Lysholt Erhverv A/S

Location: Vejle, Denmark
Size: 5000 m²
Architect: C. F. Møller Architects
Client: Lysholt Erhverv A/S

总平面图 SITE PLAN

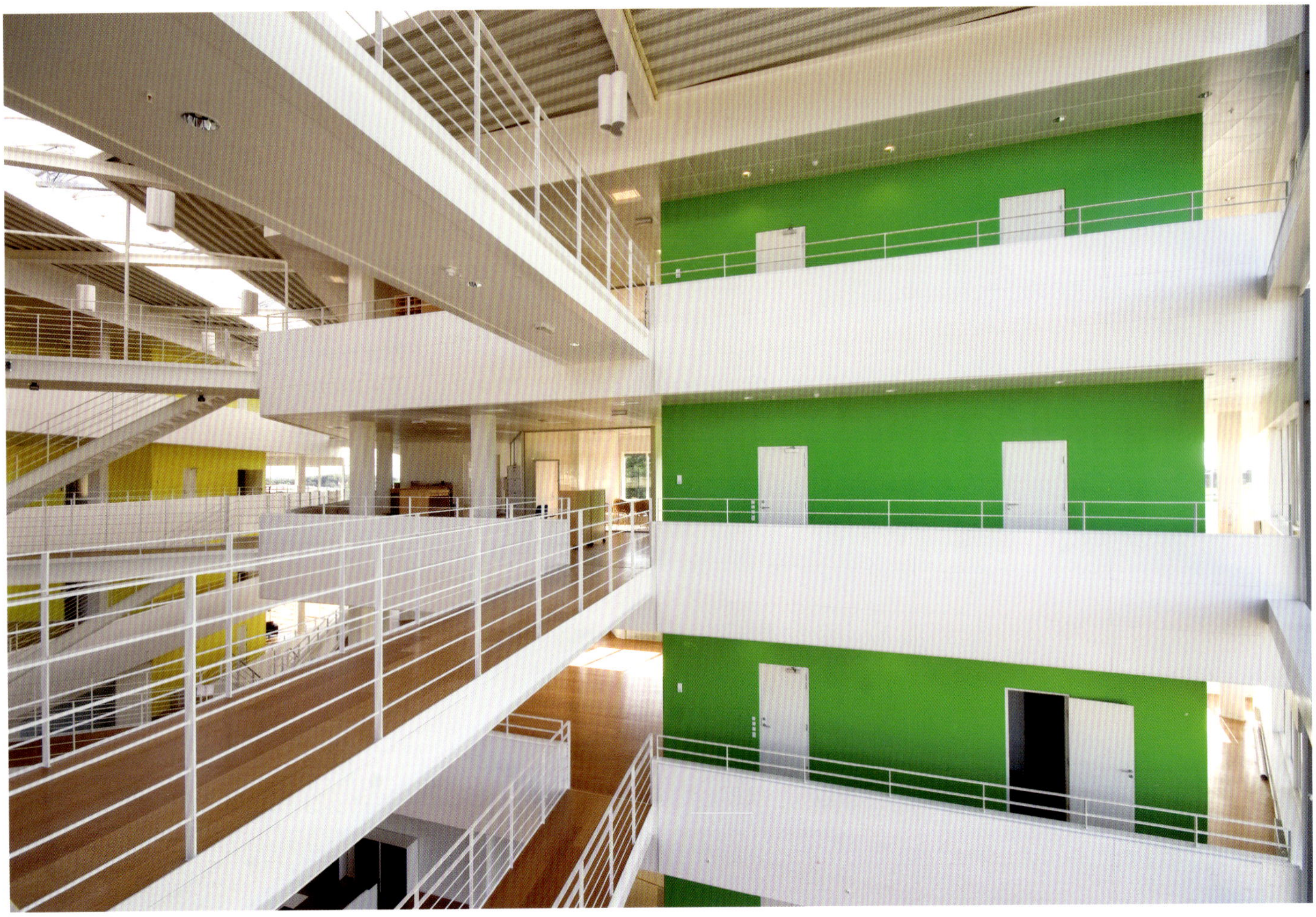

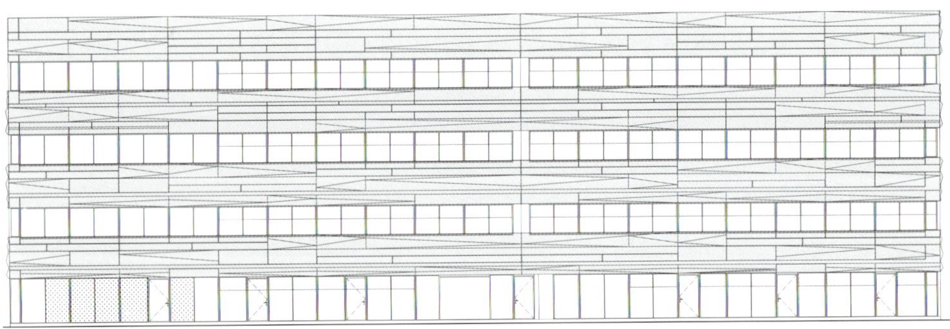

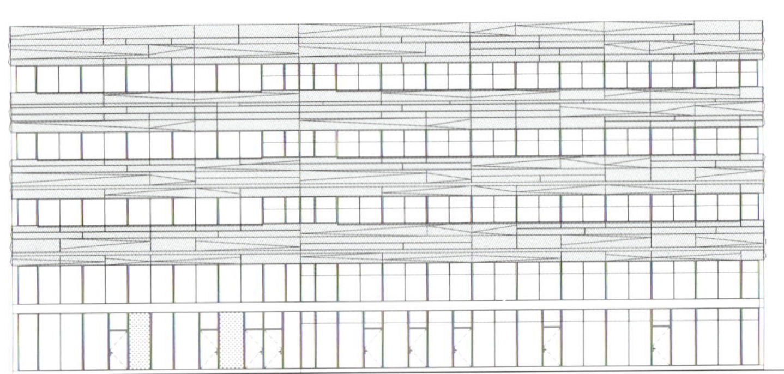

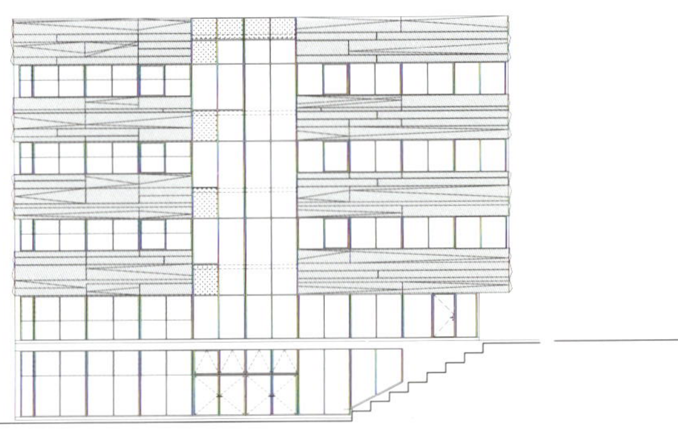

立面图 ELEVATIONS

项目概况

Advice House是瓦埃勒北部新开辟的Lysholt工业园区中第一座竣工的建筑。它的位置靠近高速公路，因此被设计成代表园区形象的引人注目的地标建筑。大楼由互成角度的两个办公翼楼组成，由一个中庭分隔，分别与两翼构成相同的角度。建筑整体类似于有一个角被推向内侧的六边形。穿过中庭的人行道连接两个翼楼，连续不断的窗子形成序列，为空间设计提供了极大方便。

建筑热环境解决方案

大楼采用了自然通风和机械通风，形成了混合通风模式。活动窗为大楼提供新鲜空气，独立办公区的隔断上留有通风口，保证空气流动。宽大开敞的中庭利用堆栈效应使用自然通风，热空气通过中庭顶部的天窗排出。内部开敞通透的空间依靠自然通风。此外，大楼还安装了机械通风系统。变风量控制空气交换系统与通风设备及热交换器等相结合，确保环境的舒适宜人。覆层条带由13个多角度的覆层面板错落有致地排列而成。传感器控制的半透明百叶窗对采光进行调节。

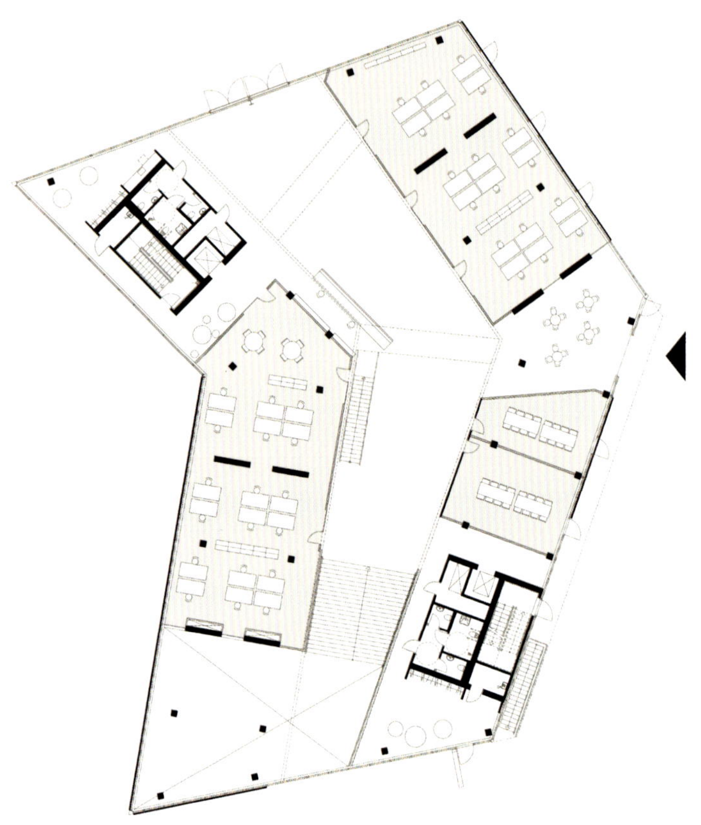

一层平面图 GROUND FLOOR PLAN

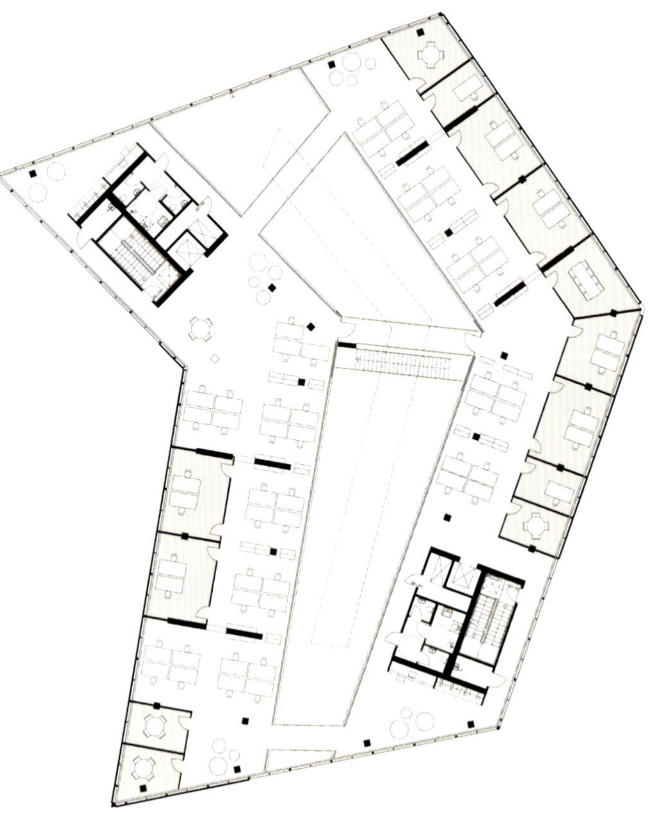

标准层平面图 TYPICAL FLOOR PLAN

OFFICE
OFFICE
OFFICE
OFFICE
ATRIUM
OFFICE
OFFICE
CONFERENCE ROOM
LOBBY
ENTRANCE
CONFERENCE ROOM
STORAGE & PLANT ROOM

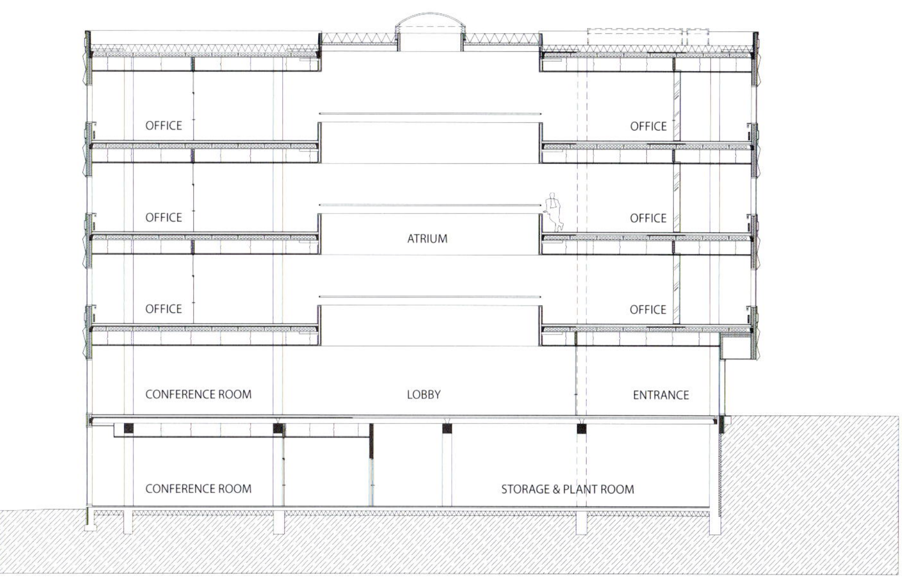

剖面图 SECTION

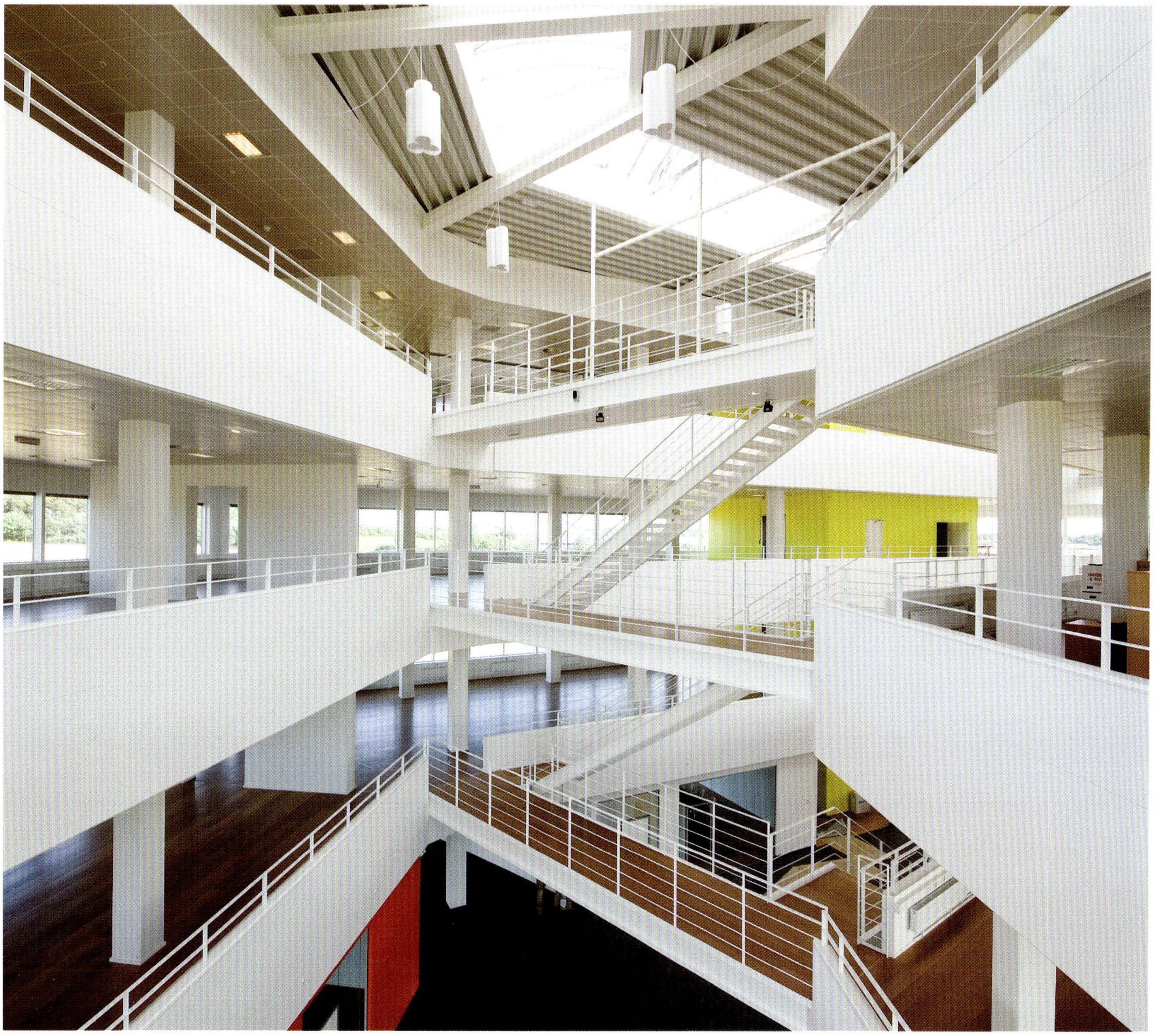

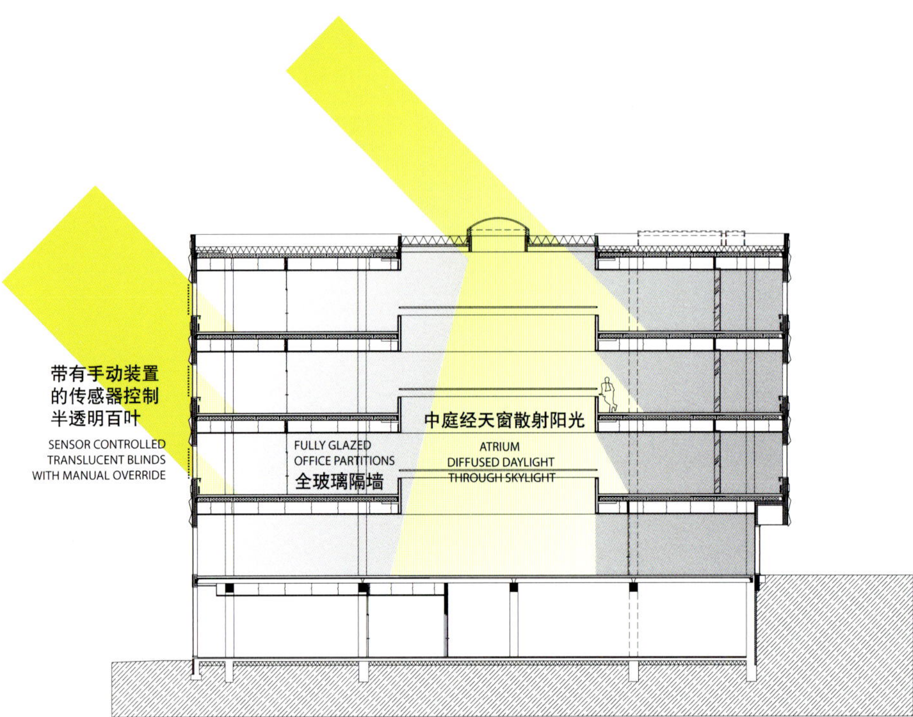

自然采光分析图 DAYLIGHT DISTRIBUTION DIAGRAM

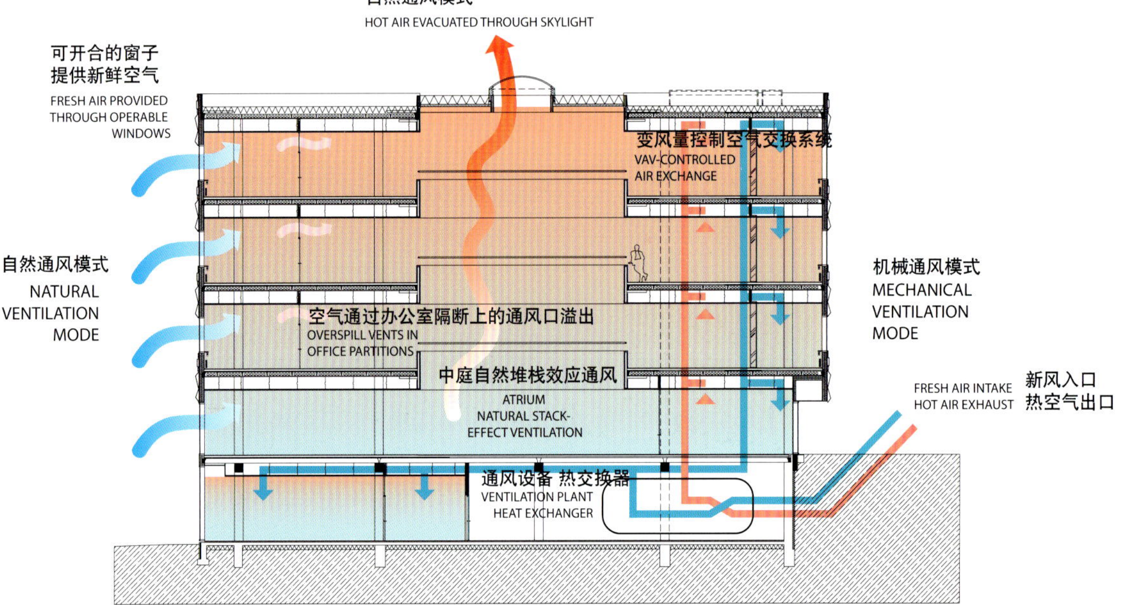

自然通风与机械通风结合的混合通风模式分析图 HYBRID VENTILATION DIAGRAM - NATURAL AND/OR MECHANICAL

Program Description

Advice House is the first completed building in the Lysholt Park, a new business-park north of Vejle, and is with its proximity to the motorway designated to act as landmark and eye-catcher for the entire development. The building is shaped around two angled office wings, separated by an equally angled atrium, resulting in a plan resembling a hexagon with one angle pushed inwards. The two wings connect by walkways across the atrium, and the floors' continuous window-bands give a high degree of freedom in the space-planning.

Building Thermal Environmental Engineering Solutions

Both natural and mechanical ventilation measures are fulfilled in the building, contributing the hybrid ventilation solution. Fresh air is provide though operable windows forming the window-bands, and there are overspill vents in office partitions. The open atrium is designed to use natural ventilation basing on the stack effect. Hot air evacuates though skylight of the atrium. The open and transparent interior is also naturally ventilated. On the other hand, the building is equipped with mechanical ventilation. VAV-controlled air exchange combines with ventilation plant and heat exchanger ensures the environment comfortable and pleasant. The cladding-strips are composed of a 'random' sequence of a total of 13 differently proportioned cladding panels. Daylight distribution are adjusted by sensor controlled translucent blinds with manual override.

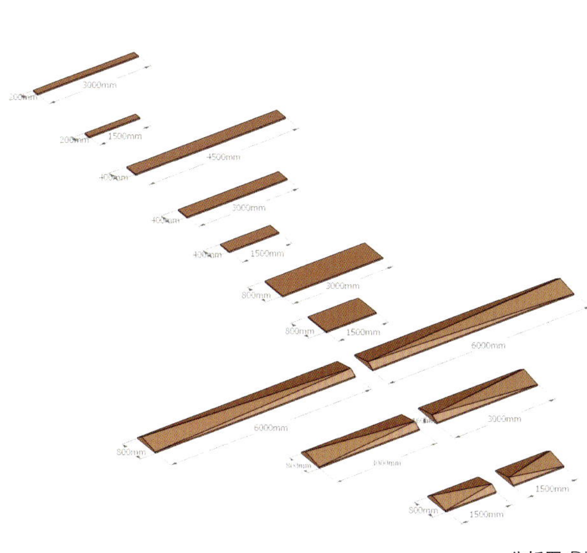

分析图 DIAGRAM

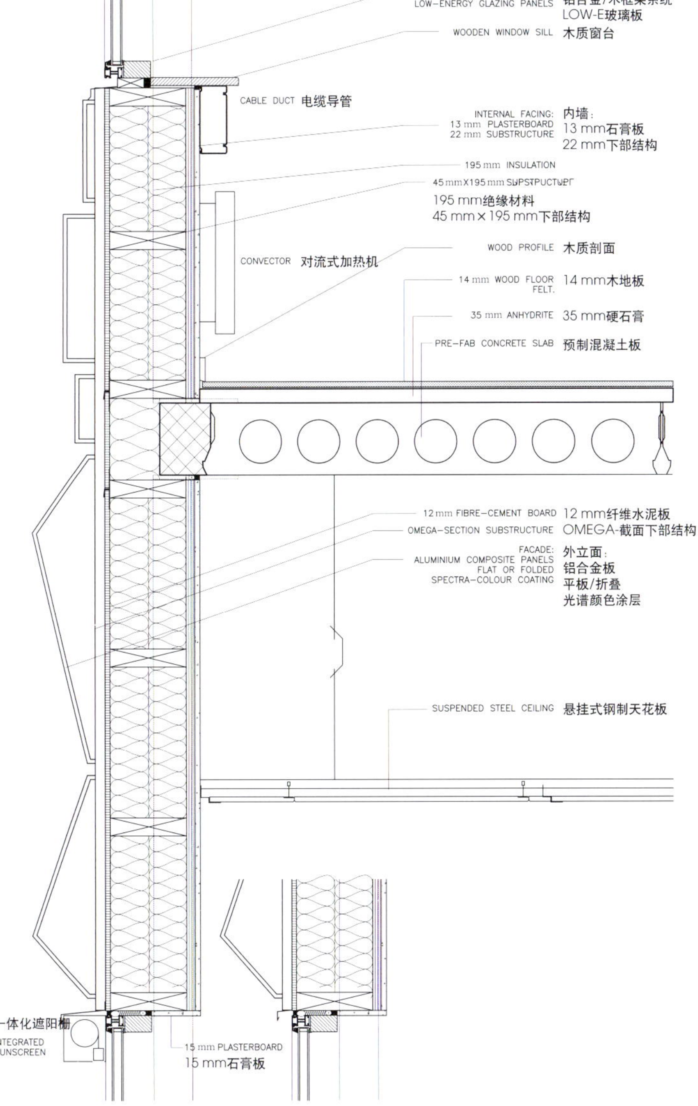

节点图 DETAIL

ENEA HEADQUARTERS
ENEA公司总部

地点：瑞士Jona
占地面积：2787 m²
建筑设计：Oppenheim Architecture+Design
所获奖项：中标方案（第一名）
摄影：Martin Rütschi

Location: Jona, Switzerland
Area: 2787 m²
Architect: Oppenheim Architecture+Design
Selected Awards: Competition Entry (First Place)
Photography: Martin Rütschi

项目概况

项目位于瑞士Obersee湖岸旁的建筑将成为世界知名景观设计公司Enea Garden Design的总部。它以木质体量拔地而起，严谨优雅，与周围景观完美融合，与自然环境相呼应，与边界线和不同的空间条件相匹配。

项目包括了温室、仓库、工作室、展示间和管理区域。这些功能区域的排列顺序有助于大楼的体量安排。所有的区域都通过一个服务走廊相连，这个走廊贯穿大楼，它不仅组织了空间的流通，还统一了公司自身的生产线。

大楼形态根据功能重要性和场地条件设计：一方面，大楼的功能和项目流程的流线性决定其体型狭长而低矮，同时需要设置可达性的点状入口；另一方面，场地自然条件比如照明和通风，服务区域的可达性以及入口和服务区域前后区分的必要性，也让长形的体量成为一个恰到好处的设计。适度设计和尊重自然元素的设计理念加强了建筑设计的简明性。

设计师对建筑材料的运用是为了将建筑和自然环境完美融合。因为节能是设计过程中的重要原则，所以项目完全是从对自然资源利用的理念出发，包括自然采光，自然隔热系统，自然能量保持系统——地热交换，绿色屋顶等。为了达到更好的建筑性能，设计中还将采用一些绿色材料，比如可回收木头、节能装置。

5 6
8
4
1
7

2
3
1

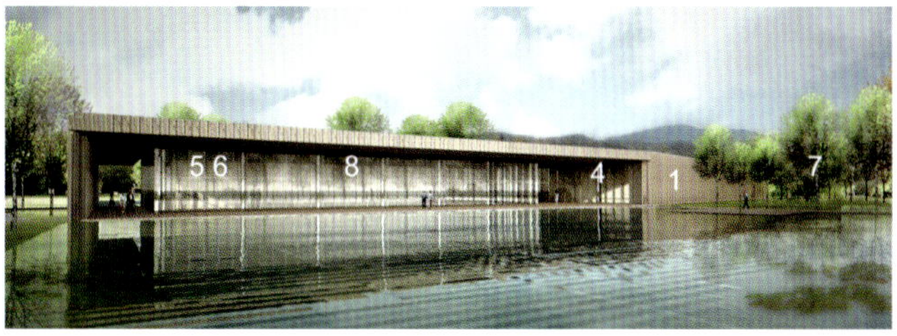

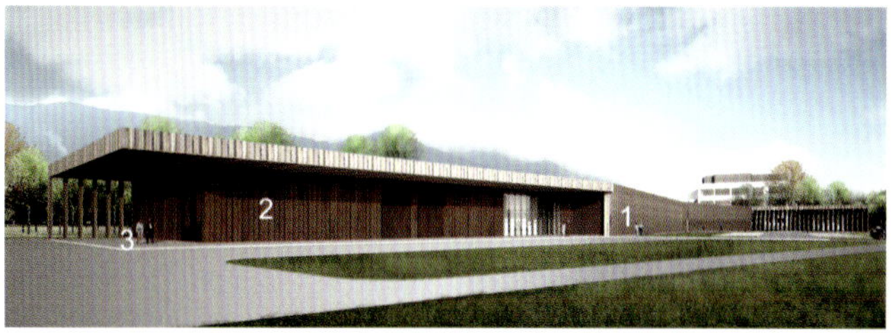

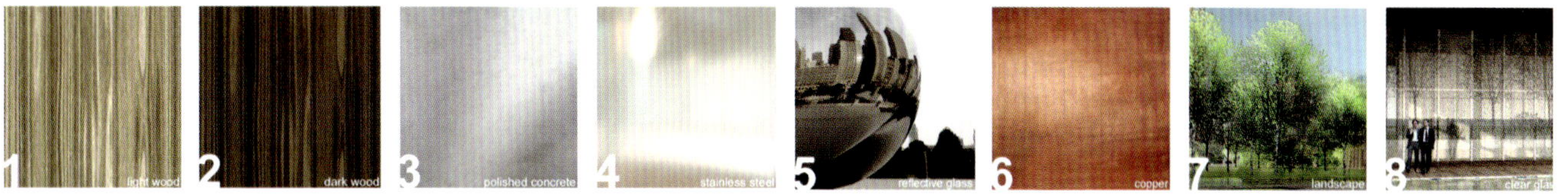

1 light wood
2 dark wood
3 polished concrete
4 stainless steel
5 reflective glass
6 copper
7 landscape
8 clear glass

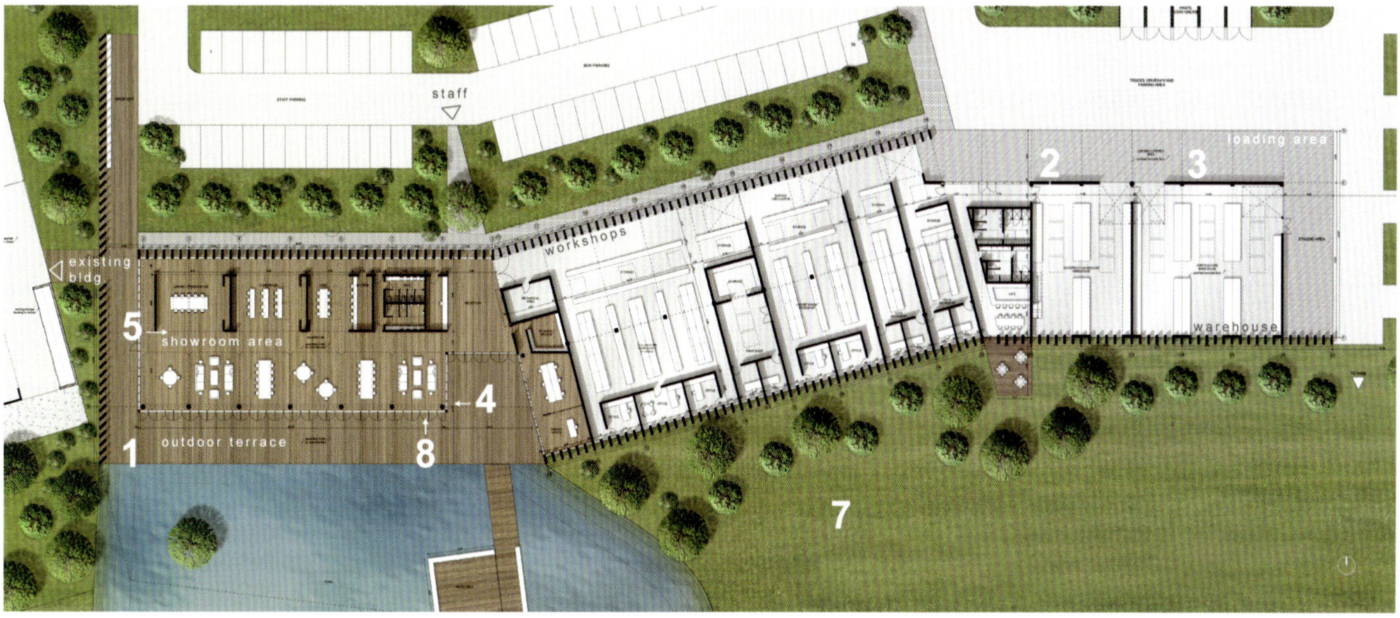

staff
loading area
2
3
workshops
existing bldg.
5
showroom area
4
outdoor terrace
1
8
warehouse
7

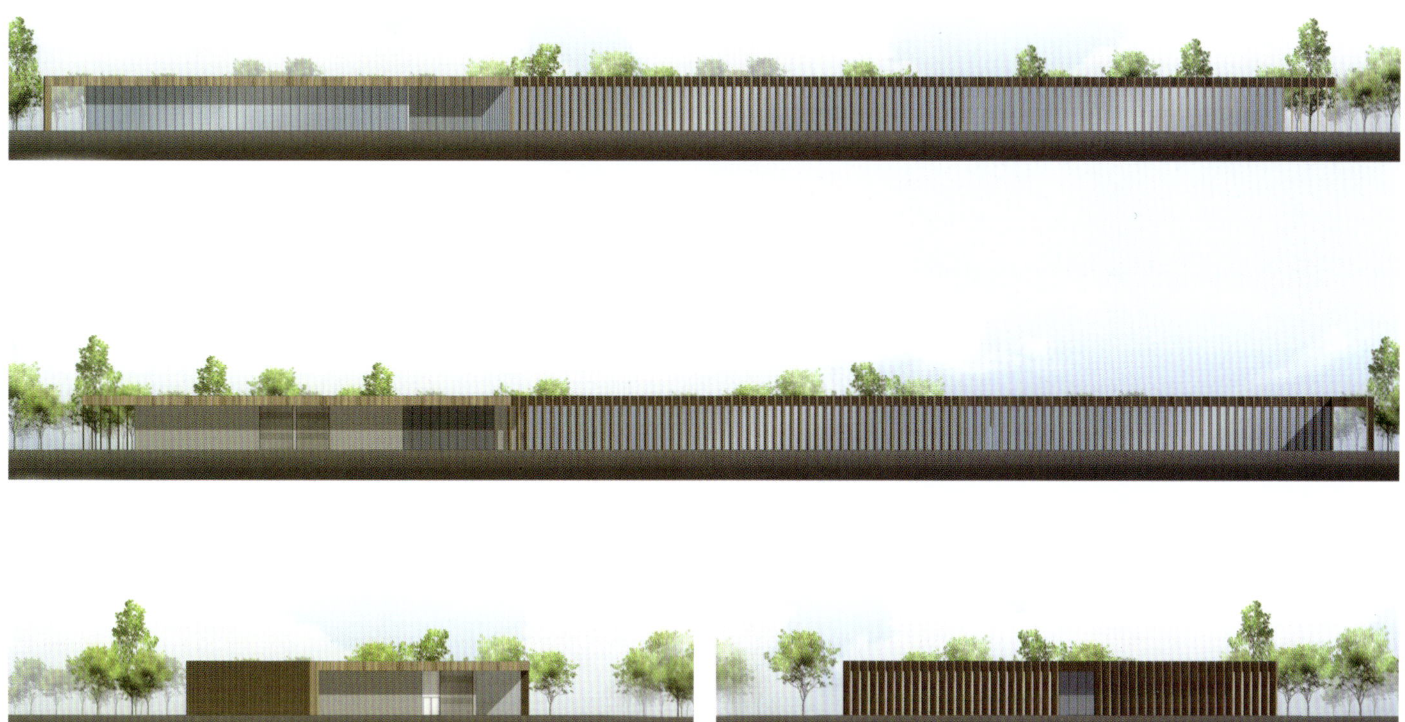

立面图 ELEVATIONS

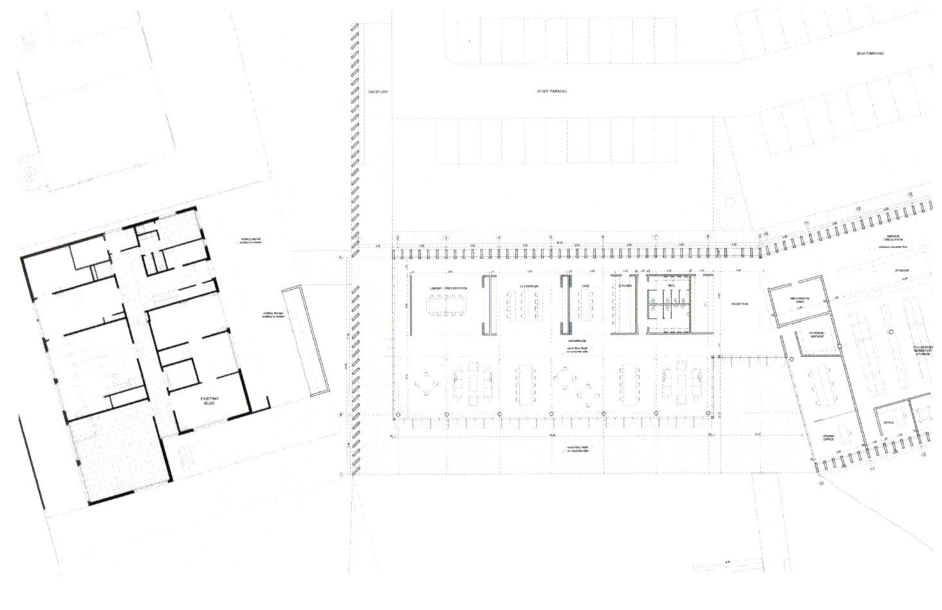

一层平面图 GROUND FLOOR PLAN

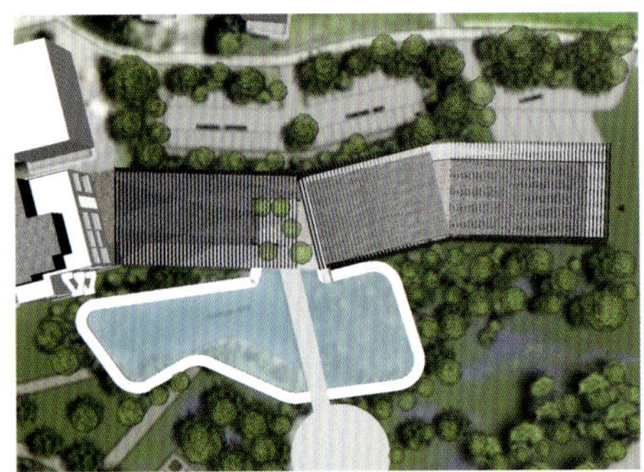

建筑热环境解决方案

1.生态墙

生态墙是一个立体花园。墙面是自然空气过滤器，当空气景观墙体进入建筑的办公空间时，过滤掉挥发性有机化合物及CO_2。生态墙是一个生物过滤器——用以过滤空气的绿植墙。这个过滤器的功效证明了生态墙能够改善空气质量，改良工作环境。生态墙可以重建与自然环境十分类似的生态环境。它的隔热效果非常好，因此降低了能量消耗，既能在冬天保温，又能在夏天提供一套降温系统。

2.储热器

在被动太阳能建筑中，热质有两种作用：第一，昼夜循环中快速吸收建筑所需太阳热能并避免热度过高；第二，当户外没有阳光时能够使储热缓慢释放。根据当地气候条件及建筑的用途，延时放热可能经历数小时或者数天。

储热策略：地面材料、太阳能吸热壁、绝缘砌体或混凝土墙体、双层石膏板、水。

3.绿色屋顶

建筑的屋顶部分或全部被植物和土壤所覆盖，或为置于防水膜之上的培养基。植物覆盖的屋顶为用户提供宜人的空间。种植水果蔬菜花草，可减少建筑的热荷载（通过增加热质和热阻系数）及冷荷载（通过蒸发降温），降低城市热岛效应，增加屋顶使用寿命，降低雨水流失率，过滤污染物及空气中的CO_2和雨水中的重金属元素，在建筑上层增加野生动物栖息场所。

4.蒸发冷却

通过蒸发降温是液体汽化时吸热降温而冷却的过程。蒸发冷却是建筑降温保证热舒适度的一种常用手段，因为较其他手段来说，该种方法成本更低，耗能较少，而且通常情况下水比建筑环境中的混凝土、灰泥、金属和玻璃等硬表面温度要低。

其他可持续设计措施

1.太阳能热水板

太阳能热水器是利用太阳能加热液体，用来向储热器传递热量。例如，对于家庭用户来说，可以将饮用水加热并存储到热水箱中。平板太阳热能收集器通常置于屋顶，具有一个附带循环管道的吸热板。吸热装置一般覆以特制表面，确保将太阳辐射转换为热能，管道中的循环液体将热量带到储存或使用它们的地方。

2.光伏建筑一体化（BIPV）

将光伏（PV）设备植入建筑围护。光伏组件起到双重作用——代替传统建筑围护材料并生成电能。因为省去了常规材料的费用，光伏设备的边际成本随之下降，而寿命周期成本上升。因此BIPV系统的总成本比单独使用PV系统更低。

3.消防用水池
4.自然采光和自动控光系统
5.雨水收集重用
6.绿肺
7.灰水处理重用

D
E
F
6.00
18.00

LOADING
parking
PARKING - BCH
PARKING - OFFICES
loading
offices
EXISTING BUILDING
showroom
workshops
f&b
warehouse
ARTIFICIAL POND

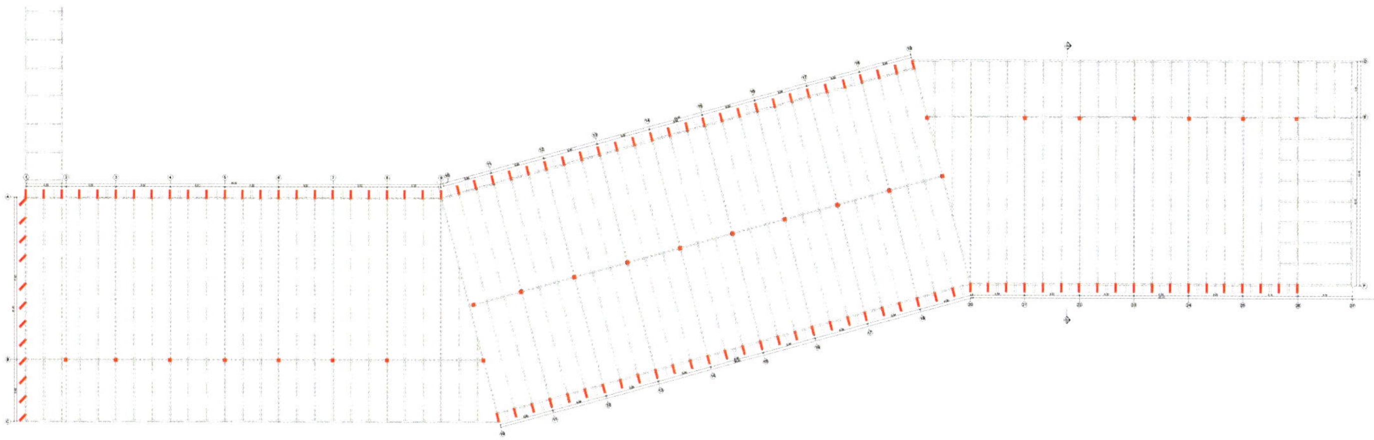

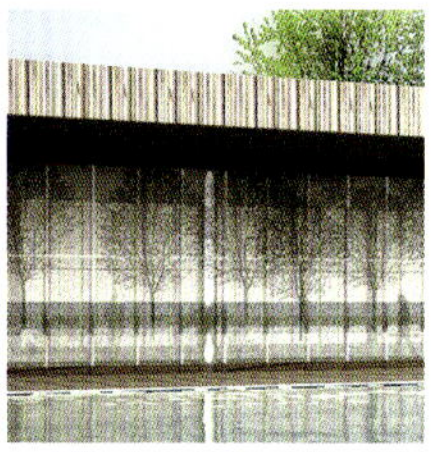

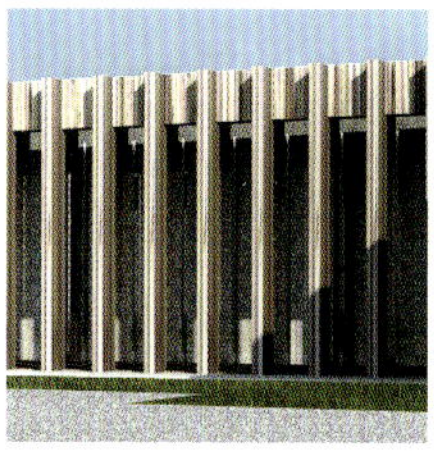

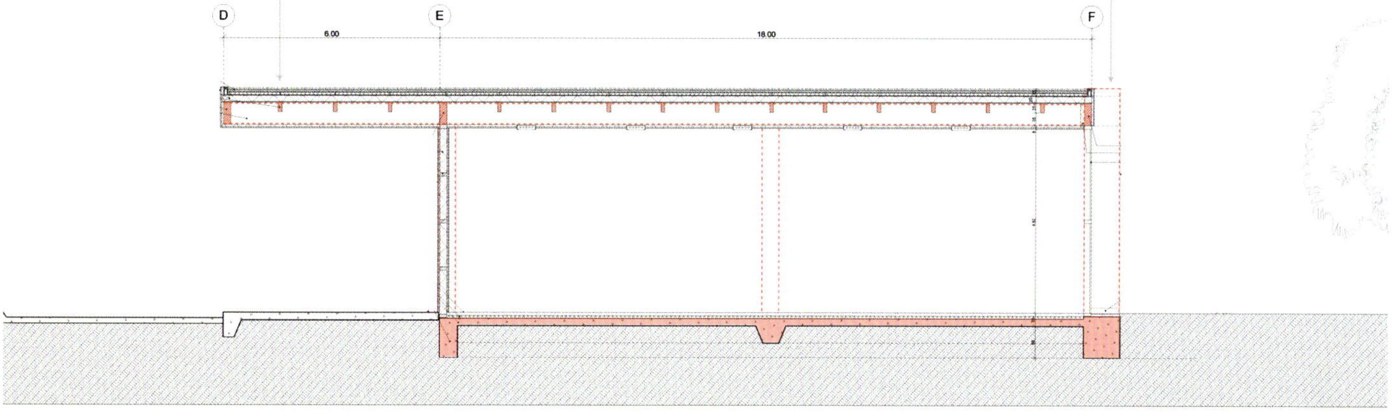

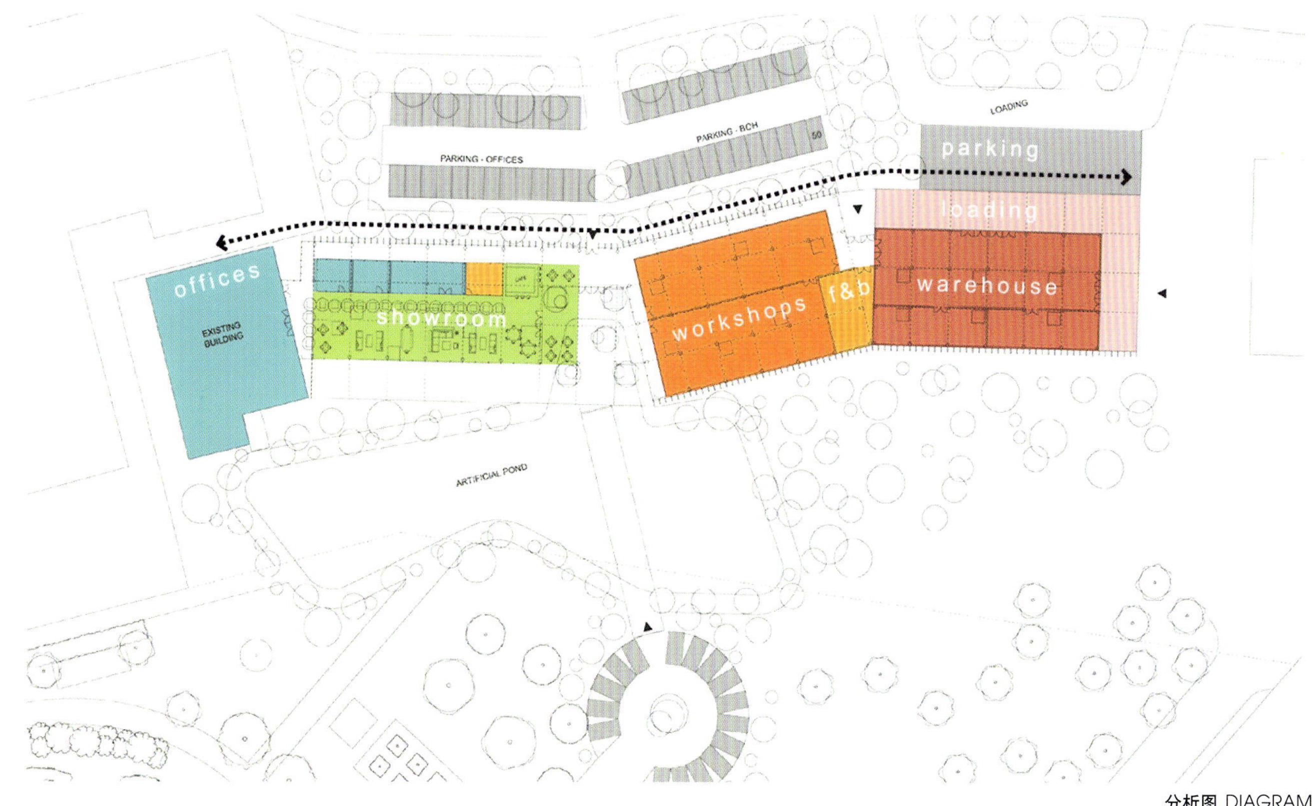

分析图 DIAGRAM

Program Description

The project, located at the shores of the Obersee in Switzerland, will house the headquarters of the internationally recognized landscape architecture firm Enea Garden Design. It emerges from the site as a wooden volume, discrete and elegant, and blends with the landscape as it responds to the natural environment, boundary lines and zoning conditions.

The program includes greenhouses, warehouses, workshops, showrooms and administrative areas. The sequence of these components contributes to the volumetric arrangement of the building - all areas are connected by a service corridor that runs along the building and organizes not only the circulation between spaces, but the production line of the company itself.

The building derives its form from programmatical and site conditions – on one hand, the flow of the processes and program required in the program and their architectural resolution generate a long and low building broken where access points are needed, on the other hand, natural site conditions such as illumination and wind, the service access and the necessity of differentiating front and back - entrance and service areas - make the long mass the appropriate response. Its architectural simplicity is reinforced by the necessity of architectural sobriety and respect towards the elements and nature at the site.

The materials used look forward to merge with the natural environment externally - in aesthetical terms, as well as internally - in functional terms. Energy efficiency is decisive in the design process, so the project, consequently, is based entirely on the idea of taking advantage of natural resources such as natural light, natural insulating systems and natural energy conservation systems such as geothermal exchange, green roofs, etc. It is also planned that 'green' materials such as reclaimed wood and energy efficient fixtures will be used for a greater building performance.

Building Thermal Environmental Engineering Solutions

1. Living Wall

Living wall is a vertical garden.The wall is a natural air filter which removes VOCs and CO_2 from the air as it passes through the wall into the building's office spaces. A Living Wall is a biofilter - a vertical wall covered in vegetation used for filtering air. The function of this biofilter is to demonstrate how living walls can enhance air quality, improve the work environment. Living wall allows to recreate a living system very similar to natural environments. It's very effective with the thermic isolation effect, thus lowering energy consumption, both in winter by protecting from the cold and in summer by providing a natural cooling system.

2. Thermal Storage

Thermal mass in a passive solar building is intended to meet two needs. It should be designed to quickly absorb solar heat for use over the diurnal cycle and to avoid

overheating. It should provide slow release of the stored heat when the sun is no longer shining. Depending upon the local climate and the use of the building, the delayed release of heat may be timed to occur a few hours later or slowly over days.
Thermal Storage Strategies: Flooring, Trombe wall, Insulated masonry or concrete walls, Double gypsum board, Water

3. Green Roof
Green roof is a roof of a building that is partially or completely covered with vegetation and soil, or a growing medium, planted over a waterproofing membrane. Vegetated roofs provide amenity space for building users, grow fruits, vegetables, and flowers, reduce heating (by adding mass and thermal resistance value) and cooling (by evaporative cooling) loads on a building, reduce the urban heat island effect, increase roof life span, reduce stormwater run off, filter pollutants and CO_2 out of the air, filter pollutants and heavy metals out of rainwater, and increase wildlife habitat in built up areas.

4. Evaporative Cooling
Cooling by evaporation is a physical phenomenon in which a liquid, typically into surrounding air, cools an object or a liquid in contact with it. Evaporative cooling is a very common form of cooling buildings for thermal comfort since it is relatively cheap and requires less energy than many other forms of cooling. Furthermore, the water will most likely always be cooler than the surrounding hard surfaces such as concrete, stucco, metals, and glass.

Other Sustainable Features

1. Solar Hot Water Panel
A solar water heater that uses the sun's energy to heat a fluid, which is used to transfer the heat to a heat storage vessel. In the home, for example, potable water would be heated and then stored in a hot water tank. Flat-plate solar-thermal collectors are usually placed on the roof, and have an absorber plate to which fluid circulation tubes are attached. The absorber, usually coated with a dark selective surface, assures the conversion of the sun's radiation into heat, while fluid circulating through the tubes carries the heat away where it can be used or stored.

2. Building-Integrated Photovoltaics (BIPV)
Building-integrated photovoltaics is the integration of photovoltaics (PV) into the building envelope. The PV modules serve the dual function of building skin-replacing conventional building envelope materials-and power generator. By avoiding the cost of conventional materials, the incremental cost of photovoltaics is reduced and its life-cycle cost is improved. That is, BIPV systems often have lower overall costs than PV systems requiring separate, dedicated, mounting systems.

3. Pool Water Use for Fire Sprinklers
4. Daylighting and Automated Lighting Control System
5. Rainwater Harvesting & Re-use
6. Green Lung
7. Gray Water Treatment and Re-use

HUAQIAO FINANCIAL PARK
昆山花桥金融园

地点：中国江苏
建筑面积：132 130 m²
建筑设计：FTA
所获奖项：国家绿色建筑三星级设计标志项目；
江苏省绿色建筑示范工程

Location: Jiangsu, China
Building Area: 132,130 m²
Architect: FTA
Selected Awards: Evaluation Standard for Green Building (Three Star) Credited by MOHURD; Green Demostration Project of Jiangsu Province

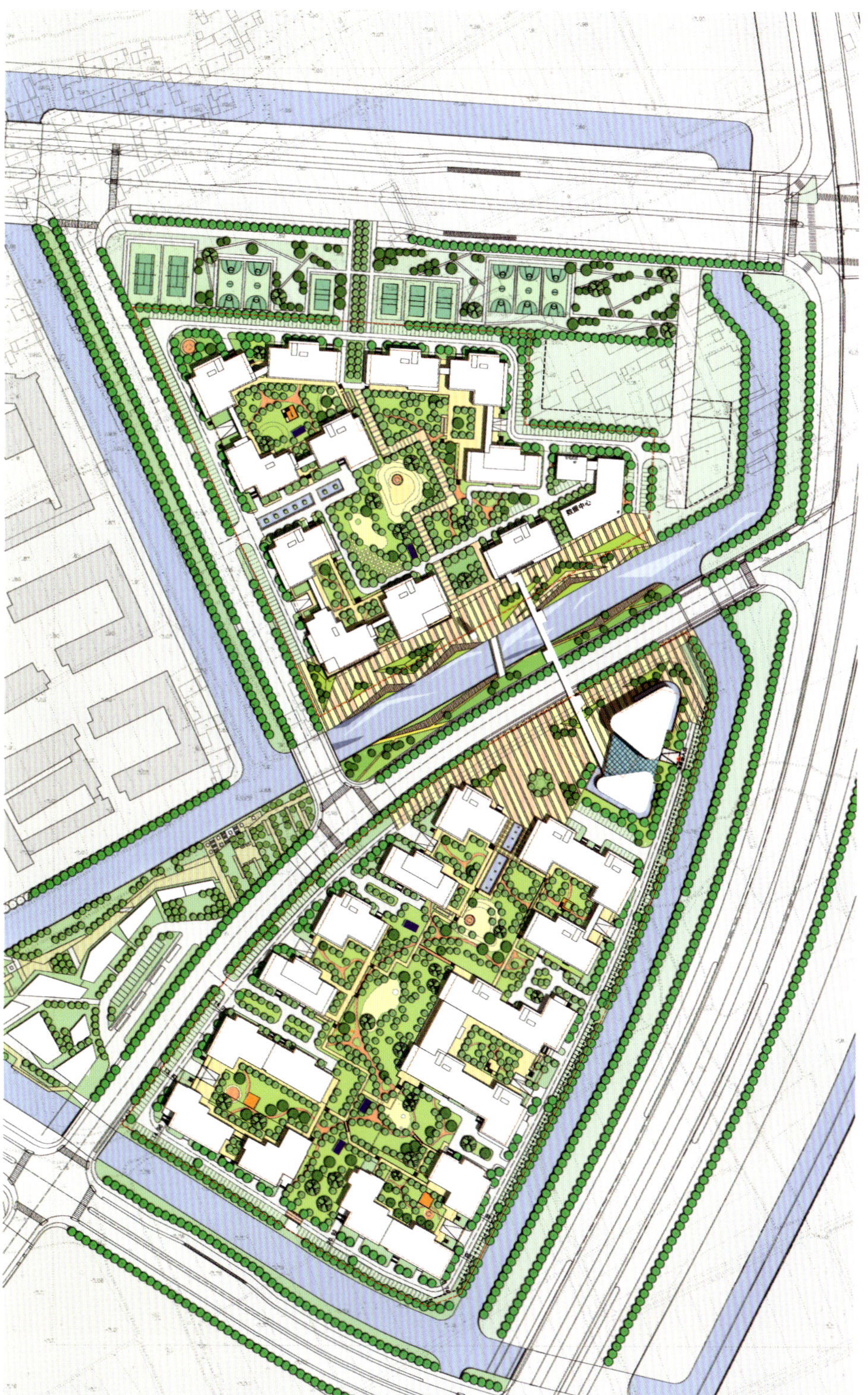

总平面图 SITE PLAN

项目概况

花桥金融园项目是2010年江苏省建筑节能和绿色建筑示范区——花桥国际金融服务外包区内的重点示范工程之一。项目共采用46项绿色建筑技术，突破了绿色建筑技术在传统单体建筑中的简单应用，实现绿色建筑技术在区域型建筑中的综合集成应用。园内全部建筑达到国家绿色建筑三星级标准，建筑节能≥65%。采用合同能源管理市场化运作机制，建设地源热泵集中式能源站，为建筑提供冷热源。推行绿色施工管理，探索绿色施工在区域性建筑中的应用。

建筑热环境解决方案

1.被动措施

项目总体规划时，按计算机日照模拟分析，确保南向办公空间均满足日照要求。适当加大外窗可开启的面积，引入穿堂风，提高自然通风的效果。建筑外窗使用固定及电动遮阳百叶，保持室内适宜温度，并使外遮阳与建筑立面一体化。

SONY
HP
IBM
SAMSUNG

SONY
IBM
HP

5.000
±0.000
-1.350
5000
1350
850
850
2400
4600
100

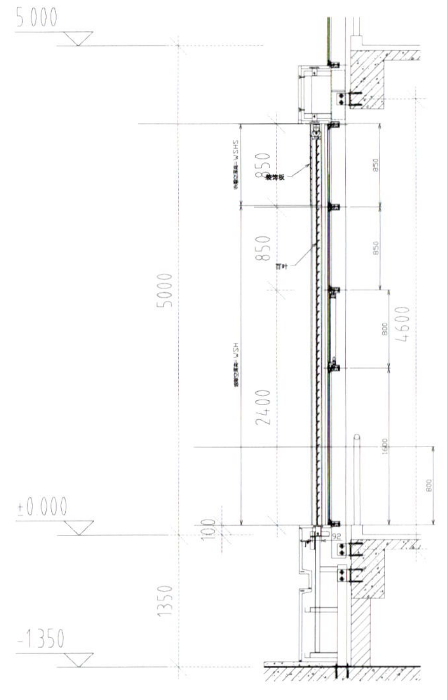

剖面图 SECTION

项目概况

花桥金融园项目是2010年江苏省建筑节能和绿色建筑示范区——花桥国际金融服务外包区内的重点示范工程之一。项目共采用46项绿色建筑技术，突破了绿色建筑技术在传统单体建筑中的简单应用，实现绿色建筑技术在区域型建筑中的综合集成应用。园内全部建筑达到国家绿色建筑三星级标准，建筑节能≥65%。采用合同能源管理市场化运作机制，建设地源热泵集中式能源站，为建筑提供冷热源。推行绿色施工管理，探索绿色施工在区域性建筑中的应用。

建筑热环境解决方案

1.被动措施

项目总体规划时，按计算机日照模拟分析，确保南向办公空间均满足日照要求。适当加大外窗可开启的面积，引入穿堂风，提高自然通风的效果。建筑外窗使用固定及电动遮阳百叶，保持室内适宜温度，并使外遮阳与建筑立面一体化。

围护结构体系：外墙保温采用30 mm厚挤塑保温板（XPS）。屋顶保温采用45 mm厚挤塑保温板（XPS），同时种植屋面的轻质种植土也有效补充和完善了屋顶保温层。窗墙面积比小于0.5，外窗选用隔热金属型材，低透光LOW-E+12空气+6透明，传热系数≤2.5 W／(m²·k)，遮阳系数≤0.40。

外遮阳系统：建筑的南向、东西向外窗设置铝合金可调节式遮阳卷帘；南面横向设置固定遮阳百叶，既达到遮阳效果，也美化建筑立面。

2. 主动措施

地源热泵集中式空调系统：采用地源热泵复合冷水机组加冷却塔作为空调系统冷热源，夏季地源热泵系统冷却水由地埋管和冷却塔共同提供，系统运行高峰或调峰时运行冷却塔。冬季由地源换热器和地源热泵机组提供机组全部所需热源水。每年可节约电量5 250 000 kWh，减排CO_2 5234 t，年节省运行费用约400万元。

全热回收新风系统：将室内污浊的空气排出,并将新鲜的室外空气引到室内；新风处理机通过热回收转轮，将排气中的能量转换给供气（新风），能量转换效率（热回收率）在75%～88%之间；新风处理机的能量转换为全热转换，即：显热（温度）和潜热（湿度）同时转换，使能耗显著降低（与同等风量的传统新风设备相比，3~12个月内即可收回新风处理机组的投资）；拥有两条独立气流回路，保证排风和新风气流完全隔离，不会产生交叉污染和泄漏。

其他可持续设计措施

1.节水与水资源利用

采用节水器具。所有用水器具均满足CJ164《节水型生活用水器具》及GB/T18870《节水型产品技术条件与管理通则》的要求。选用揿钮式和感应式龙头、冲洗阀、陶瓷芯水龙头等，节水效果良好。

2.雨水回用系统

分别在A区与B区设置雨水处理池，收集所在片区屋面、地下室顶板、绿地以及道路等雨水，进入雨水处理池前的整流井，多余雨水经过溢流管排至市政雨水管。收集的雨水经过处理后，供项目范围内的绿化、冲洗、洗车等使用。系统由PLC控制自动运行。

3.节材与材料资源利用

本项目通过设计采用资源消耗和环境影响小的建筑结构体系，最大范围应用可循环与可再生材料来实现节材与材料资源利用。

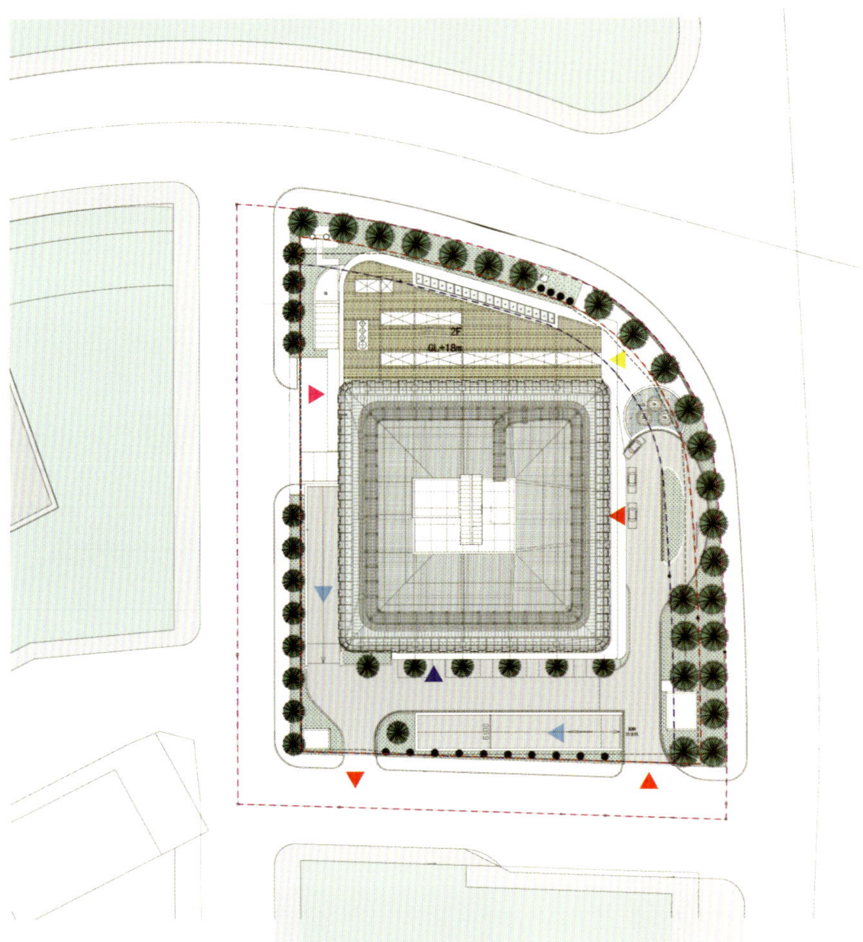

项目概况

上海太平金融大厦是中国太平保险集团在上海的总部大楼，大楼规模110 000 m²，高208 m，是一座集总部和出租为一体的商务办公建筑。大楼面向上海陆家嘴金融贸易区中心绿地，毗邻金茂大厦、环球金融中心以及即将建成的上海中心等建筑，成为象征上海国际金融中心地位的又一标志性建筑。同时也是集节能环保技术与外观个性于一体的典型案例。

建筑热环境解决方案

覆盖于主楼立面之上的幕墙采用了LOW-E玻璃。光影变幻的幕墙不仅为建筑创造了丰富多彩的表情，同时也是提高防雨安全性能、实现建筑绿色节能的重要装置。建筑采用了中空核心自然换气系统和风向压力自然换气系统。引入室内的自然风又通过预设在核心筒内的通风管井的重力拔风，可以增强数倍的换气功能，以达到春秋季节大幅度节约空调用电的节能效果。

利用幕墙单元凸出板块的室内空间，在窗台处每隔3 m设置了自然通风的机械开关装置，通过幕墙的竖框形接缝导入外部新鲜的空气，并在横框的腔体内转换后从窗台上排出，避免了外墙上出现通风用的开口。

为实现带有立体编织肌理的全工厂组装的单元式幕墙，在建筑上大胆采用了宽的巨型铝合金型材和应用在汽车制造上的高分子保温材料等工业用材料产品。

总平面图 SITE PLAN

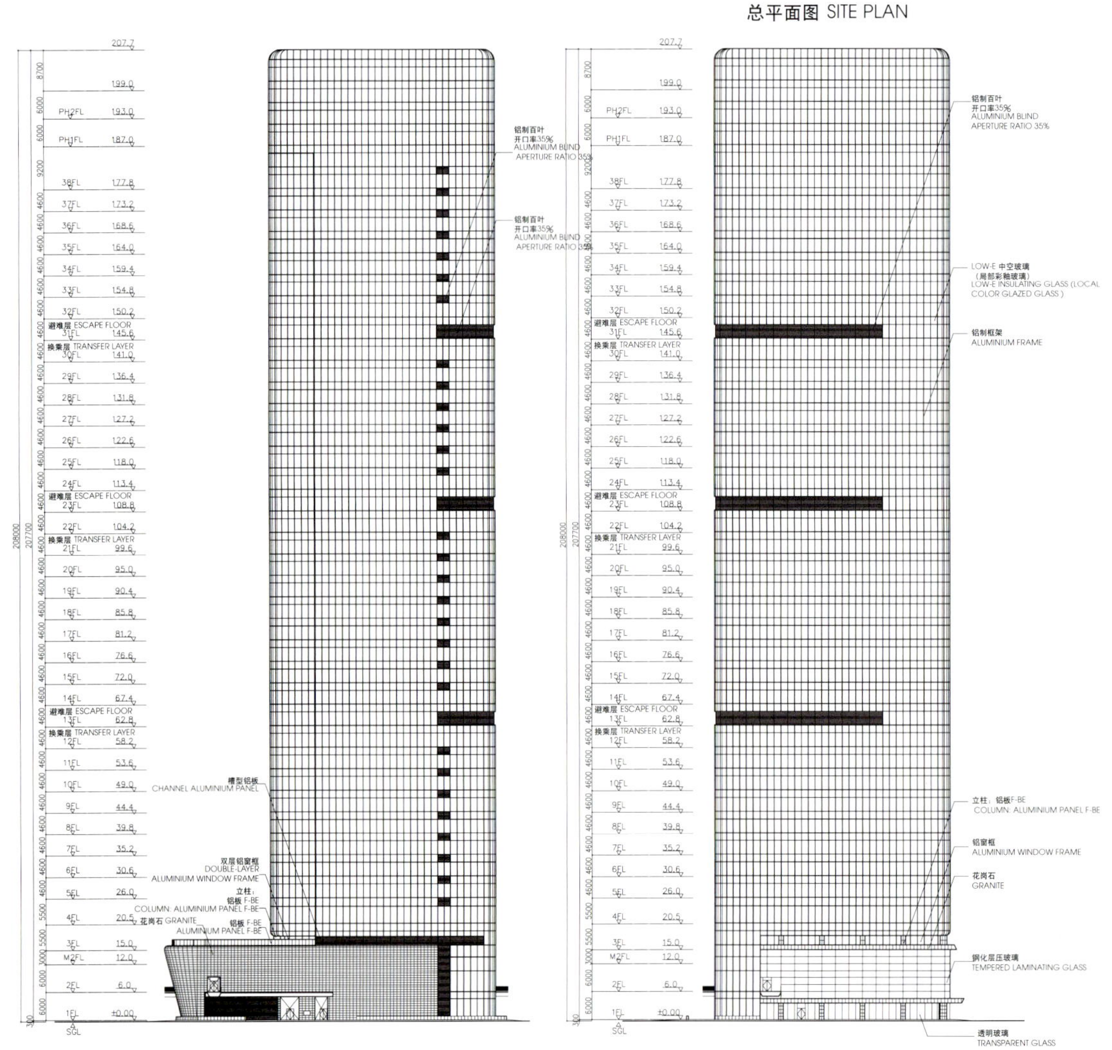

立面图 ELEVATIONS

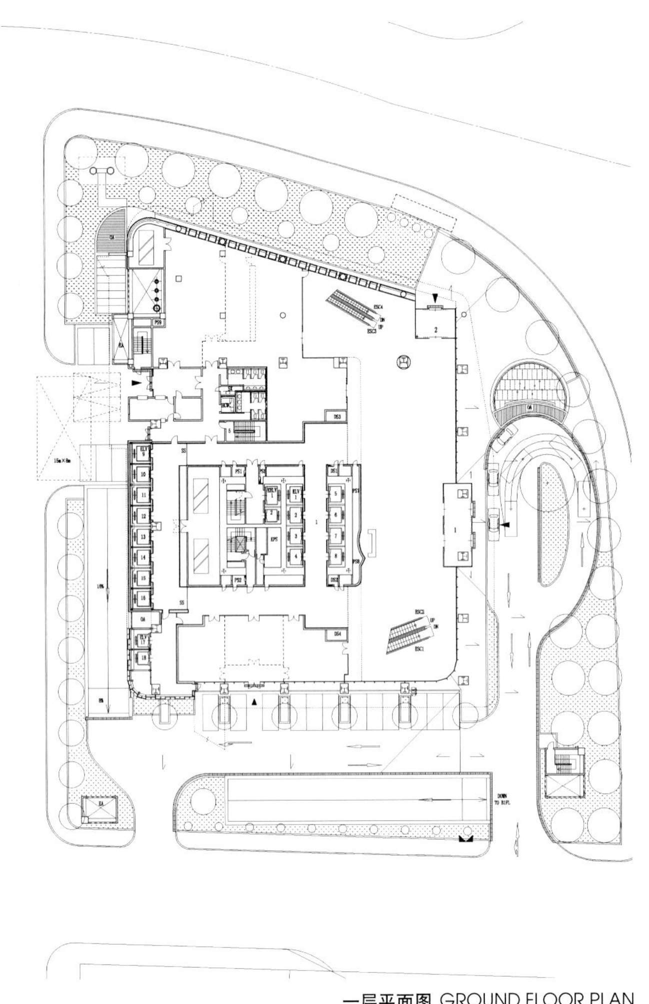

一层平面图 GROUND FLOOR PLAN

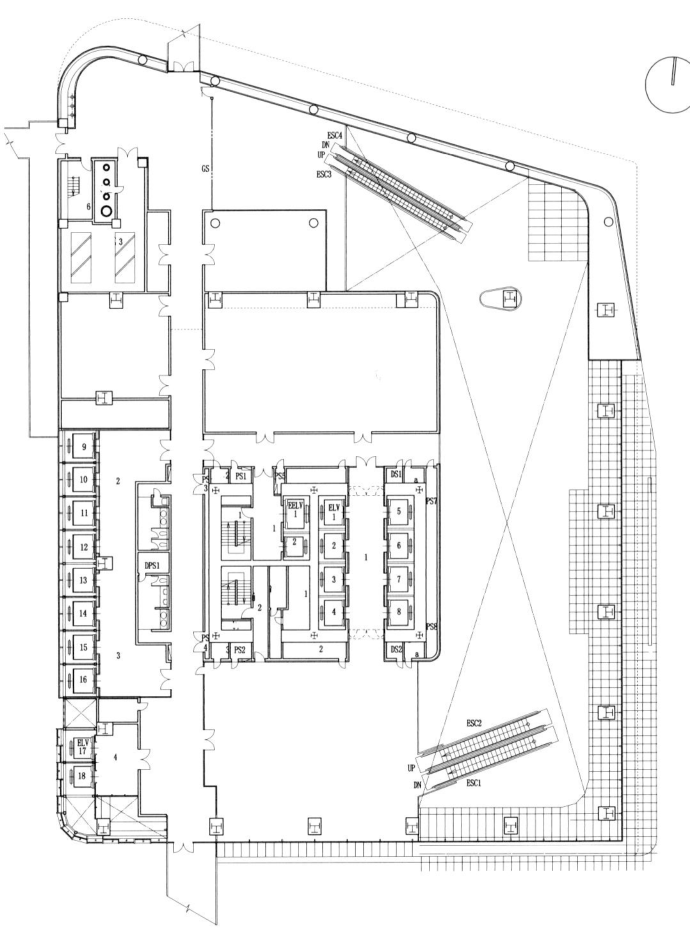

二层平面图 SECOND FLOOR PLAN

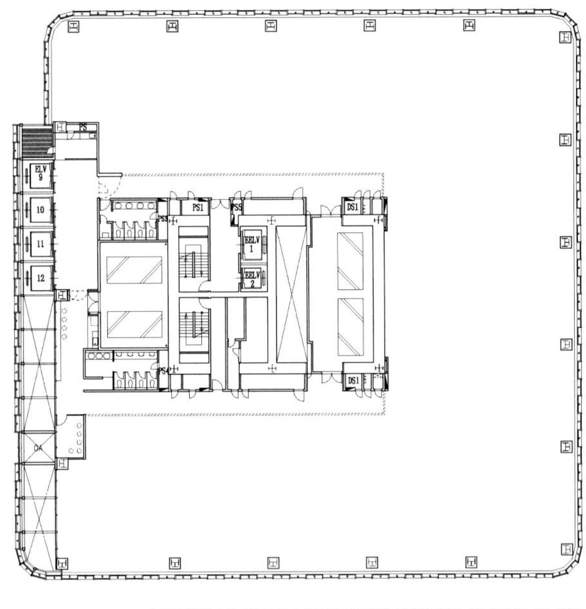

高区标准层平面图 HIGH ZONE TYPICAL FLOOR PLAN

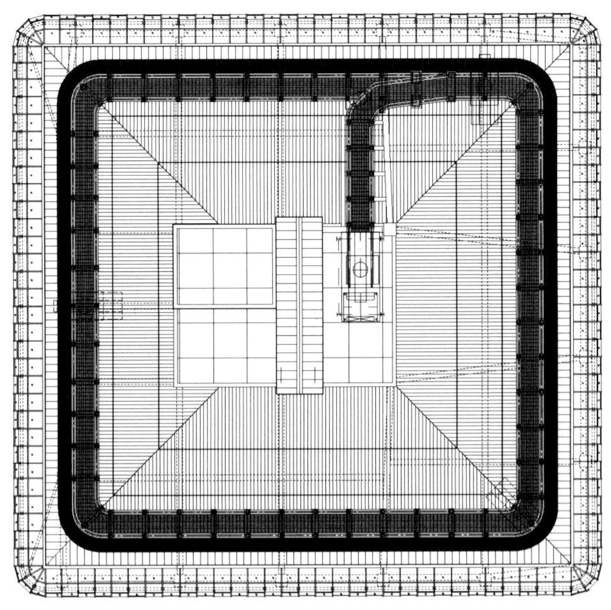

屋顶平面图 ROOF PLAN

MOKUZAI KAIKAN
木材会馆

©Harunori

地点：日本东京
建筑面积：1011 m²
建筑设计：日建设计
所获奖项：环保和建筑服务系统设计奖，2011；AIJ年度建筑设计奖，2011；JIA奖，2011
摄影：Harunori Noda；Nacasa & Partners Inc.

Location: Tokyo, Japan
Building Area: 1011 m²
Architect: Nikken Sekkei
Selected Awards: Environment and Building Service System Design Award 2011: AIJ Annual Architectural Design Commendation 2011: JIA Award 2011
Photography: Harunori Noda; Nacasa & Partners Inc.

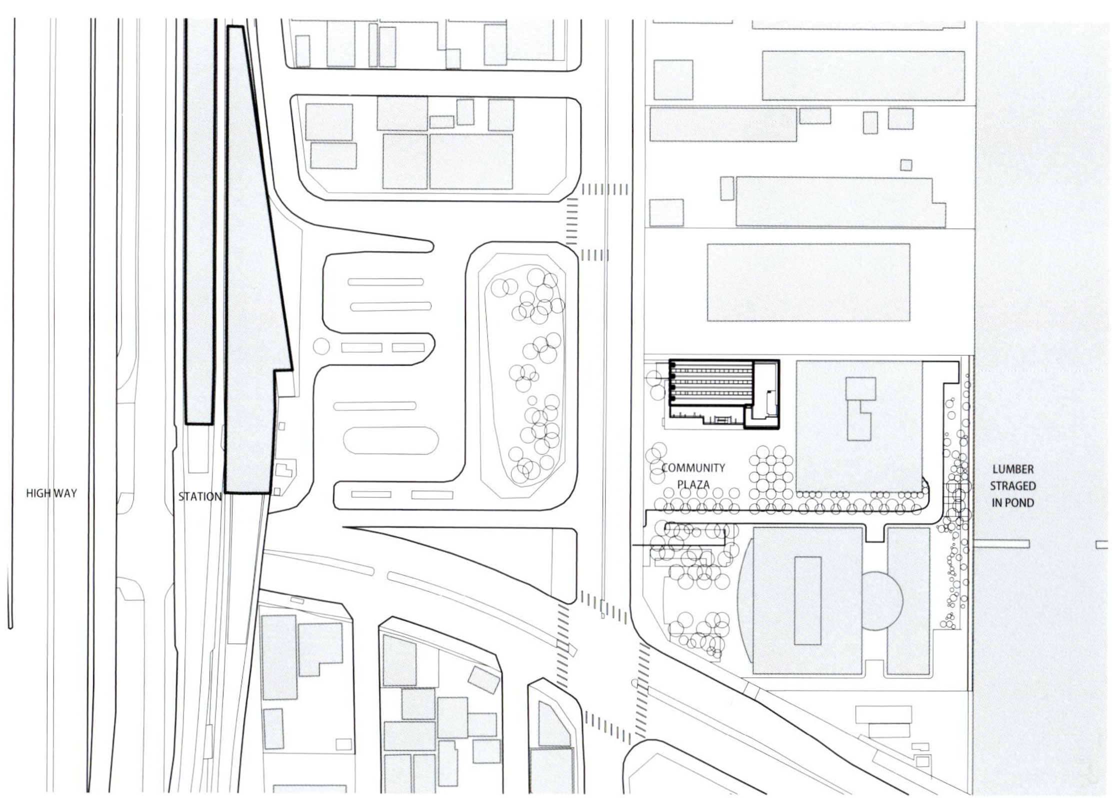

总平面图 SITE PLAN

©Nacasa & Partners Inc

©Nacasa & Partners Inc

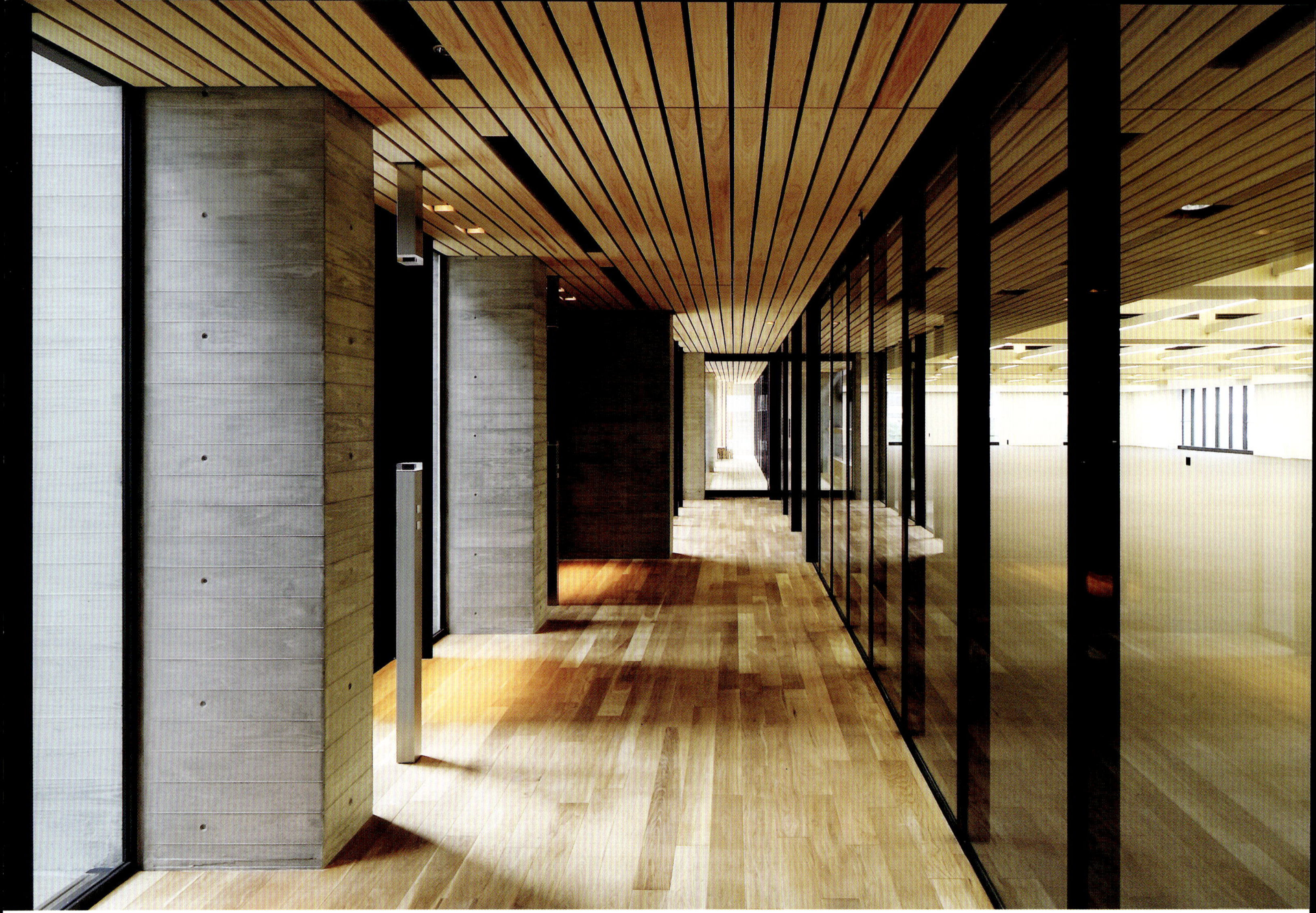

©Nacasa & Partner

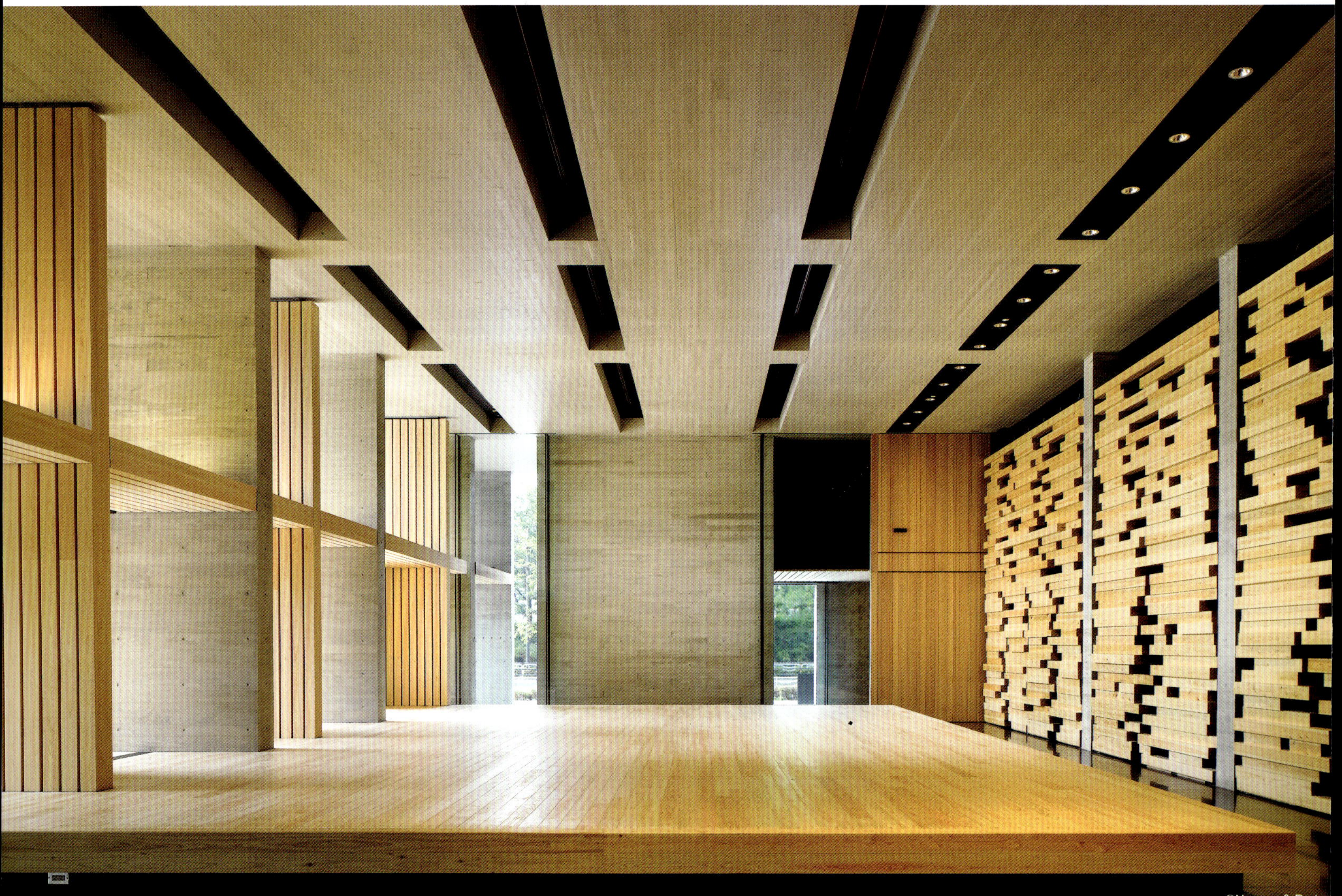

©Nacasa & Partne

项目概况

该项目是东京木材批发商协会原总部的改迁项目，展示了木材作为城市建材的潜力。一方面，木材很好地融入建筑物结构之中，同时建筑外露混凝土在松木模板中浇铸。由于该建筑物大量使用木材，因而设计过程中特别考虑到了各种消防措施。此外，分层设计和自然光线的利用，使得整个空间结构有机地连接在一起。

建筑热环境解决方案——被动措施

被称为缘侧的外部半圆形空间，是日本传统建筑的保留元素，它能够实现自然通风，同时能够阻挡强烈的阳光直射，从而在炎热和潮湿的恶劣环境中营造一个舒适的室内环境。传统的缘侧形成了一种新的空间形态。在这个配置中，办公室的元素，例如其结构、循环系统及设施，在半圆环境中获得重建。这个配置使户外空间尽量与通用办公区域相连，同时又作为一个缓冲器来对抗严酷的外部环境。一系列露台共同构成一种可渗透的表层。深凹的屋檐将强烈阳光和雨水阻隔在外，同时提供一个舒适的外部环境，人们能感受到自然光线、微风和其他自然元素，仿佛生活在树林中一般。此外还采用了电子活动遮光板和建筑内部遮光设备等。设计师通过安装的固定装置使连接空间给人以抽象之感。如此一来，各个空间相互流动。房间、墙壁和屋顶分散布局，地板之下有风管通气道，安静、清洁地调节室内温度。

©Nacasa & Partners Inc

©Nacasa & Partners Inc

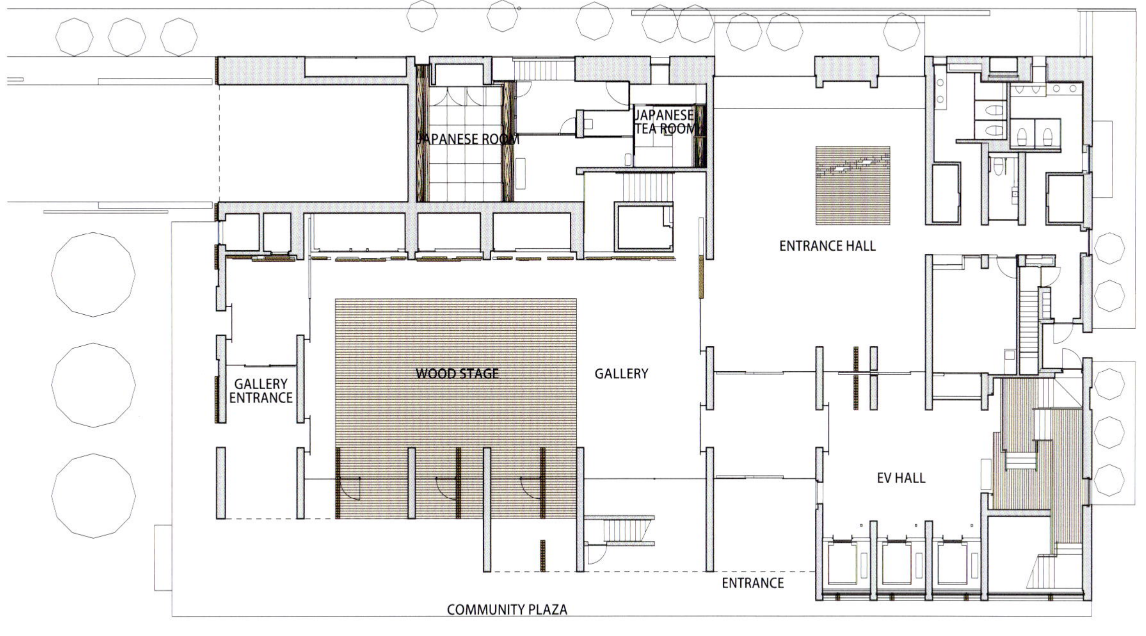

一层平面图 GROUND FLOOR PLAN

©Nacasa & Partners Inc

Program Description

This project involved the relocation of the offices of the Association of Wood Wholesalers in Tokyo. It serves as a showcase to demonstrate the possibilities of wood as an urban construction material. Lumber were integrated into the building's structure, and architectural exposed concrete was cast in cedar formwork. Since the building uses a large amount of wood, great attention was given to fire safety measures. The design focused on creating spatial continuity with the use of layering and natural light.

Building Thermal Environmental Engineering Solutions—Passive measures

A semi-outdoor space called engawa, existing in traditional Japanese homes, allows a natural breeze to enter the shelter while shutting out strong sunlight, achieving a comfortable indoor environment in the midst of harsh heat and humidity. From the traditional engawa, a new space configuration was realized. In this configuration, elements of an office, such as its structure, circulation system, and facilities, are reconstructed in a semi-outdoor environment. This configuration strives to connect the outdoors with the universal office, while simultaneously acting as a buffer against the harsh side of the external environment. Sets of terraces come together to make a porous skin. Deep eaves shut out direct sunlight and rain as well as offer a comfortable outdoor area where one can experience natural light, breezes, and other aspects of nature, as if living in the middle of the woods. Other passive measures, like electric rolling blind and internal blind, are also adopted in the building. Fixtures are installed to give an abstracted impression of the adjacent space. In this way, spaces flow from one into other. Rooms, walls, and the roof are dispersedly located. Under the floor there is an air duct space, conditioning the interior air cleanly and quietly.

A 柏木面板 115 mm × 115 mm
B 柏木露台 120 mm × 30 mm
C 钢框窗
D 内部遮光
E 铝制遮光板
F 钢制滑动门（2.7 m宽）
G 送风管区域
H 侧光
I 2 mm铝制屋顶，t-2
J 顶光
K 电子活动遮光板
L 1.61 m梁，由115 mm × 115 mm的日本柏木构成（最长4 m）

A cypress elevation panels, 115 mm x 115 mm
B terrace deck in cypress, 120 mm x 30 mm
C steel sash window
D internal blind
E aluminum eaves panels
F steel sliding door (2.7 m wide)
G air duct space
H side light
I 2 mm aluminum roof, t-2
J top light
K electric rolling blind
L 1.61m- deep beams, composite of Japanese cypress in 115 mm x 115 mm sections (maximum length 4 m)

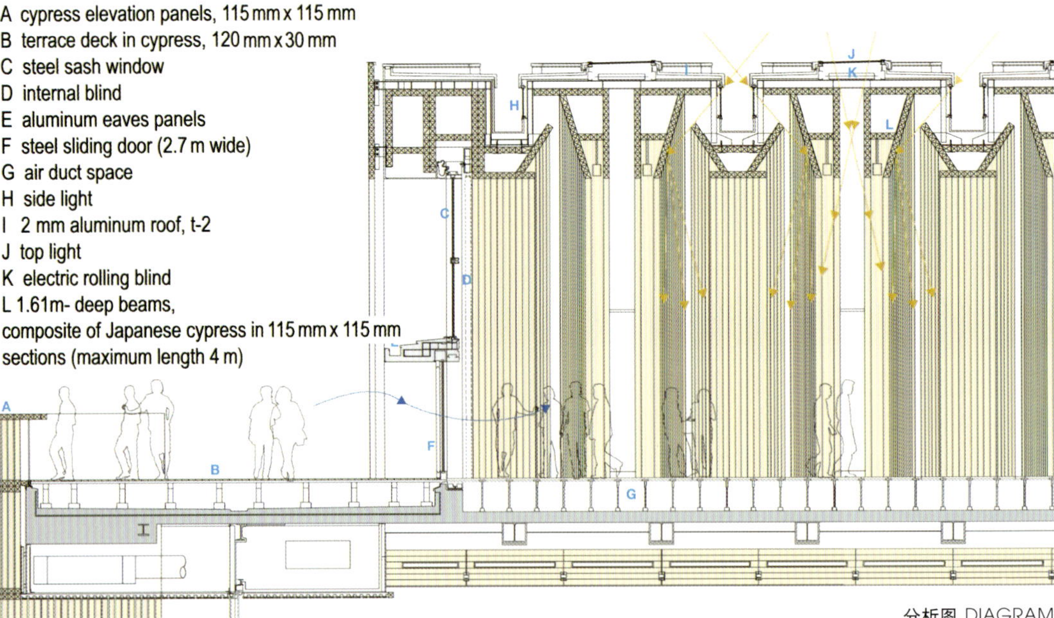

分析图 DIAGRAM

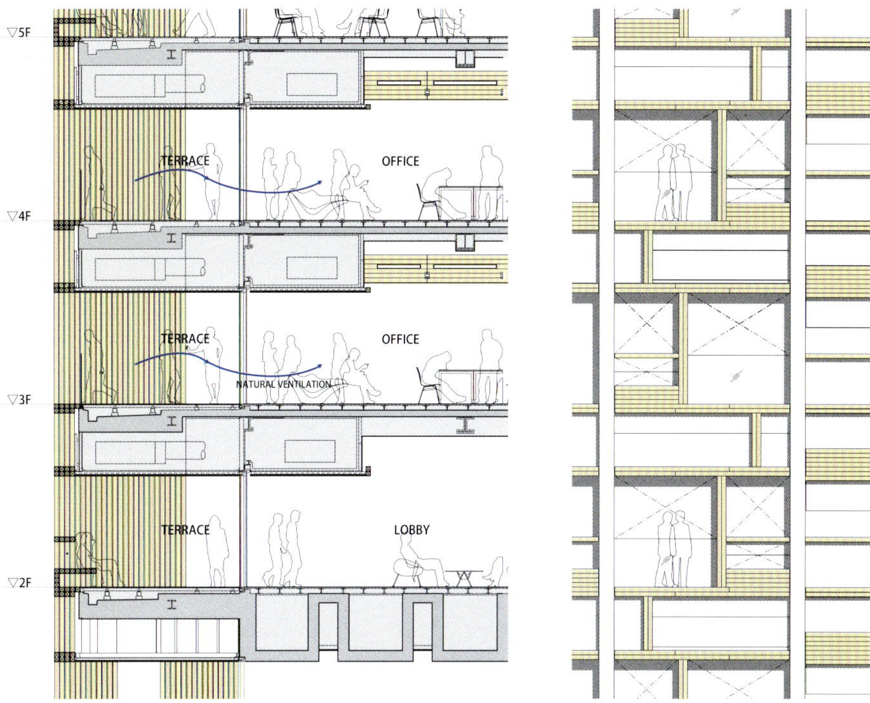

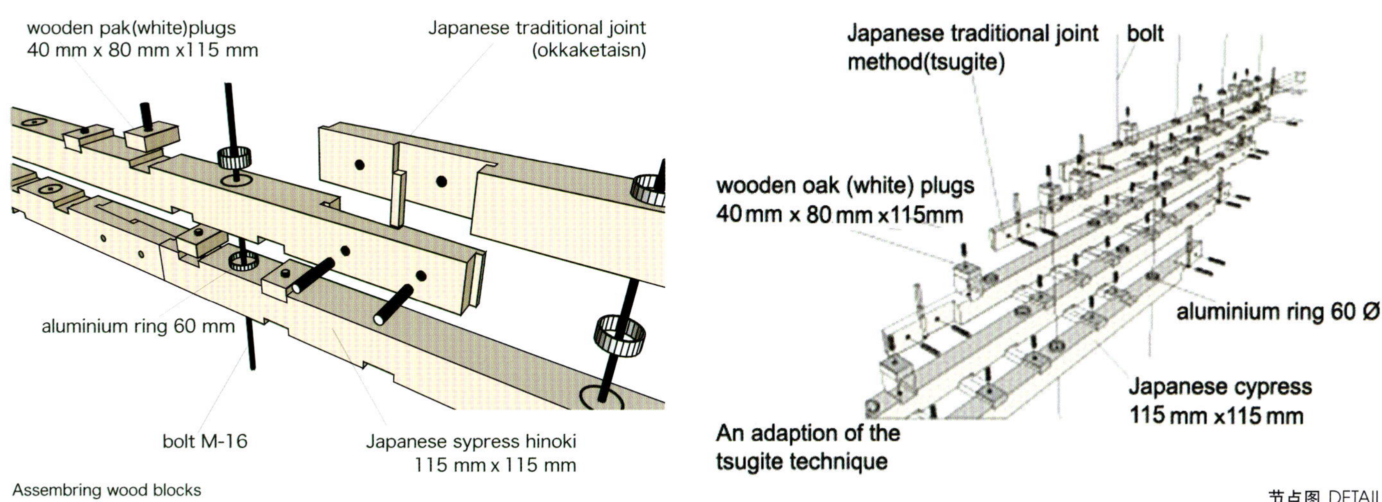

Assembring wood blocks

节点图 DETAIL

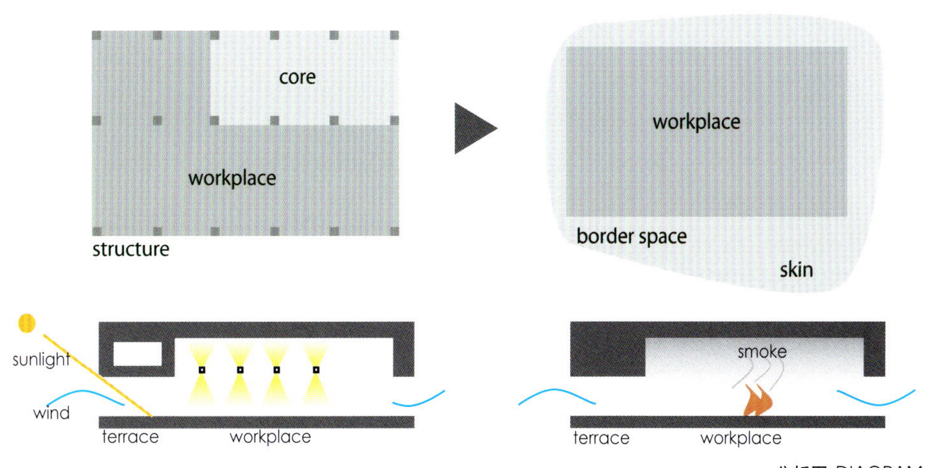

分析图 DIAGRAM

COCOON EXCLUSIVE OFFICE HEADQURTER
COCOON 办公总部

©Camenzind Evolution, Zü

位置：瑞士苏黎世
建筑面积: 1900 m²
建筑设计: Camenzind Evolution
委托方：Swiss Life
摄影: Ferit Kuyas; Nick Brändli; Camenzind Evolution

Location: Zurich, Switzerland
Building Area: 1900 m²
Architect: Camenzind Evolution
Client: Swiss Life
Photography: Ferit Kuyas; Nick Brandli; Camenzind Evolution

©Camenzind Evolution, Zürich

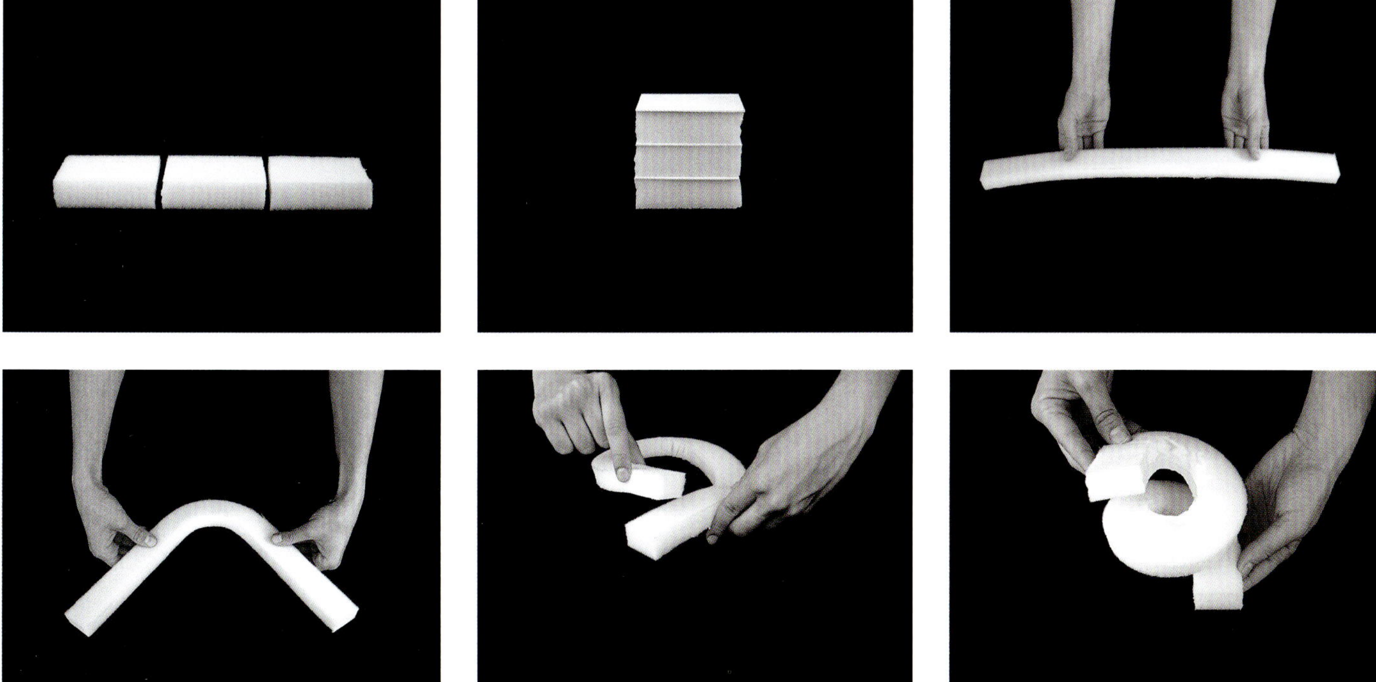

©Camenzind Evolution, Z

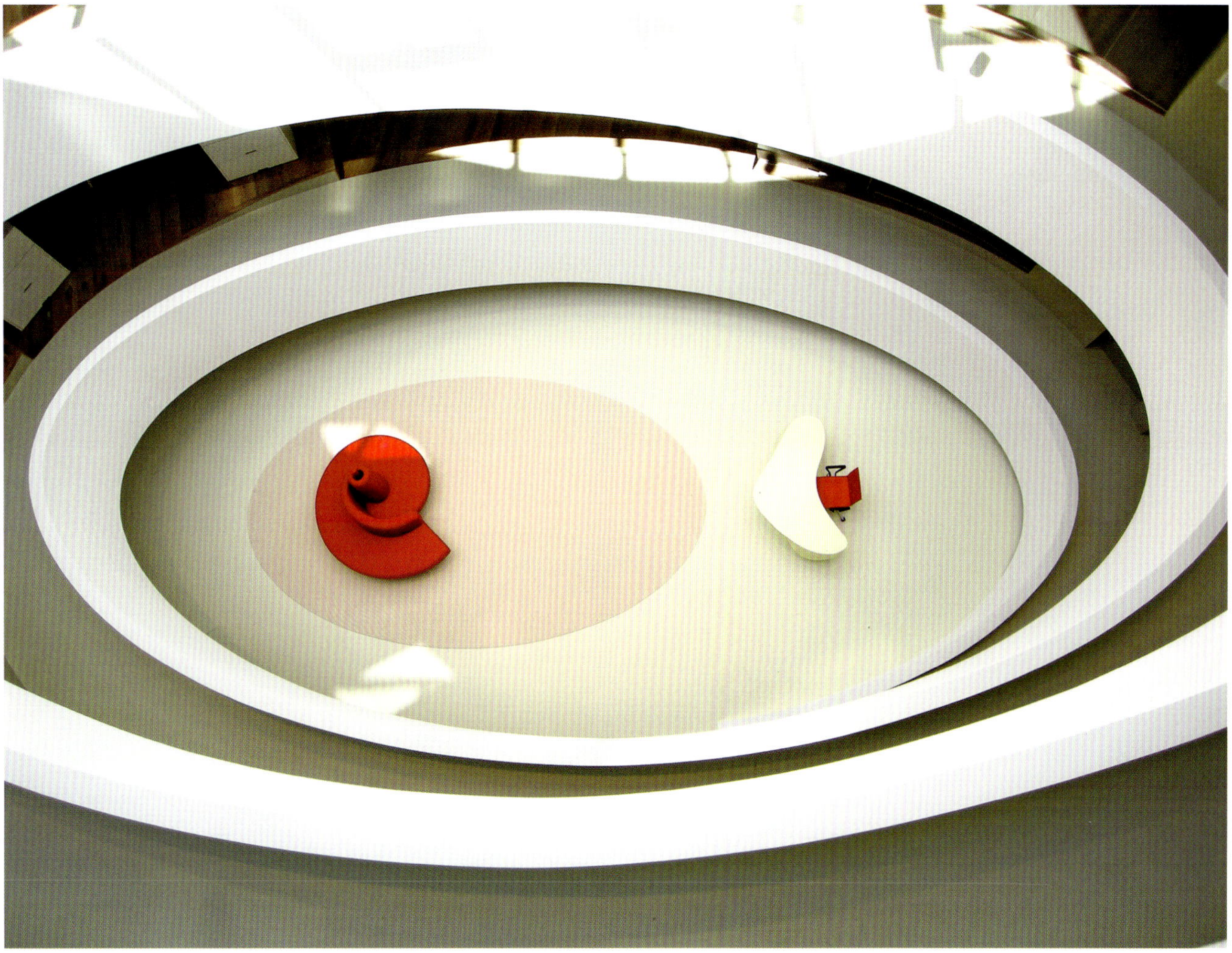

©Camenzind Evolution, Zürich

©Camenzind Evolution, Zürich

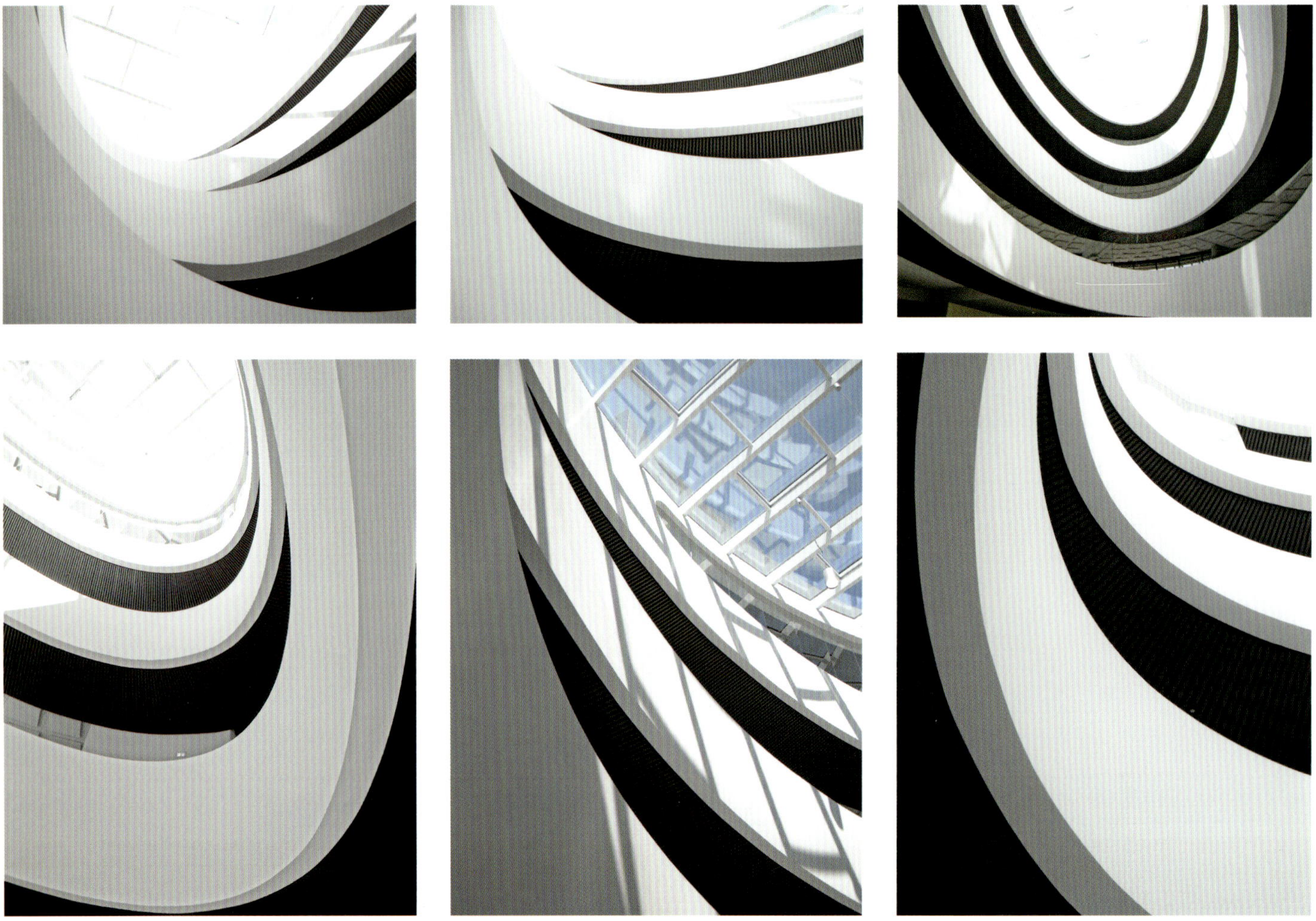

©Camenzind Evolution, Zürich

©Nick Brändli, Zürich

©Nick Brändli, Zürich

Program Description

Cocoon - Zurich Switzerland Cocoon is located in Zurich's Seefeld district on a beautiful hillside, which enjoys excellent lake and mountain views. The location's distinctive flair stems from the exceptional park-like setting - a green oasis into which Cocoon snugly nestles. The stainless steel mesh enveloping the building combines visual privacy with restrained elegance. The bold stand-alone building embodies an innovative conception of interior spatial organization and interaction with the surrounding environment. It caters for a wide variety of workplace and occupancy concepts. With its spiral massing, Cocoon may be conceived as a sort of "communication landscape" that creates a unique spatial configuration and working environment in a matchless setting.

©Ferit Kuyas, Wädenswil

©Ferit Kuyas, Wädenswil

Building Thermal Environmental Engineering Solutions

Internally, a light-flooded, upwardly widening atrium forms the centrepiece of Cocoon. As the ellipses expand with each turn of the spiral, the skylight void opens up in a stunning spectacle. Externally, the building adopts the guise of a dynamic, upward-reaching sculpture. The dramatic atrium, with its wealth of internal visual links and providing natural daylight and ventilation, generates a natural ambience conducive to communication and a sense of community. The shrouded, sculptural stand-alone building, introverted during the daytime as it looks inwards towards the atrium, is recast in the evening hours as a transparent shining beacon. Cocoon uses a Air-source heat pump system for environmentally-friendly heating and cooling.

©Ferit Kuyas, Wädenswil

Program Description

Located among a low-rise residential neighborhood, a new location for a software developer--SUNONE presents designers with a great challenge dealing with quality of living for existing residence and working condition for the new comer. In order to preserve serenity of an area, respect privacy for each residence and simultaneously create pleasing working environment for staffs, pattern of horizontal screening was introduced to the project. Not only does it create visual barrier between inside and outside; allowing no direct sight from inside to surrounding residence, but does it serve as light filter generating array of light quality for interior space.

Interior design strategy was designed to promote flexibility and transparency in work environment. In contrary to the traditional office space, SUNONE provides a range of work settings including bench desks completed with task lights for concentrated work, and informal places such as lounge/bar setting for group meeting and chats. This socially oriented space is intended to improve the possibilities for social interaction among staffs rather than previously tended electronic communication.

Truly the most delightful feature of an interior space is the design of interior screen that scattered throughout space. While exterior screen reflects technological character of company's field of business, interior screen, on the opposite, directly reflects adventurous personality of the owner and staffs. Industrial material like colorful PVC strip curtain has been cleverly incorporated through the building to reflect energy and liveliness of these extreme-sport lovers. Besides operating as space divider and enclosure, this screen, with its transparency property, also functions as an indicator of an area's privacy level. By using PVC strip curtains, openness and transparency of office space are maintained.

With deliberate architectural strategy, the two building typologies, office and home, can perfectly fuse in one environment regardless of their architectural style. By creating a work space in the low corporate presence, SUNONE has successfully provides a place that encourages both collaboration and relaxation among its staffs.

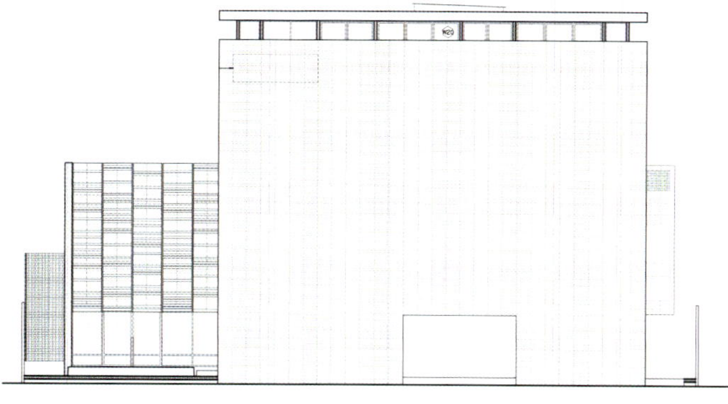

立面图 ELEVATIONS

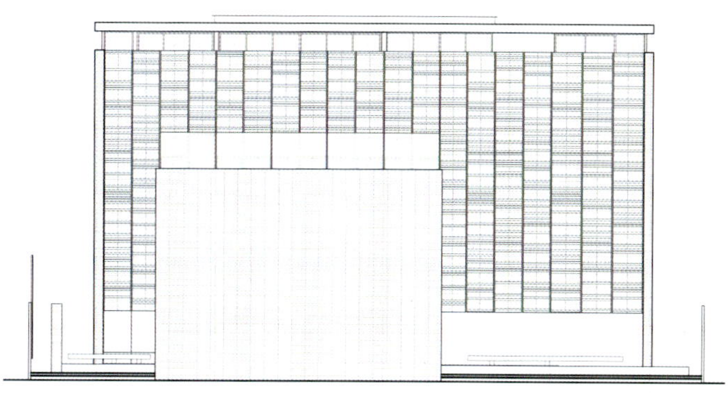

剖面图 SECTION

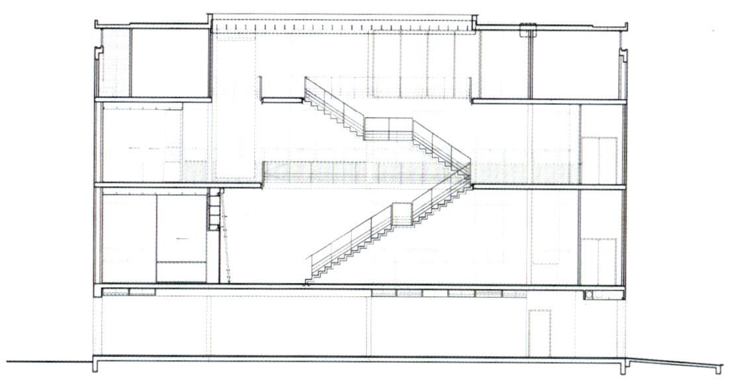

剖面图 SECTION

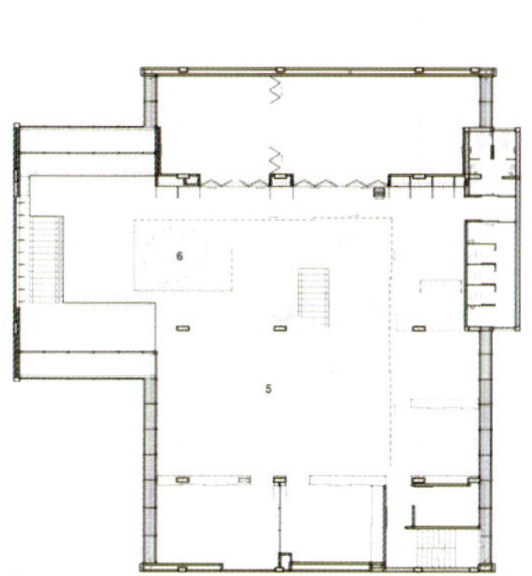

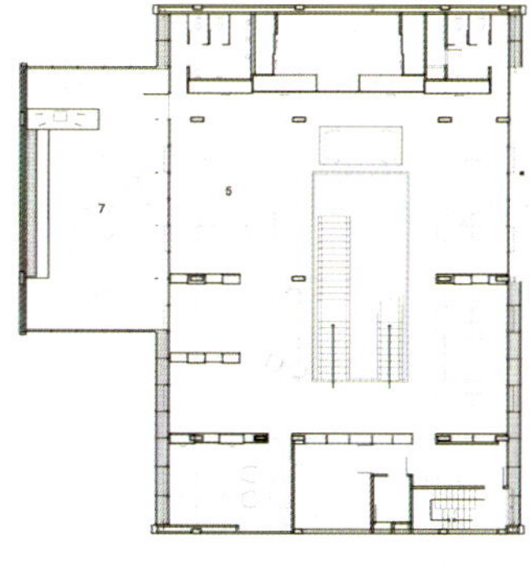

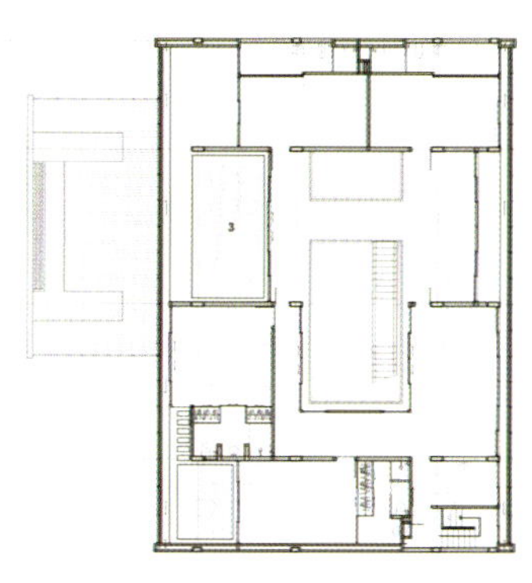

1停车场
2花园
3生活区
4餐厅
5工作区
6会议室
7户外露台

1 Parking
2 Garden
3 Living Area
4 Ganteen
5 Working Area
6 Meeting Room
7 Outdoor Deck

平面图 FLOOR PLANS

B&Q STORE SUPPORT OFFICE
B&Q 后勤办公大楼

©David Ba

©David Ba

地点：英国汉普郡伊斯特利
建筑设计：BDP
结构工程：BDP
委托方：B&Q
摄影：David Barbour

Location: Eastleigh, Hampshire, UK
Architect: BDP
Structural Engineering: BDP
Client: B&Q
Photography: David Barbour

1 中心广场 1 Heart Square
2 大厅 2 Foyer
3 饭店 3 Restaurant
4 办公室 4 Offices
5 楼梯间/卫生间 5 Cores/WC
6 会议区 6 Meeting Pods
7 休憩区 7 Break Out Area
8 茶座、复印处 8 Tea+Copy Points
9 商业中心 9 Business Centre
10 庭院 10 Courtyard
11 研发部门 11 Research+Development
12 体育馆 12 Gym

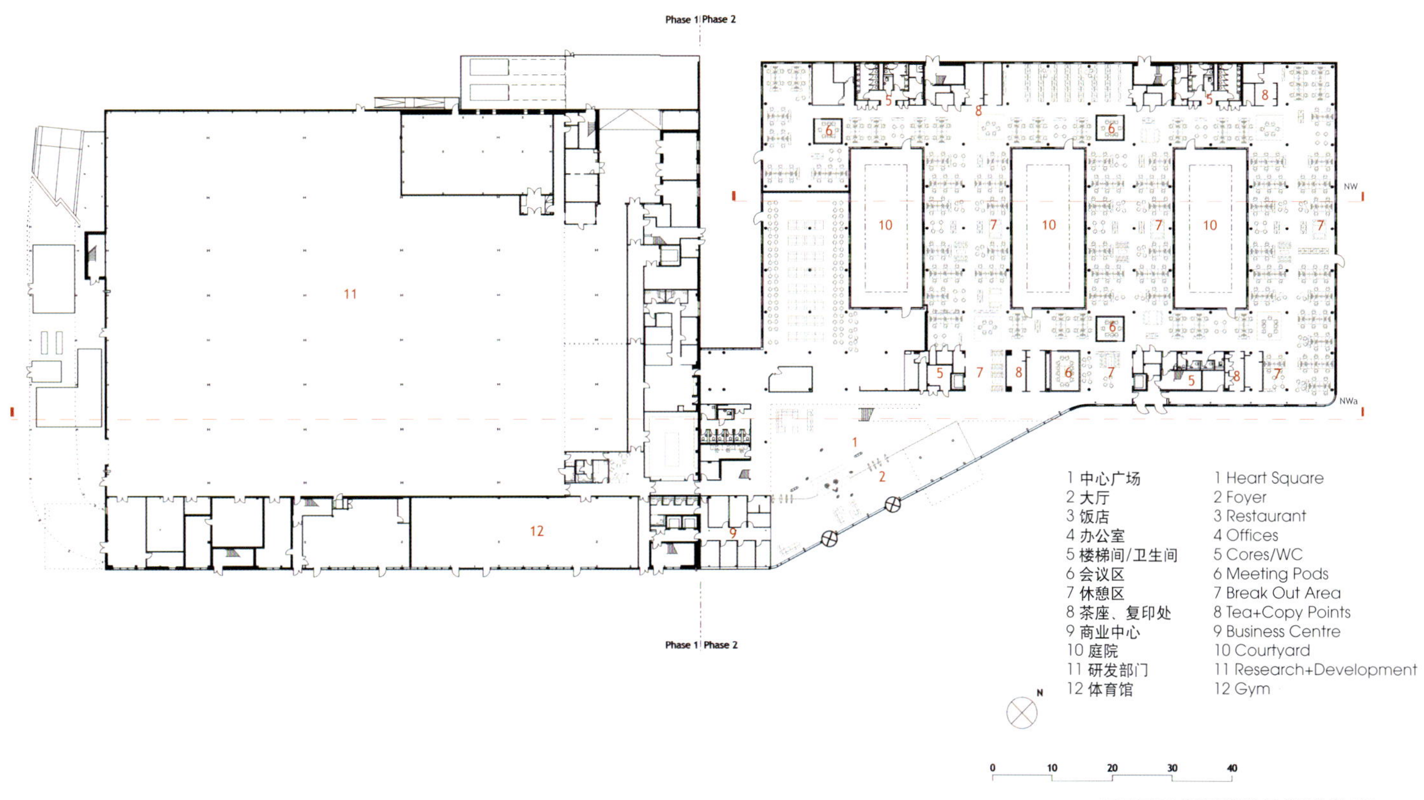

一层平面图 GROUND FLOOR PLAN

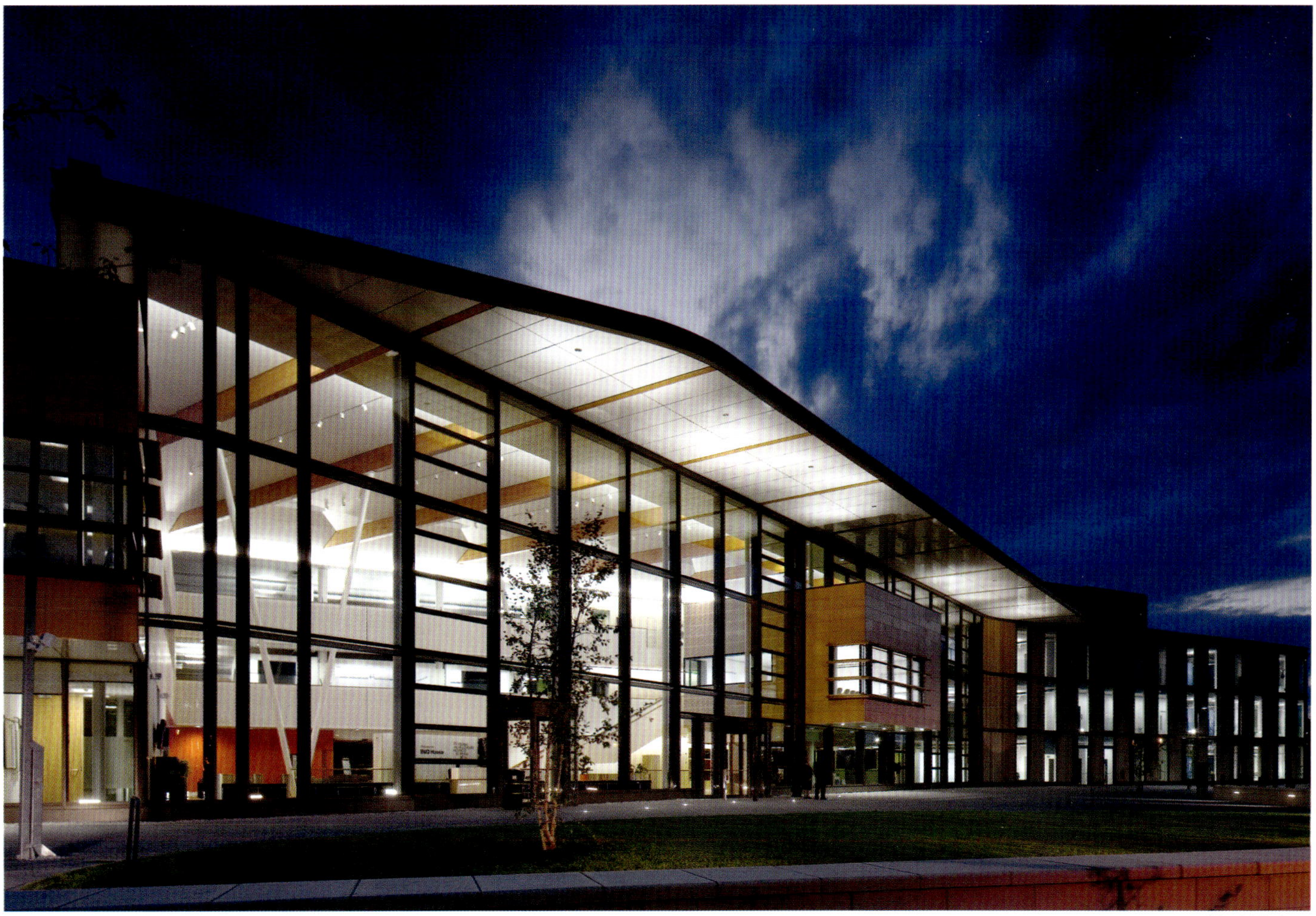

©David Barbour

©David Barbour

项目概况

B&Q（百安居）公司1969年成立于英国的南安普敦，是一家大型DIY与家庭园艺工具材料连锁卖场。B&Q新建的店铺支持中心大楼（Store Support Office）位于英国汉普郡的伊斯特利。来自BDP建筑设计公司的不同学科背景的专家组成项目组，遵循“为人而创造”的理念展开设计。这座建筑将原来位于三个不同办公地点的1400个员工集中到一个办公楼中，并为各个商店提供产品销售方面的基础支持。

设计的要旨在于：根据B&Q的整体理念，创造一个高效协作的工作环境；吸引并留住B&Q的优秀员工；展示品牌形象、企业文化及价值观，并表明他们在环境可持续问题上的关注和作为。建筑位于一个商贸集散地的南端，建筑设计师努力使新建筑对周边已有建筑的影响降到最低，在此之上，尽全力为企业员工提供舒适的环境。办公区域设置在大楼的正面，使人在Chestnut大街的街面上就可以对大楼的功能和性质形成鲜明的第一印象。

建筑主要由研发实验室和办公区两个主要部分组成。建设分两个阶段完成，实验室将办公空间与先期建造完成的部分连接起来。这样的设计强调了建筑物内不同分区的不同功能。大楼的东北及东南部是公共空间，设有步道和入口；中部是半开敞空间，包括办公空间和贯通三层楼的中庭；西北部和西南部是私密空间。大楼围护结构的设计也是对上述空间组织形式的反映，东南和东北部的立面较为开放，而西北和西南部的立面更适用于营造私密空间。

建筑热环境解决方案

项目通过节能、节水、节约自然能源来减少浪费和污染，防止环境退化。建筑主要依靠自然通风，在极端气候条件下采用低能耗的机械设备进行调节。建筑使用了混凝土框架等高效保温材料，还采用了木制外覆面等低能耗材料，以便在保持良好的室内环境的同时，减少冬季热损耗和夏季热量吸收。整座大楼的核心是橙色大厅，来自全国各地的商店经理主要在这里和店铺支持中心的人员沟通交流。挑空的开敞空间使大楼内部的空气流通和热交换得以顺畅进行。室外植有柏树，为相关设施遮蔽阳光，减少热量吸收。

大楼共三层。具有高度适用性，并且在视觉上相互连通的空间分布于各个楼层，并由一个面向大门的中庭统领。中庭位于大楼入口处，是一个大型的过渡空间，将实验室和办公区联系在一起。中庭是人流通过的主要空间，还设有可供召开非正式会议、休憩交流的区域。大楼核心区的体量很大，相当于三层楼高度的挑高与办公楼其他区域形成鲜明对比，旨在强调其在整个建筑中的重要作用。大厅为各办公室提供自然采光和自然通风。

主要的办公空间设置在13.5 m间距的柱网结构内，每层设有一个室内庭院。建筑设计师对办公区的设计原则是最大限度地利用自然采光和自然通风，同时为B&Q提供一个高效的办公环境。室内庭院的设计使办公环境与外部景观相互交融，同时在每层充分利用自然采光。办公空间在设计上比较开放灵活，以适应不同的使用要求。每层的庭院设计都各具特色。

大楼的办公区域装有玻璃幕墙，并以隔板间隔。这样的设计使得建筑具有良好的通透性，并且可以合理控制进光量。日间，大楼外墙使一部分阳光射入室内；晚上，室内的灯光照亮夜空，整座建筑闪闪发光。

在中庭旁边还设有一个餐厅、一个咖啡厅和一家店铺，它们也在公共活动区域和办公空间之间起到了缓冲过渡的作用，同时也为办公环境提供了社交和互动空间。这座建筑是B&Q业务的一座可持续展示中心，体现了其品牌理念和文化价值观念。这是一座具有很强适应性的建筑，传达了设计者对环境可持续发展的认同和遵从；这也是一座有效利用自然条件的建筑，最大限度地采用了自然光照和自然通风。

Program Description

B&Q plc is a multinational DIY and home improvement retailer headquartered in Eastleigh, United Kingdom. B&Q's new store support office building in Eastleigh, Hampshire, was designed by an interdisciplinary BDP team following its principle of 'creating places for people'. The building brings together 1400 staff from three separate sites into one main office, and provides the stores with technical support for product sales.

The brief was to create an inspiring, efficient and sociable working environment which stands comparison locally and nationally, helping the business to attract and retain the best staff and to be a showcase for B&Q, reflecting its brand, culture and values, and demonstrating its commitment towards environmental sustainability. The building is located on the southern edge of a mixed commercial retail area; it maximises amenity space for staff while minimising impact on the adjacent buildings. The offices areas are placed towards the front of the site to create a strong visual foreground and identity to the building's setting when viewed on the approach from Chestnut Avenue.

©David Barbour

©David Barbour

SOLAR POWER OFFICES
太阳能办公楼

地点：斯洛文尼亚卢布尔雅那
占地面积：32 500 m²
建筑设计：OFIS Architects

Location: Ljubljana, Slovenia
Area: 32,500 m²
Architect: OFIS Architects

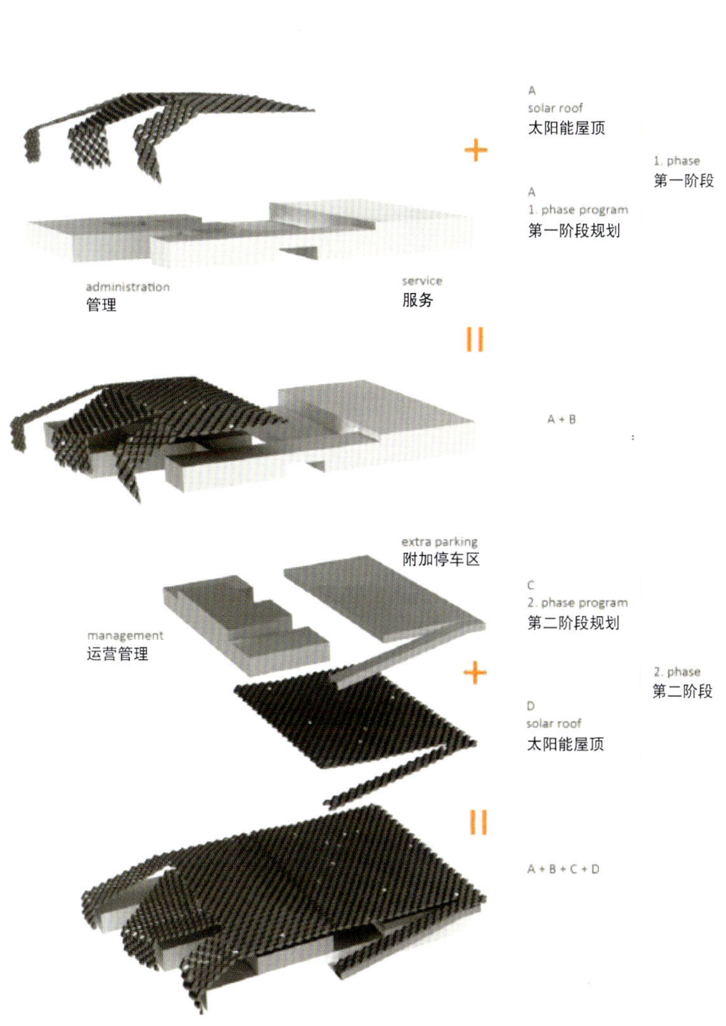

分析图 DIAGRAM

一层平面图 GROUND FLOOR PLAN

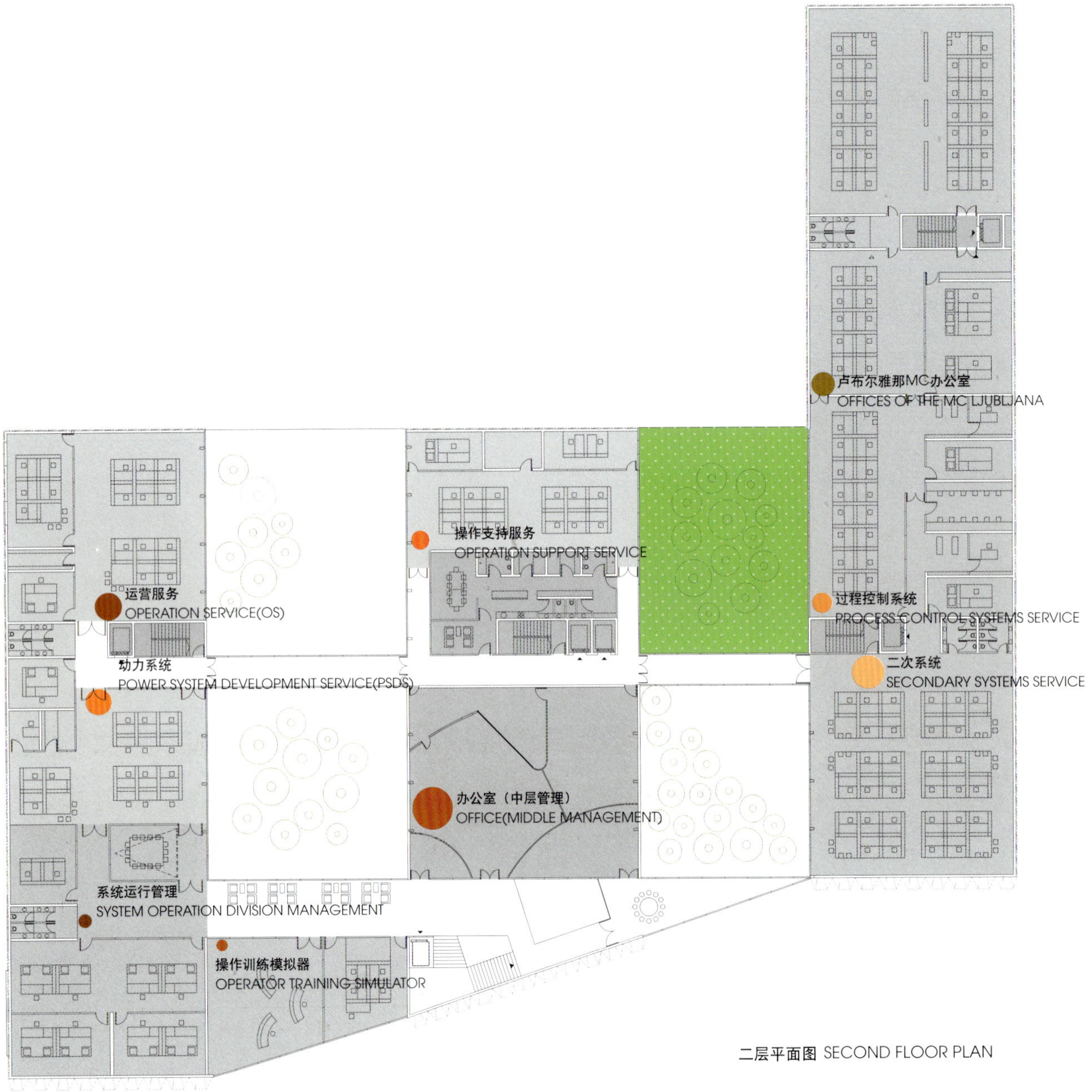

卢布尔雅那MC办公室
OFFICES OF THE MC LJUBLJANA

操作支持服务
OPERATION SUPPORT SERVICE

运营服务
OPERATION SERVICE(OS)

过程控制系统
PROCESS CONTROL SYSTEMS SERVICE

动力系统
POWER SYSTEM DEVELOPMENT SERVICE(PSDS)

二次系统
SECONDARY SYSTEMS SERVICE

办公室（中层管理）
OFFICE(MIDDLE MANAGEMENT)

系统运行管理
SYSTEM OPERATION DIVISION MANAGEMENT

操作训练模拟器
OPERATOR TRAINING SIMULATOR

二层平面图 SECOND FLOOR PLAN

项目概况

这座建筑的布局和外观体现了ELES公司的目标和愿景。大楼被组织成一个园区或者说一个可以清晰反映公司目标和活动的微缩城市，同时它也创造了一个积极的工作环境。建筑设计和装置设计都是为了更有效地使用燃料和可替代能源，从而形成一个完全的碳平衡设计。

建筑热环境解决方案

数据测算显示，当地全年中187天是晴天，这反映了使用太阳能的潜力，同时也说明热水供应和加热系统可大量有效使用太阳能。设计考虑了对当地资源的利用，比如用地下水直接加热和制冷，促进大楼自然通风，注重自然采光和太阳能的使用。所有的通风系统都配置了房间的自动控制和管理设备，这些设备的安排取决于房间的占有率和检测到的空气质量。通风系统被连接到一个中央控制系统，一个可以让我们通过操作进而优化通风的系统。

场地的每个功能区被分为独立的单元或者说区域，这些单元通过一个环绕的天井连接在一起，并通过交叉连接的桥梁使单元不同的部分连接在一起。天井起到了内部循环的作用。它将光照和新鲜空气带入内部空间，同时将自然和绿意带进了办公区，给大楼营造了宜人的工作氛围。

其他可持续特性：

基于设计对象和目标以及全年1402 mm的平均降雨量的考虑，项目拟采用供公共卫生、灌溉、洗车使用的雨水收集装置。大楼室内将使用混凝土和自然材料，如木头、纺织品和铝等。工作室将会被设计为预制混凝土大厅。

这个项目的能源系统对一些可再生能源（如太阳能、风能），进行了优先利用。此设计同时还优先考虑利用当地资源从而减少（在交通运输过程中的）能源消耗和CO_2排放量。此外还对雨水以及废水进行回收。

能量的使用率取决于表面太阳能屋顶的使用和室内绿色植被与天井的综合使用。外部太阳能薄膜在产生电力的同时也代表了大楼和公司的设计个性。因为有机材料的使用，设施达到了现代的、理性的设计效果。太阳能板结合隔热板和铝板构成了关键性的薄膜结构。使用的太阳能板的数量已经足以产生充足的能量，这些能量可以解决第二阶段建造的能量需要。

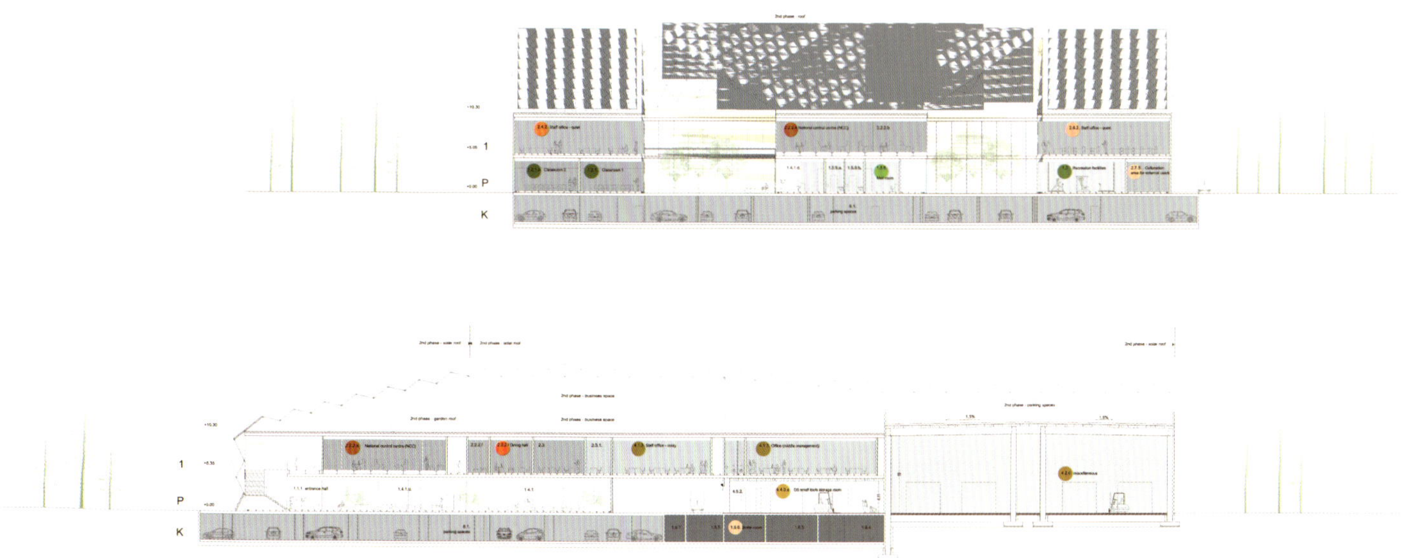

剖面图 SECTION

PAS PÅ DØRENE

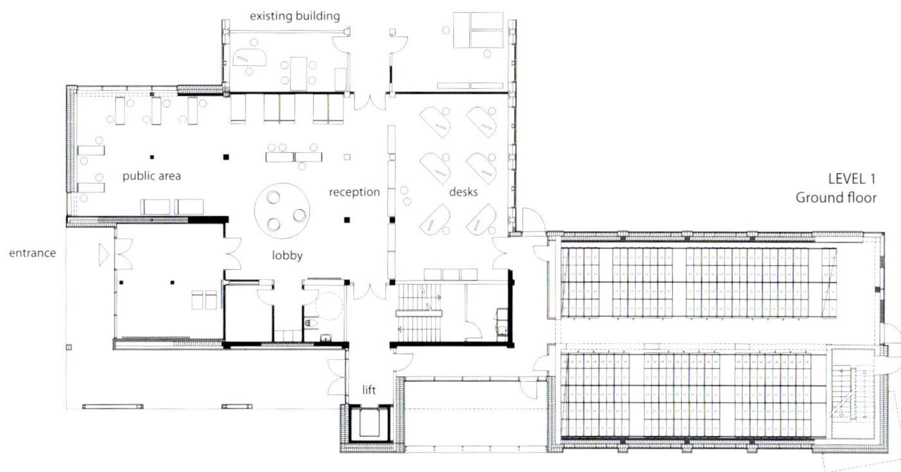

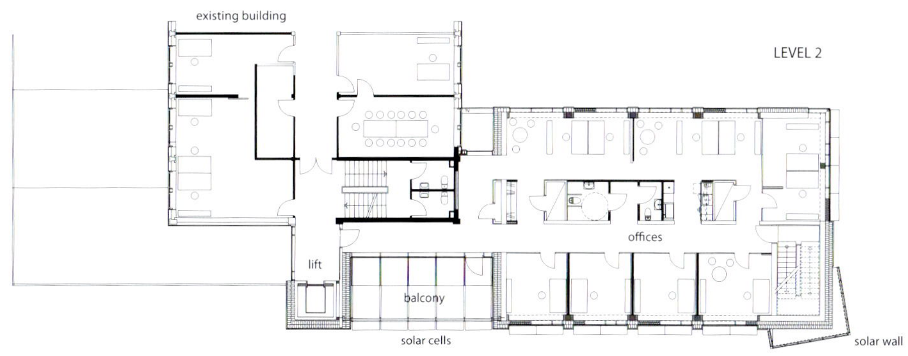

平面图 FLOOR PLANS

PALMAS ALTAS CAMPUS
西班牙帕尔马阿尔塔斯园区

地点：西班牙塞维利亚
占地面积：100 000 m²
Abengoa总部办公区建筑面积：27 800 m²
其他办公区建筑面积：19 200 m²
建筑设计：Roger Strik Harbour + Partners, Vidal And Asociados Arquitectos
结构工程：奥雅纳
所获奖项：2010年英国皇家建筑师协会（Riba）杰出建筑奖；2010年美国建筑师协会（Aia）杰出环境设计奖；2010年欧洲Prime Property Award优秀可持续实践；LEED铂金预认证
委托方：Centro Tecnologico Palmas Altas (Abengoa)

Location: Seville, Spain
Area: 100,000 m²
Office Building Area (Abengoa Headquarters): 27,800 m²
Office Building Area (Speculative): 19,200 m²
Architect: Roger Strik Harbour + Partners, Vidal And Asociados Arquitectos
Structural Engineering: Arup
Selected Awards: 2010 Riba Award For Architectural Excellence: 2010 Aia Award For Environmental Excellence: 2010 Prime Property Award By Union Investment: LEED Platinum Pre-Certification
Client: Centro Tecnologico Palmas Altas (Abengoa)

©Arup

©Arup

项目概况

西班牙塞维利亚市的帕尔马Altas科技园区的主要特色在于，设计师以可持续建筑理念为原则，在修建过程中采用了先进的环境技术和节能措施。该项目荣获了多个奖项，包括LEED铂金预认证（Platinum LEED Pre-certification）。

1. 采取的措施

借鉴安大路西亚地区的传统，在建筑中设置天井。这项设计的优点在于，它可以创造稳定的微气候，而且在夏季可以通过多种途径使天井降温，例如设置池塘，产生蒸发作用；栽植植物，水分通过植物蒸发；设置喷雾（喷泉）设备。采用热质并利用热辐射，建筑外表面得到充分利用。热质可以有效降低热量的吸收和散失，有利于建筑在夜间保温，混凝土结构以及直接与外界接触的混凝土板有助于达到这项效果。天井中设计了一个水滴系统，用来促进天井内空气的冷却，即空气被压入一个冷却管道，其中布满水滴，通过蒸发降温，为天井提供冷湿空气，同时这些冷空气可以输送到各个办公空间。太阳能烟囱可以促进建筑的空气流动，将天井中的冷气传送到其他楼层；南向的玻璃通风管道将屋顶通过热交换形成的热空气通过风力释放出去；利用塞维利亚当地较大的日夜温差（大于15 ℃），通过对流设计措施使空间的通风/冷却不需要任何能耗；利用热质吸收的日间热量可通过建筑的通风系统在夜间将热量释放；利用地热能/生态能。

62 COUNCIL FLATS

62号公寓

©Sergio G

规模：5120 m²
建筑设计：Hamonic+Masson (Gaelle hamonic+Jean - Christophe Masson and Marie-Agnès de Bailliencourt)
委托方：Paris Habitat
摄影：Sergio Grazia; Frederic Delangle

Size: 5120 m²
Architect: Hamonic+Masson (Gaelle hamonic+Jean - Christophe Masson and Marie-Agnès de Bailliencourt)
Client: Paris Habitat
Photography: Sergio Grazia; Frederic Delangle

©Sergio Grazia

项目概况

项目包括两座公寓楼，一座11层，另一座为8层。每层公寓地板的形状都不完全相同，可完成不同的功能和作用。在内部，公寓围绕一个中央核心筒展开，管线、楼梯和电梯据此分布。每隔三四层设置一个楼梯平台。大楼只有核心筒和外立面承重，其他的楼板可以变动。

建筑热环境解决方案

每层楼都有凉廊（或者露台），较之传统的阳台而言，它们完全可见，允许自然风在公寓外侧流动，使用户感觉自己在户外居住。这个"浇铸的花园"将建筑与外界环境紧密联系起来。凉廊的天花板可以起到为室内遮阳的作用，不封闭的阳台适合于巴黎当地温和的海洋性气候，有利于空气流动和自然的通风。

建筑师还进行了气候设计和隔声设计，从而独立于公寓室内部分形成了一个独立的户外空间，创造了层层叠叠开敞或封闭的露台和凉廊，环绕于塔楼的四周，形状如同蛇形线一般。住户可以绕公寓四周行走，例如走出卧室经露台到达起居室，因此有多种可选的步行路线。

从公寓的所有房间都可以进入蛇形凉廊，使之成为房间的延伸，提供了一片私人的户外空间。这些空间用金属栅条和树脂玻璃与建筑边缘相隔离，可以避免夏日阳光直射，对风和阳光起到缓冲作用。同时还使用了不锈钢和生铝等材料，使楼宇更加通透而富于动感。

©Frederic Delangle

©Frederic Delangle

©Frederic Delangle

©Sergio Grazia

©Sergio Grazia

Project Description

The project involves two blocks of flats, one of 11 storeys above the ground floor and the other of 8 storeys above ground floor. Each level and each flat has a different floor lending itself to different practices and uses. Inside, the flats are arranged around a central structural core that houses the flows, stairs and lifts. Each landing serves three to four flats. Only these cores and the facades are weight-bearing, which means that the decks can be opened up and the floor-plan reversed.

Building Thermal Environmental Engineering Solutions

Rather than being like a balcony, a loggia (or a terrace), which can be seen and used on a daily basis, winds its way around the outside of the flats and gives residents the feeling that they live outdoors. This "poured garden" creates close ties to the building's external environment. The ceilling formed by the loggias can shade the sunlight for the interior spaces, and the open terraces are adapt to the local mild maritime climate of Paris, which are propitious to air flow and natural ventilation.

Climate planning and sound-proofing have also left their mark, and permitted a system of "truly outdoor spaces" that are therefore independent of the internal floor-plan, creating a stack of more or less closed terraces and more or less open loggias wrapped around the four sides of each tower like a "serpentine". One can stroll around a flat, walk out of the bedroom and into the living room; there are many paths to choose from and many surprises in store.

The internal space at each level of the two towers, is well designed and structured as an open space around a central core housing the stairs and lifts, but instead of the expected openings and the hoped-for transparency, there is a game of mirrors. Flowing balconies snake around the outside of the flats, some of them so large that they are like loggias. They are accessible from all the rooms in the flat, forming an extension of the indoor areas as well as a private space outdoors. These spaces are protected from the buildings opposite by varying arrangements of wire and plexiglass, thus they can prevent the direct sunlight in summer and be a buffer area to wind and daylight. Added to the alternating combination of stainless steel and untreated aluminium, this composition produces rich and kinetic patterns of transparency.

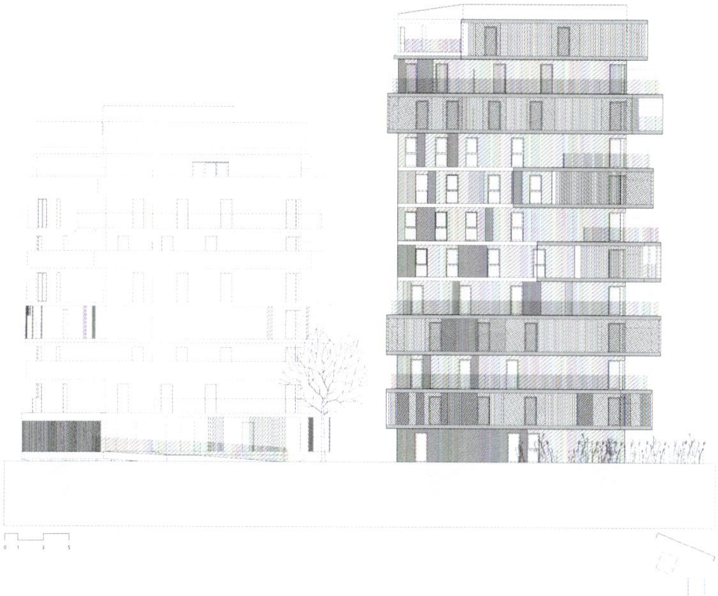

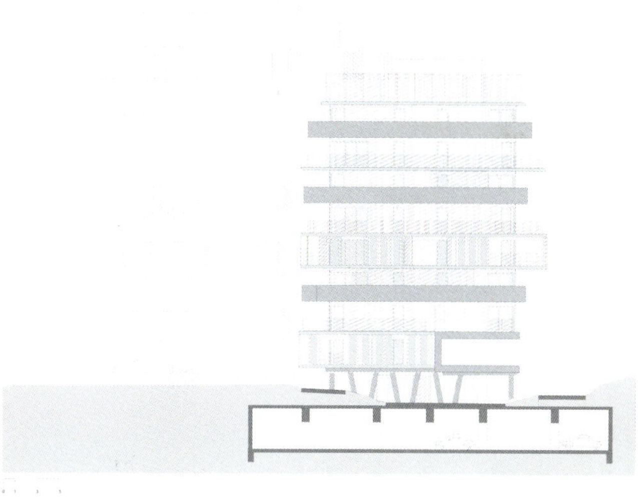

立面图 ELEVATIONS

©Sergio Grazia

©Sergio Grazia

©Sergio Grazia

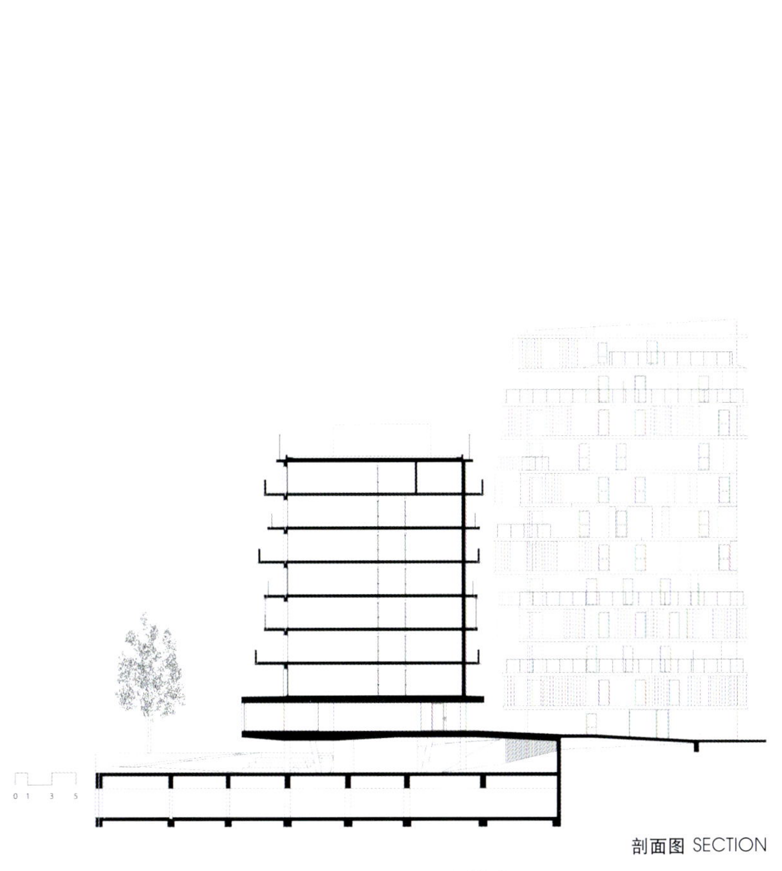

剖面图 SECTION

©Jean-Christophe Masson

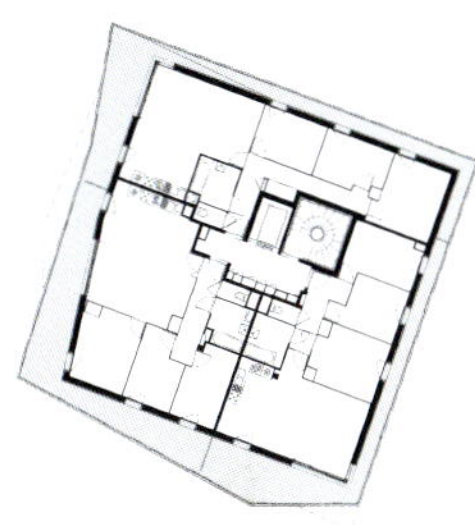

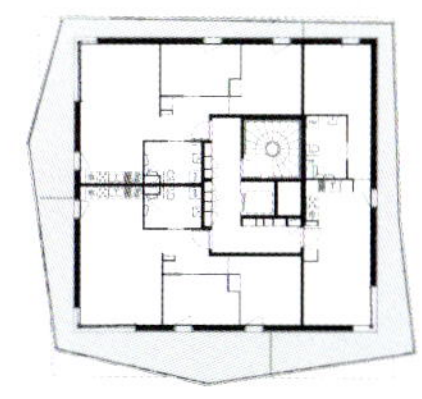

二至五层平面图 LEVEL 02~05 FLOOR PLAN

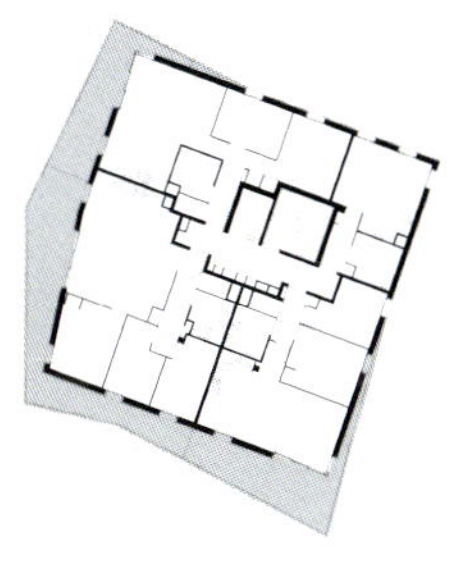

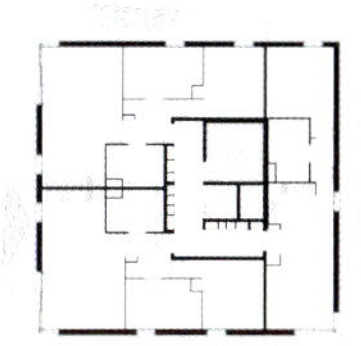

一层平面图 GROUND FLOOR PLAN

六至十层平面图 LEVEL 06~10 FLOOR PLAN

VILA ALSTRUP
VILA ALSTRUP别墅

地点：丹麦埃斯比约Hjerting
规模：311 m²
建筑设计：C. F. Møller Architects
委托方：Nis Alstrup
摄影：Helene Hoyer Mikkelsen

Location: Hjerting, Esbjerg, Denmark
Size: 311 m²
Architect: C. F. Møller Architects
Client: Nis Alstrup
Photography: Helene Hoyer Mikkelsen

项目概况

这栋位于Wadden Sea的海景别墅是一座"储能建筑"，它产生的电能和热能大于其消耗量。同时公寓的各种使用功能没有受到影响，还设有全方位的海景房。新颖的外形与沉静、自然的细节设计相结合，并运用了大量高质量的材料。

建筑热环境解决方案

1.被动措施

（1）控制太阳辐射

建筑外形简洁、轮廓清晰，面向大海的一侧开敞通透，面向邻里的一侧封闭私密。整栋建筑根据被动式节能房屋的原则设计，形制紧凑，面向西南方向设计了观景大窗，这一朝向也符合被动利用阳光热量的最佳角度。这一角度也遵守了海岸线保护区的规定，因此设计出了三角形的平面图，成为该建筑的一项特色。另一项特色是倾斜的屋顶，设置的角度和朝向都是为了使太阳能加热和太阳能电池达到最佳效果。

（2）控制外墙内表面温度

以被动方式获得的太阳能热量主要是通过内部混凝土墙和水泥板吸收聚集，同时西南向的阳台和挑檐为立面遮光，并根据季节的变化控制进光量。阳台由一块独立的混凝土板构成，完全避免了与室内发生热桥效应。

2.主动措施

别墅通过应用太阳能和地热资源可以实现电能和热能的自给。它可以有效地产生比自身消费量更多的能量。

其他可持续设计措施

根据项目的设计构想，别墅采用了高效铝的预制构件以及环境可持续的材料，例如纸制微粒，是一种循环再利用产品，同时得到了生态标志认证。按照设计，别墅的维护需求降到了最低。例如外立面覆以纤维水泥，以抵御当地的海洋气候。这也将增加建筑的耐久性。

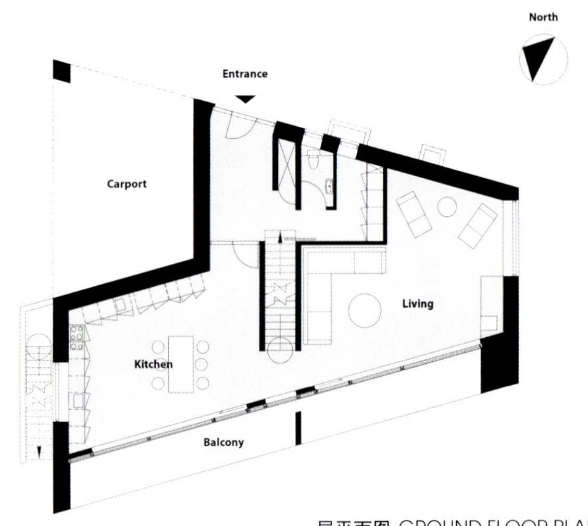

一层平面图 GROUND FLOOR PLAN

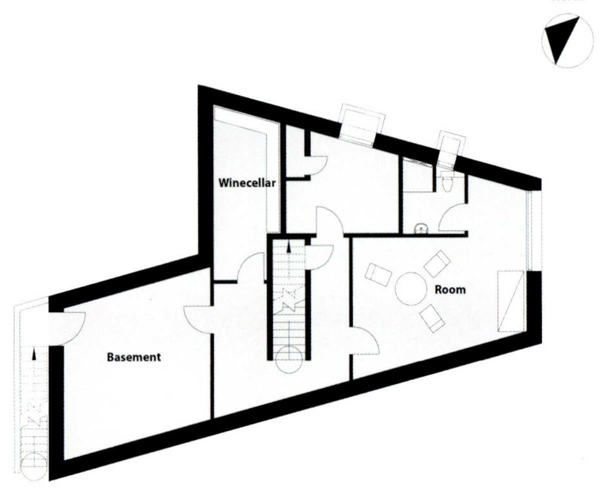

地下一层平面图 –1F PLAN

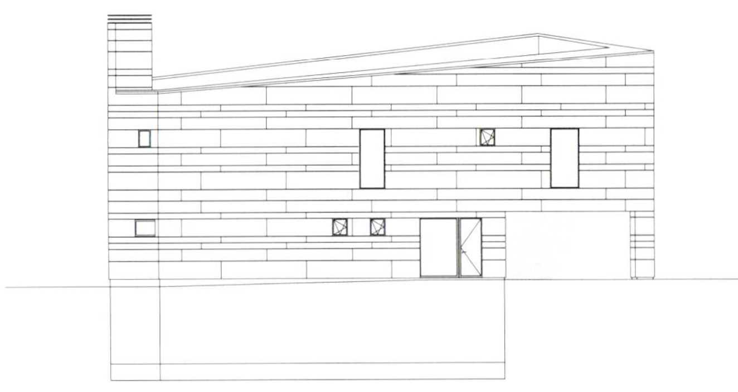

东立面图 EAST ELEVATION

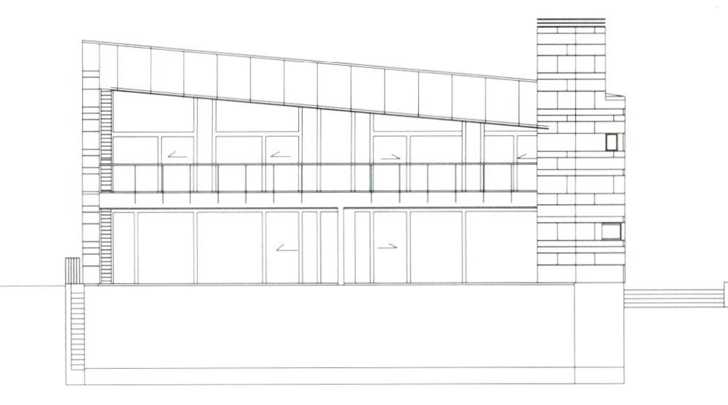

西立面图 WEST ELEVATION

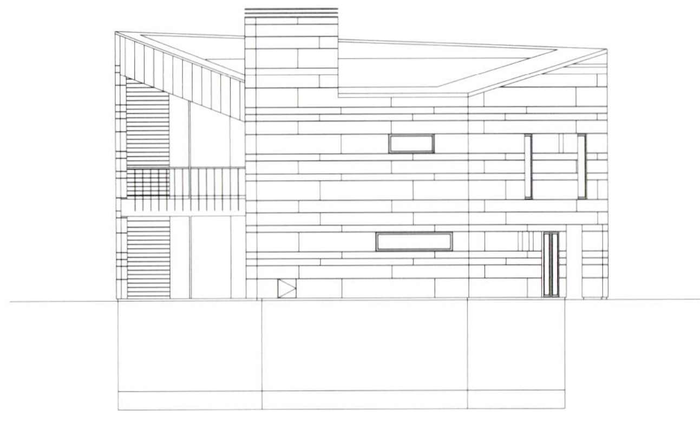

南立面图 SOUTH ELEVATION

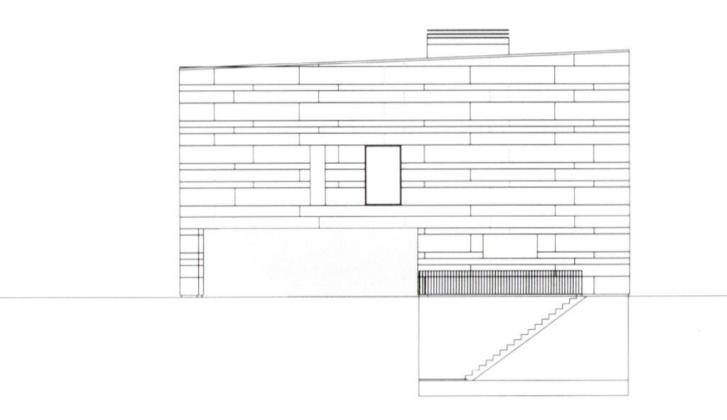

北立面图 NORTH ELEVATION

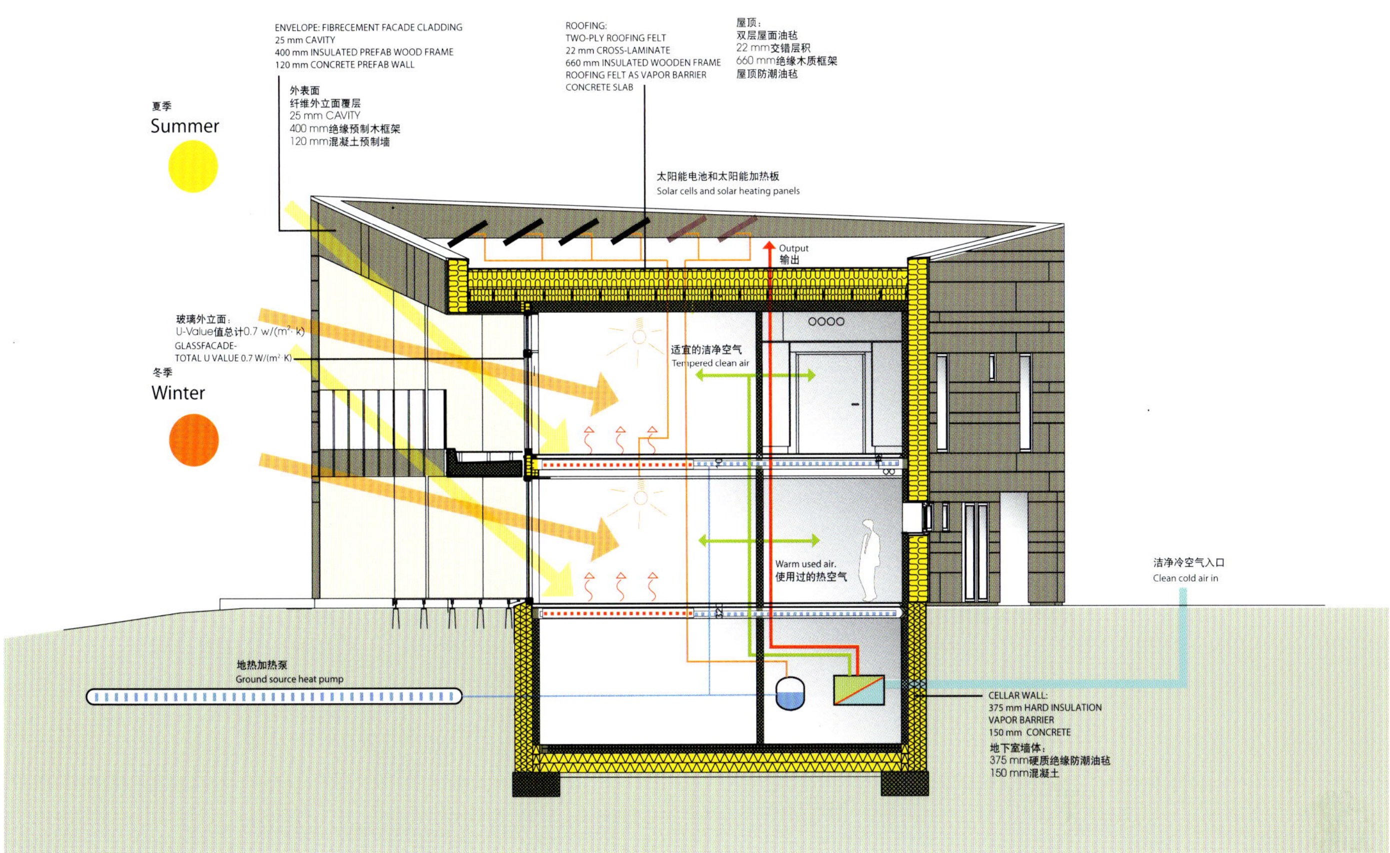

分析图 DIRGRAM

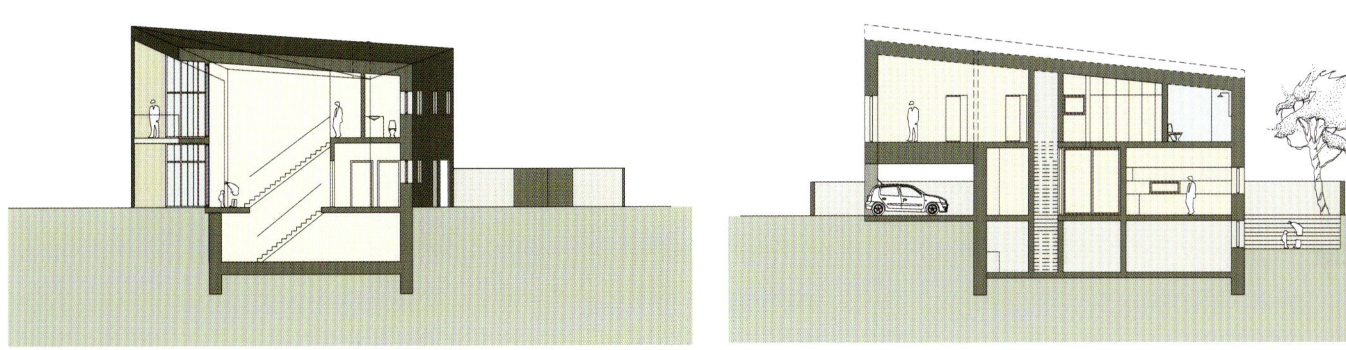

剖面图 SECTION

The house on the shore with a view of the Wadden Sea is an energy-plus house, which means that it produces more electricity and heat than it uses. This was achieved without compromising on the exclusive qualities of a large home, including large panoramic sea-views.

The architecture has a clear and simple expression, open and transparent to the sea and more closed and private towards the neighbours. The unusual geometry of the volume is combined with a calm and unpretentious detailing, and a restrained palette of high-quality materials.

The building, which is designed on the basis of the 'passive house' principle, is compact in form, with large windows facing the view to the south-west, which is also the best angle at which to make optimal passive use of the sun's heat. The angle also respects the shoreline protection zone, and out of this derives the triangular floor plan layout, which is one of the buildings characteristics. Another is the sloping roof, which is angled and orientated to optimize the performance of the solar heating and solar cells.

Passive solar heat gain is absorbed and accumulated in the interior concrete walls and floor slabs, while the south-west facing balcony and overhangs shade the facades and control the amount of solar energy in correlation to the seasons. The balcony is designed as a free-standing concrete slab completely eliminating any cold-bridging to the interior.

The house is self-sufficient in electricity and heat through this use of solar and geothermal energy, meaning in effect that it produces more energy than it consumes. By law, the building must still be connected to the grid, but the meters will simply be running backwards in periods.

In the spirit of the project, it is built using efficient pre-fabrication and environmentally-sustainable materials such as paper granules, which is a recycled product, and eco-labelled products. The house is designed to require minimal maintenance; the facade, for example, is clad in fibre cement, to withstand the local Wadden Sea climate. This will help to increase the building's sustainability over time.

HOUSING+ ZERO-ENERGY COLLECTIVE HOUSING IN AALBORG

HOUSING+奥尔堡零能耗集成住宅

地点：丹麦奥尔堡
规模：7400 m²
建筑设计：C. F. Møller Architects
景观设计：Vogt Landscape
委托方：Enggaard A/S

Location: Aalborg, Denmark
Size: 7400 m²
Architect: C. F. Møller Architects
Landscape: Vogt Landscape
Client: Enggaard A/S

World Green Buildings——Thermal Environmental Engineering Solutions

总平面图 SITE PLAN

剖面图 SECTIONS

项目概况

HOUSEING+ 概念设计将目标设定于建设零能耗别墅，其中也包括用户的日常能耗部分。建筑将100%依靠可再生能源。项目的核心是利用综合能源设计措施设计出未来房屋，产能大于耗能。将主体建筑被动能量的获得最大化，同时利用地势条件，并开发主动太阳能收集系统，由此零能耗的目标得以实现。

建筑热环境解决方案

住宅按照被动式房屋的标准建设，也可与太阳能电池阵和海湾水温地源热泵相结合，保证了取暖和热水供应过程中节约能源。一个3 m宽12 m长的高度隔热水箱用来存储日间生成的能量。用浅色落地窗保证最佳光照。夏季采用混合通风，冬季采用热回收平衡通风。混合通风是太阳能烟囱中的一个太阳能推进装置。新风通过墙面机械通风口吸入，并从烟囱排出。外露混凝土板为热质，可在夜间冷却。规格为400 mm 的高度隔热气密性围护材料结合三层节能玻璃窗，保证环境隔热性。

可再生能源利用

高度在4层至12层之间的60套单元的大楼外形为斜坡式，形成了一个巨大的南向屋顶平面，是利用太阳能的理想空间。这个出众的外形使其成为海滨令人瞩目的建筑。利用太阳能电池、太阳能加热以及二者的结合，整个屋顶表面成为建筑的发电厂。以每户年均用电量1740 kWh计算，大楼用电总量为104 400 kWh，1200 m^2的太阳能阵列可以满足其能量需求。建筑不需要外部CHP。

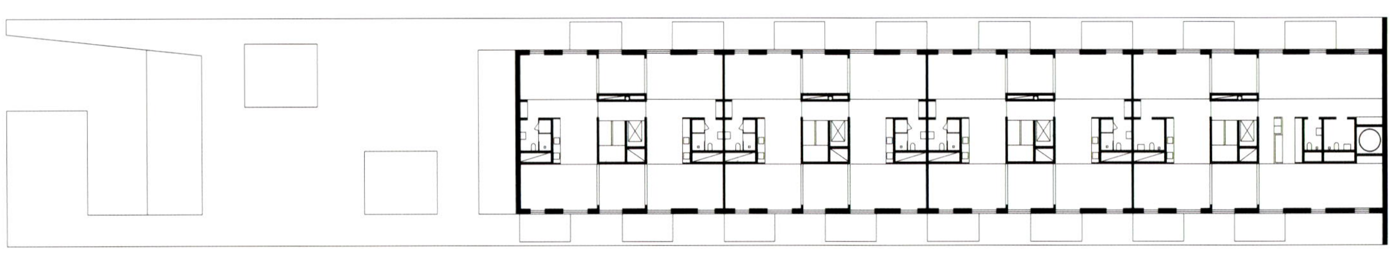

平面图 FLOOR PLANS

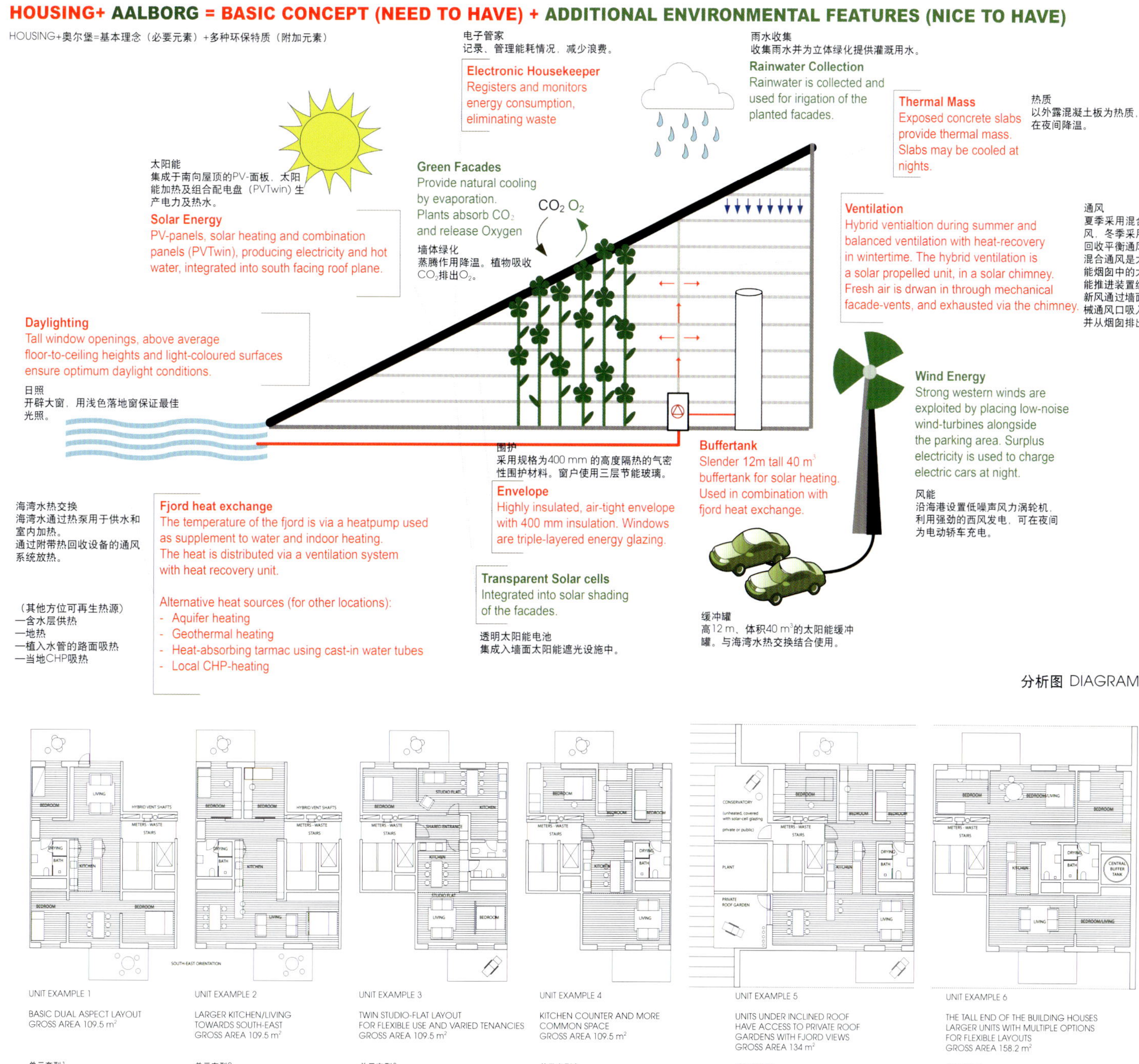

分析图 DIAGRAM

UNIT EXAMPLE 1

BASIC DUAL ASPECT LAYOUT
GROSS AREA 109.5 m^2

单元套型1
基本二层布局
总面积：109.5 m^2

UNIT EXAMPLE 2

LARGER KITCHEN/LIVING
TOWARDS SOUTH-EAST
GROSS AREA 109.5 m^2

单元套型2
大厨房/起居室
朝向东南
总面积109.5 m^2

UNIT EXAMPLE 3

TWIN STUDIO-FLAT LAYOUT
FOR FLEXIBLE USE AND VARIED TENANCIES
GROSS AREA 109.5 m^2

单元套型3
套房含两个一居室公寓，使用及出租方式灵活多样。
总面积：109.5 m^2

UNIT EXAMPLE 4

KITCHEN COUNTER AND MORE
COMMON SPACE
GROSS AREA 109.5 m^2

单元套型4
反向厨房，更多公共空间
总面积 109.5 m^2

UNIT EXAMPLE 5

UNITS UNDER INCLINED ROOF
HAVE ACCESS TO PRIVATE ROOF
GARDENS WITH FJORD VIEWS
GROSS AREA 134 m^2

单元套型5
位于斜屋顶下的单元，可进入视野良好的私人屋顶花园。
总面积134 m^2

UNIT EXAMPLE 6

THE TALL END OF THE BUILDING HOUSES
LARGER UNITS WITH MULTIPLE OPTIONS
FOR FLEXIBLE LAYOUTS
GROSS AREA 158.2 m^2

单元套型6
大楼较高的一端设置了更大的户型，可供多种用途灵活使用
总面积 1582 m^2

单元类型 UNIT TYPES

Project Description

HOUSING+ is a pilot-project for 60 zero-energy housing units on the Aalborg Waterfront. The Housing+ concept sets the ambitious target of a zero-energy housing scheme, which also includes the tenant's primary household energy consumption. The complex will thus be 100% relying on renewable. Central to the project is the use of integrated energy-design to generate the concept of tomorrow's housing, producing more energy than it consumes. This is achieved by optimizing the inherent passive gains of the main volume, and shaping it to take advantage of the orientation and potential for active solar energy-collection.

Building Thermal Environmental Engineering Solutions

The housing is built to passive-house standards, ensuring reduced energy consumption for heating and hot water supply, which can thus be covered by the solar array and heat pumps operating on fjord water temperatures. A 3 meter wide by 12 meter tall highly insulated water-tank is integrated to store the generated energy during daytimes. Floor-to-ceiling windows with light-coloured surfaces are used to ensure optimum daylight conditions. The complex uses hybrid ventilation during summer and balanced ventilation with heat-recovery in wintertime. The hybrid ventilation is a solar propelled unit, in a solar chimney. Fresh air is drawn in through mechanical facade-vents, and exhausted via the chimney. Exposed concrete slabs provide thermal mass. Slabs may be cooled at nights. The highly insulated air-tight envelope with 400 mm insulation coordinating with triple-layers energy glazing windows ensure an well-insulated environment.

Using Renewable Energy

The 60 units take the form of a sloped volume, from 12 to 4 storeys, creating a large south-facing roof-plane, ideal for solar energy. This optimized shape also creates a landmark silhouette of the park. The entire surface of the roof becomes the building's power plant using both solar cells, solar heating and a combination of the two. The 1200 m^2 solar array produces sufficient power to cover the annual 1740 kWh electricity demand of each unit, a total of 104,400 kWh. The building need not be connected to an external CHP.

COMFORT HOUSE

宜居别墅

地点：丹麦瓦埃勒Stenagervænget
规模：208 m²
建筑设计：C. F. Møller 建筑事务所
所获奖项：2007年竞标第一名
委托方：Komforthusene (by Saint-Gobain Isover and Zeta Invest)

Location: Stenagervænget, Vejle, Denmark
Size: 208 m²
Architect: C. F. Møller Architects
Selected Awards: 1st Prize in competition, 2007
Client: Komforthusene (by Saint-Gobain Isover and Zeta Invest)

项目概况

本项目由C. F. Møller建筑事务所应邀为参加在丹麦瓦埃勒附近举行的主题为"宜居别墅——生态友好及低能耗别墅展览"而设计并建造。宜居别墅传达了设计师对独户别墅的未来设计理念，致力于减少CO_2排放，将热量支出降到最低，并创造适宜的内部气候。

建筑热环境解决方案

在建筑设计方面，宜居别墅的设计灵感源于北欧现代主义明亮、简约的传统，经过深入思考，采用可持续材料，例如用方砖装饰外立面（brick for the exterior）。在朝北的方向，别墅完全封闭；在朝南的一侧，门窗可开启，并置以大型玻璃面板构成的落地窗；同时，起居室也设计在朝南的方向。在别墅前方设有一个木制的露台。玻璃面板的材料中加入了致密的元素，以降低热损耗，屏蔽阳光。别墅外立面为深灰色砖，内墙为木质，两种材料都有利于室内气候的自然调节。别墅的室内外都采用这种墙体材料建造，从而将室内和户外的设计巧妙地连接贯通起来。别墅的墙体和房顶均为预制构件，由起重机设备置于项目选址之上。

大屋檐由屋顶延伸而出，遮住朝南的落地窗，使得室内除了冬天之外，都不会受到阳光直射的影响。特别设计和制造的窗口框架同样起到保温隔热的作用。别墅采用的空心墙隔热材料比普通隔热材料更厚：墙内厚度300 mm，地面和屋顶均为450 mm，从而避免了热桥效应。可进行热量回收的暖通空调设备使建筑内的热能得以循环利用，为了使其高效运行，整栋建筑采取全封闭设计。

立面图 ELEVATIONS

总平面图 SITE PLAN

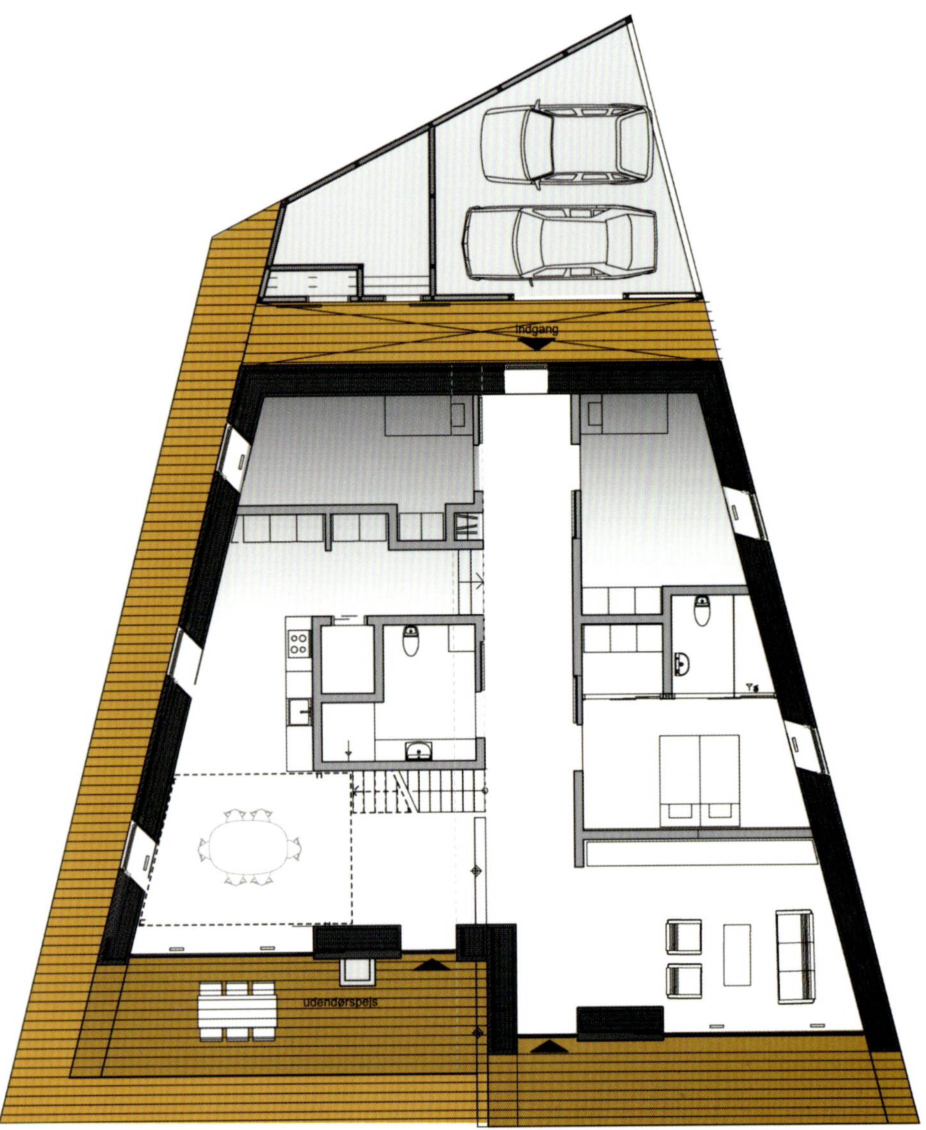

平面图 FLOOR PLANS

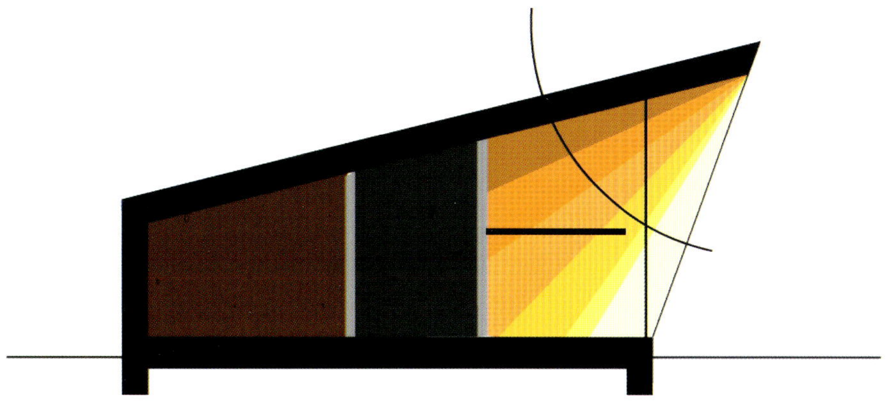

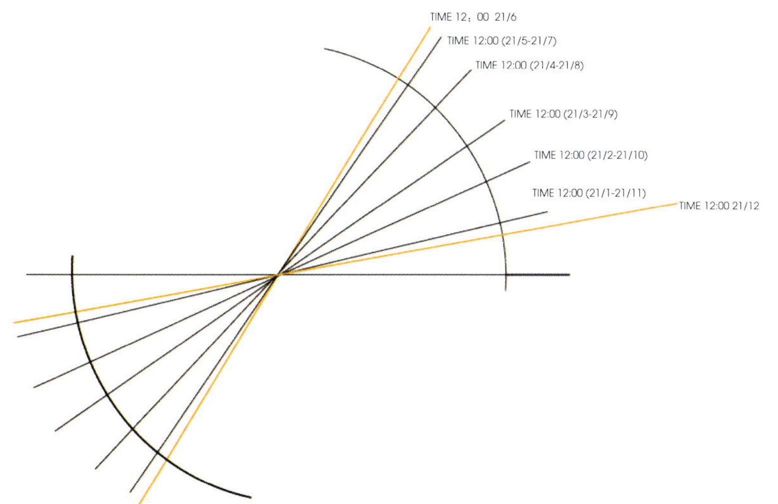

Project Description

The project is C. F. Møller Architects' contribution (as one of a total of ten consortia) to the eco-friendly low-energy housing exhibition 'Comfort Houses' near Vejle, Denmark, a project initiative by insulation manufacturer Saint-Gobain Isover. It is developed in cooperation with Tækker Consultant Engineers, Fiberline (GRP-composites manufacturer) and contractors Buch Andersen, and is currently under construction.

Building Thermal Environmental Engineering Solutions

The Comfort Houses represent suggestions for the single-family houses of the future, which will contribute to reducing CO_2 emissions, as well as having almost no heating expenses and an optimum inner climate. Architecturally, C. F. Møller's Comfort House takes its cue from the bright and simple tradition of Nordic modernism, with a deliberate use of reliable materials such as brick for the exterior.
The house is closed off towards the north and open, with large glass panels, towards the south, which is also where the living-rooms are located. A wooden deck in front of the house provides a terrace. The glass panels have integrated solid elements to reduce heat loss and screen off the sun. The house is made of dark grey brick with inner walls of wood; both materials help to create a natural regulation of the inner climate. The materials are used consistently both inside and outside the house, thereby creating a smart continuity between the indoor and outdoor design. The walls and roof of the house have been pre-fabricated, and craned into position on the site.
The large glass panels face south, but are drawn in under a roof eave, so that they are struck only by the relatively low winter sun.
The specially developed and manufactured window frames also help to provide heat insulation. Cavity insulation is thicker than normal: around 300 mm in the walls and 450 mm in the floors and roof, and thermal bridges have been avoided. The heat in the house is recycled with the help of heat recovery and controlled ventilation. In order to be able to exploit this principle fully, the building is completely sealed.

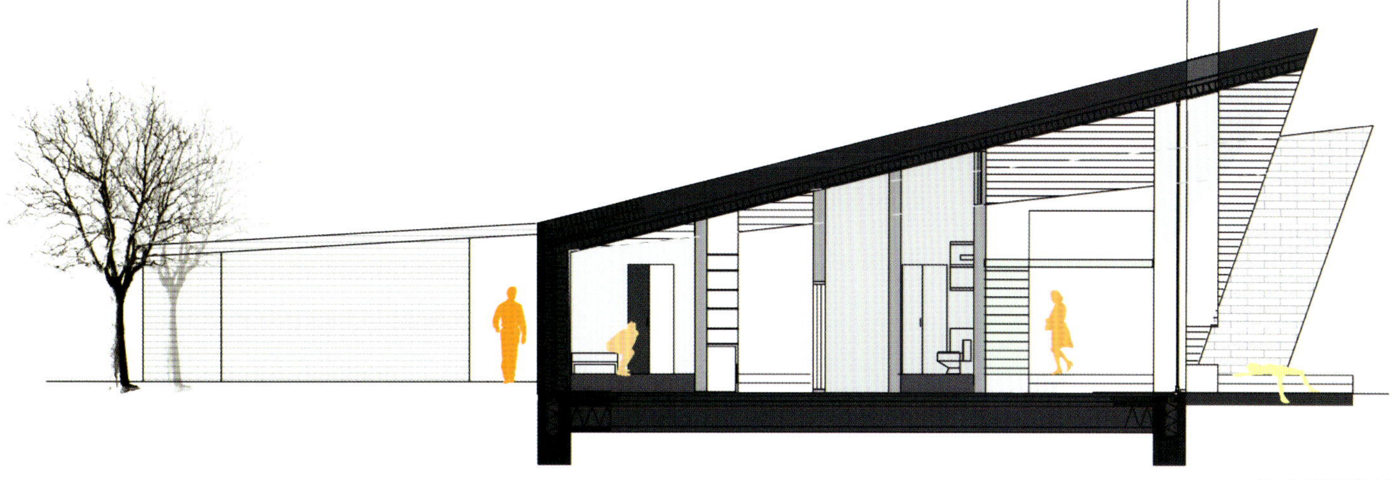

剖面图 SECTION

DA VINCI, RESIDENCIAL TOWER

地点：墨西哥
建筑面积：16 649.82 m²
建筑设计：Pascal Arquitectos
摄影：Sófocles Hernández

Location: Mexico
Building Area: 16,649.82 m²
Architect: Pascal Arquitectos
Photography: Sófocles Hernández

World Green Buildings——Thermal Environmental Engineering Solutions

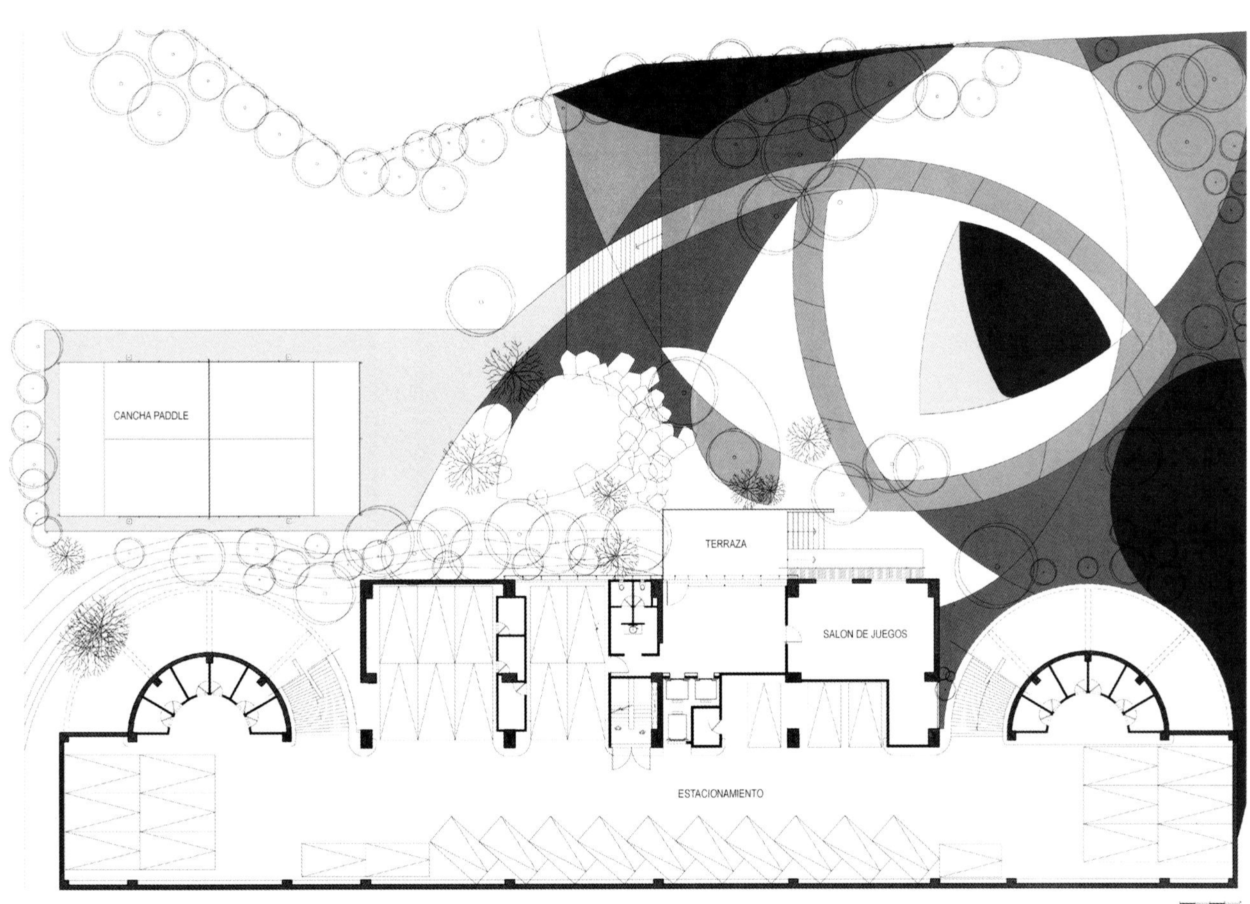

总平面图 SITE PLAN

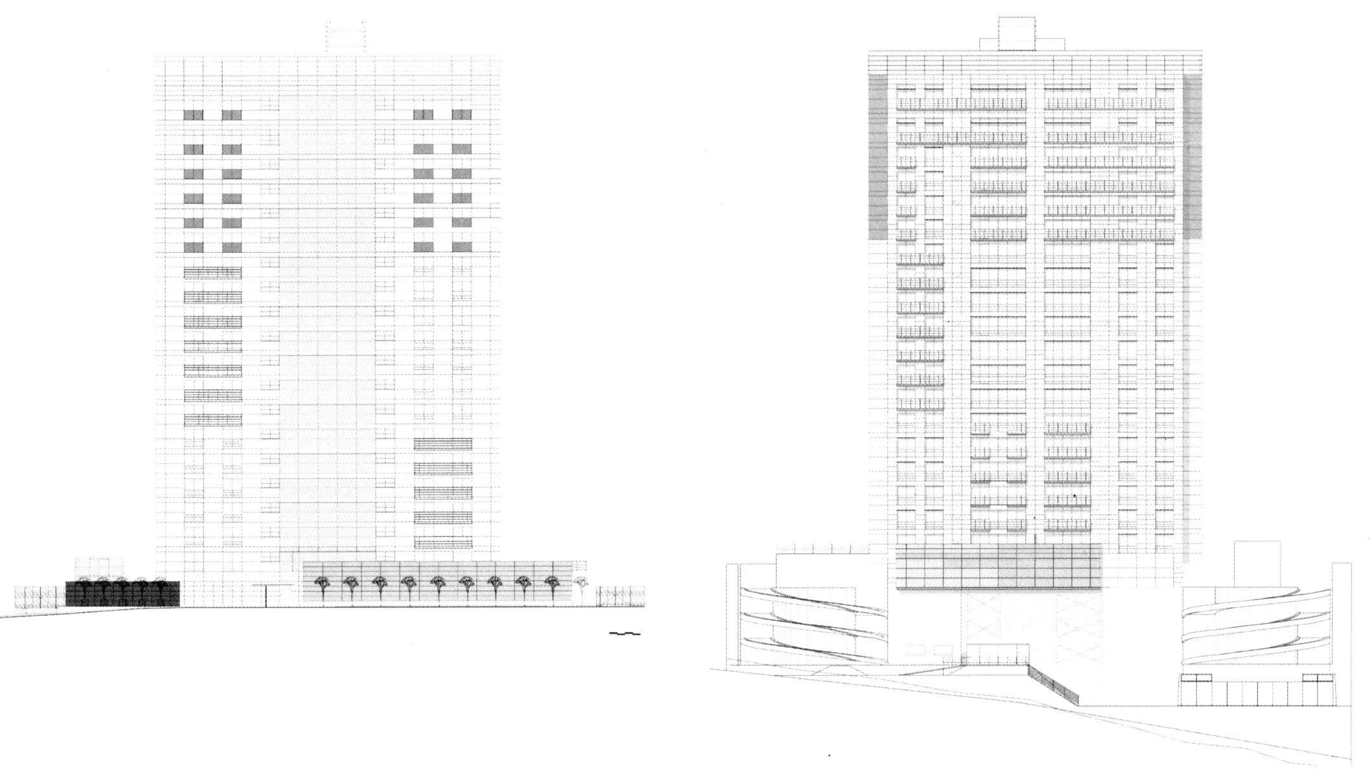

立面图 ELEVATIONS

MOUNTAIN DWELLINGS

山居住宅

地点：丹麦哥本哈根
规模：33 000 m²，80个单元，480停车位
建筑设计：BIG with JDS
所获奖项：2009年AID论坛大奖；2008年密斯·凡·德·罗奖提名奖；
2008年丹麦伍德奖；2008年WAF世界建筑节最佳住宅项目奖
委托方：Hopfner NS, Danish Oli Company

Location: Copenhagen, Denmark
Size: 33,000 m², 80 Units, 480 Parking Spaces
Architect: BIG with JDS
Selected Awards: Forum AID Award 2009; Mies van der Rohe Prize 2008 / Nomination: Danish Wood Award "Traprisen" 2008; WAF World Architecture Festival Award 2008 / Best Housing Project
Client: Hopfner NS, Danish Oli Company

©JDS

©JDS

项目概况

本项目中，场地面积的2/3用于停车，1/3为住宅。以停车场为基础建造带有露台的住宅，就像由混凝土山上建起的一层层简练的住宅，并从第11层起逐渐迭退，一直延伸到公路边。

建筑热环境解决方案

停车场需要临街，住宅需要开敞空间。汽车要靠近地面，不受风雨侵蚀，不受太阳直射。而另一方面，住宅需要南向采光、新鲜的空气以及良好的景观。建筑师将这两种需求叠加在一起，从而构思出这座建筑的形态，停车场占据北侧大部分空间，住宅在上层，坐北朝南，全部向阳。因此，将两个计划相结合，设计方案中所有的公寓都有阳光充盈的屋顶花园，宜人的观景效果。整座建筑如同在一个10 m米高的大楼之上漂浮着的郊区花园别墅。

建筑的北立面和西立面由铝制多孔板覆盖，使日光和空气进入停车区，保证日照和通风。建筑内外，每层走廊都用金属材料封装起来，然而从前门进入住宅的感觉完全不同。对流通风、开阔敞亮、阳光充溢的房间以玻璃幕墙围护；屋外是面积90 m^2的露台，种植着人工草坪。白色外墙、原木地板和木框玻璃窗显示出了沉静的丹麦现代风格。每一户都可得到同等的美景，屋顶花园开敞而私密，植物随四季不同而变换。大楼设置了灌溉系统，保证屋顶花园的维护运营。公寓和花园之间只用玻璃幕墙和推拉门隔开，室内可以得到充足的阳光和新鲜空气。

Moe & Brødsgaardg公司负责了所有的工程设计工作。包括地面和排水工程，钢铁和混凝土结构，并达到多种严格的环境分级准则要求。

©JDS

©JDS

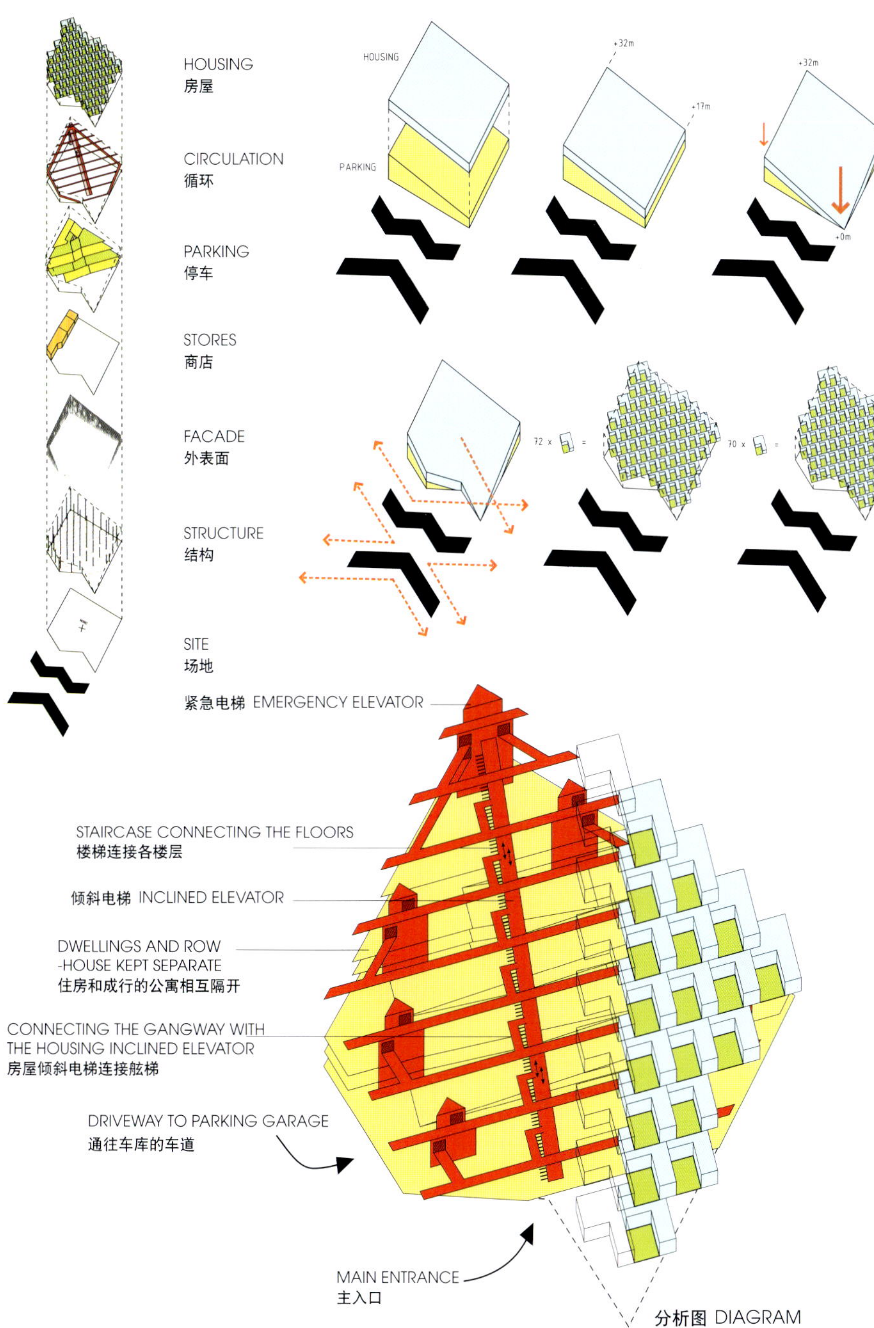

分析图 DIAGRAM

Project Description

The program consist of l/3 living and 2/3 parking. The parking area became the base upon which to place a terraced housing development - like a concrete hillside covered by a thin layer of housing, cascading from the eleventh floor to the street edge.

Building Thermal Environmental Engineering Solutions

The parking area wants to be connected to the street, and the residences want open spaces. Cars, are near the ground, sheltered, and kept away from direct sunlight. Housing, on the other hand, "wants" southern exposure, fresh air, and a view. Gradually, when mixing the needs of cars and housing —they'd gravitate to this form, with the parking in the deep space in the north and the houses on top, facing south. So by combining the two programs, all apartments can have a sun filled roof garden, amazing views and parking on the same level as the apartment. The Mountain appears as a suburban neighborhood of garden homes floating over a 10-story building.

The north and west facades are covered by perforated aluminum plates, which let in air and light to the parking area. Each residence is provided with equal amenities and the gardens are grouped in a way that is large yet intimate, with plants that change character according to the season. The building also has an integrated watering system, which helps to maintain the roof gardens.

Moe & Brødsgaard is doing the project design on all engineering works. This includes ground- and sewerage works, and steel- and concrete structures in several environmental classifications, among these aggressive environmental classification.

©JDS_FELIX LUONG

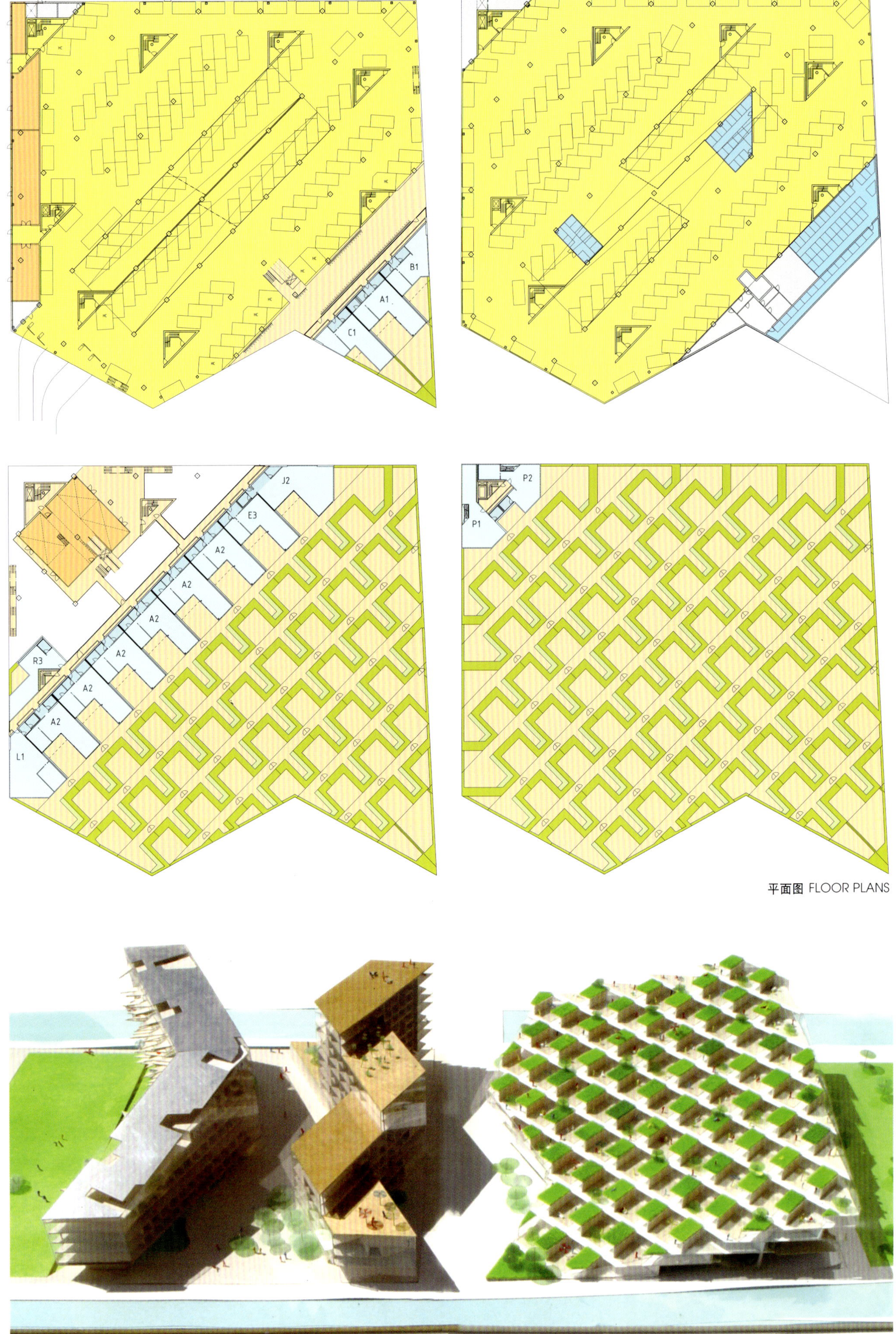

平面图 FLOOR PLANS

模型图 MODEL DIAGRAM

©JDS

©JDS

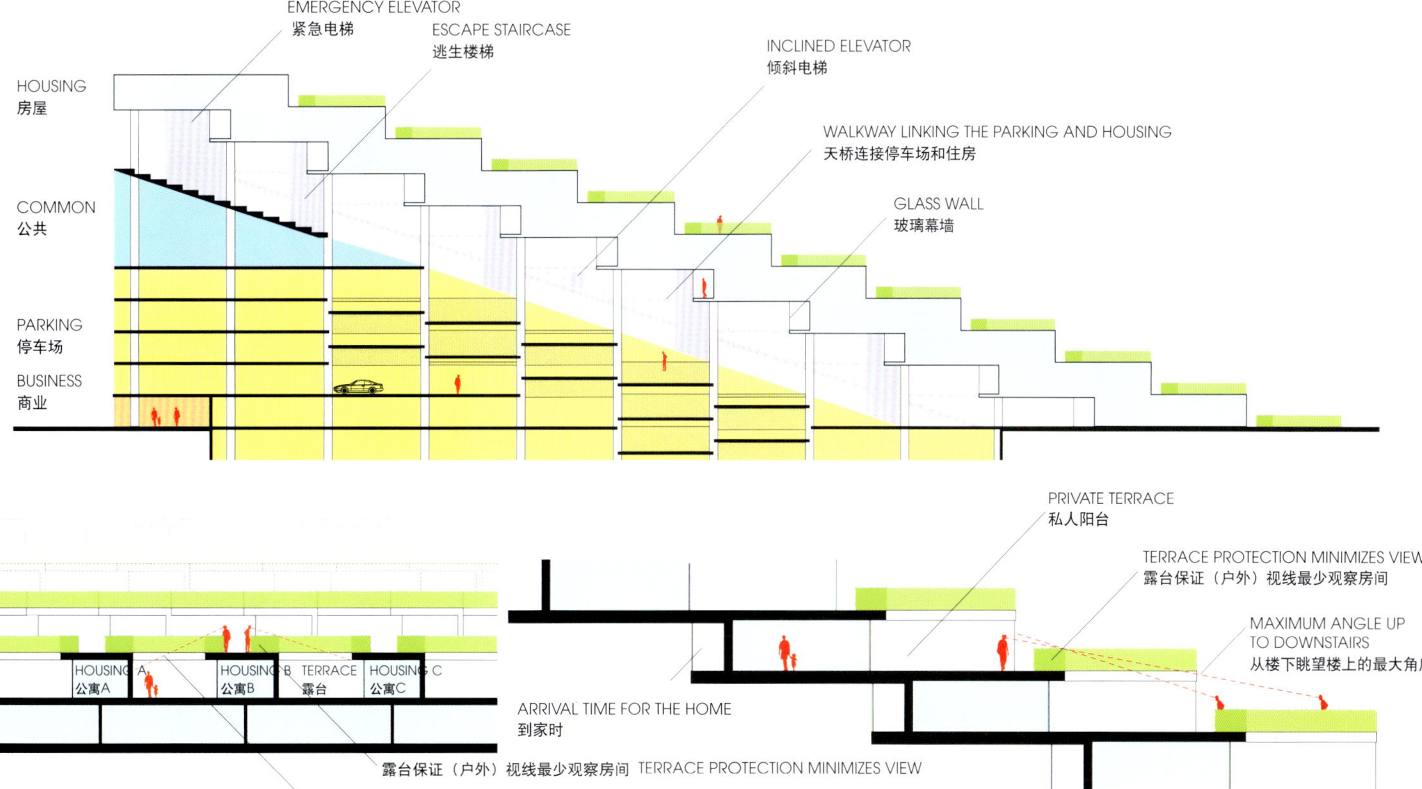

分析图 DIAGRAM

HOUSE FOR MR.R

R氏别墅

地点：俄罗斯莫斯科
占地面积：560 m²
建筑面积：300 m²
建筑设计：Za Bor architects
摄影：Zinur Razutdinov；Peter Zaytsev

Location: Moscow, Russia
Architect: Za Bor architects
Area: 560 m²
Building Area: 300 m²
Photography: Zinur Razutdinov; Peter Zaytsev

不久前莫斯科出现了一座壮观的绿色别墅。建造过程中没有破坏树木，在别墅内部只使用了自然材料。业主提出的唯一要求是在较小的范围内（1800 m^2）保留所有树木。因此建筑面积只用了300 m^2（住宅的占地面积为560 m^2）。在建设过程中，树木也带来了不少麻烦：打地基时，他们发现了一个巨大的树根。因此设计了一个混凝土特种管植入地基，这样就保存了树根。起居室、家庭影院、餐厅以及一些配套设施设在一楼，主人书房、卧室和儿童房等位于二楼。

房子有坚实的钢筋混凝土框架结构，支柱、桥接和砖墙。一面墙是倾斜的（堆高砌坯）。公寓有两层，地下室向地下加深1.2 m。设施包括水泵、吹风井、紧急柴油发电机和燃料储罐。清洁管道、天然气和电力供应也被安排在地下室。一体化屋顶最大的斜坡角度为12°。坡屋顶有利于建筑屋面的防水、隔热、保温、节能等。屋顶开有照亮楼道的天窗以及老虎窗，在侧厅之间屋顶区域可供日光浴或聚餐使用。二楼有两个阳台和一个凹阳台。凹阳台有金属框架支撑的玻璃棚架，用以防雨。细不锈钢管制成的排水管放置在屋顶上方。

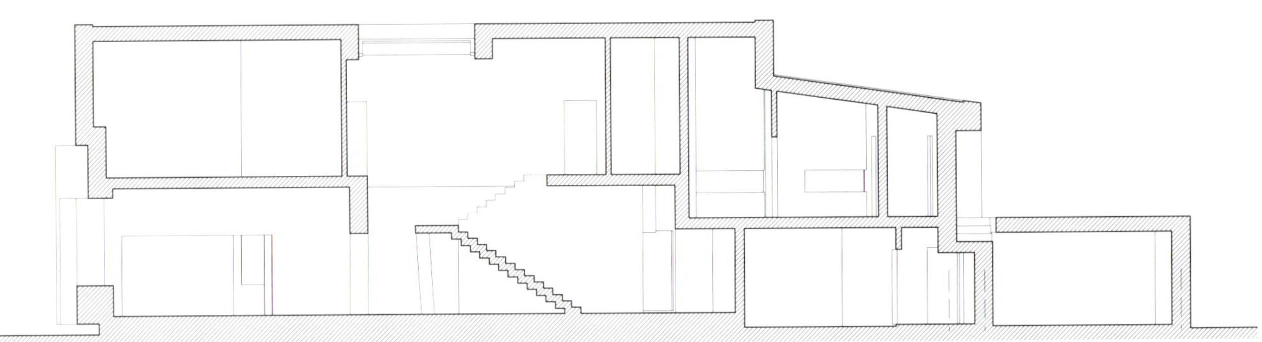

剖面图 SECTION

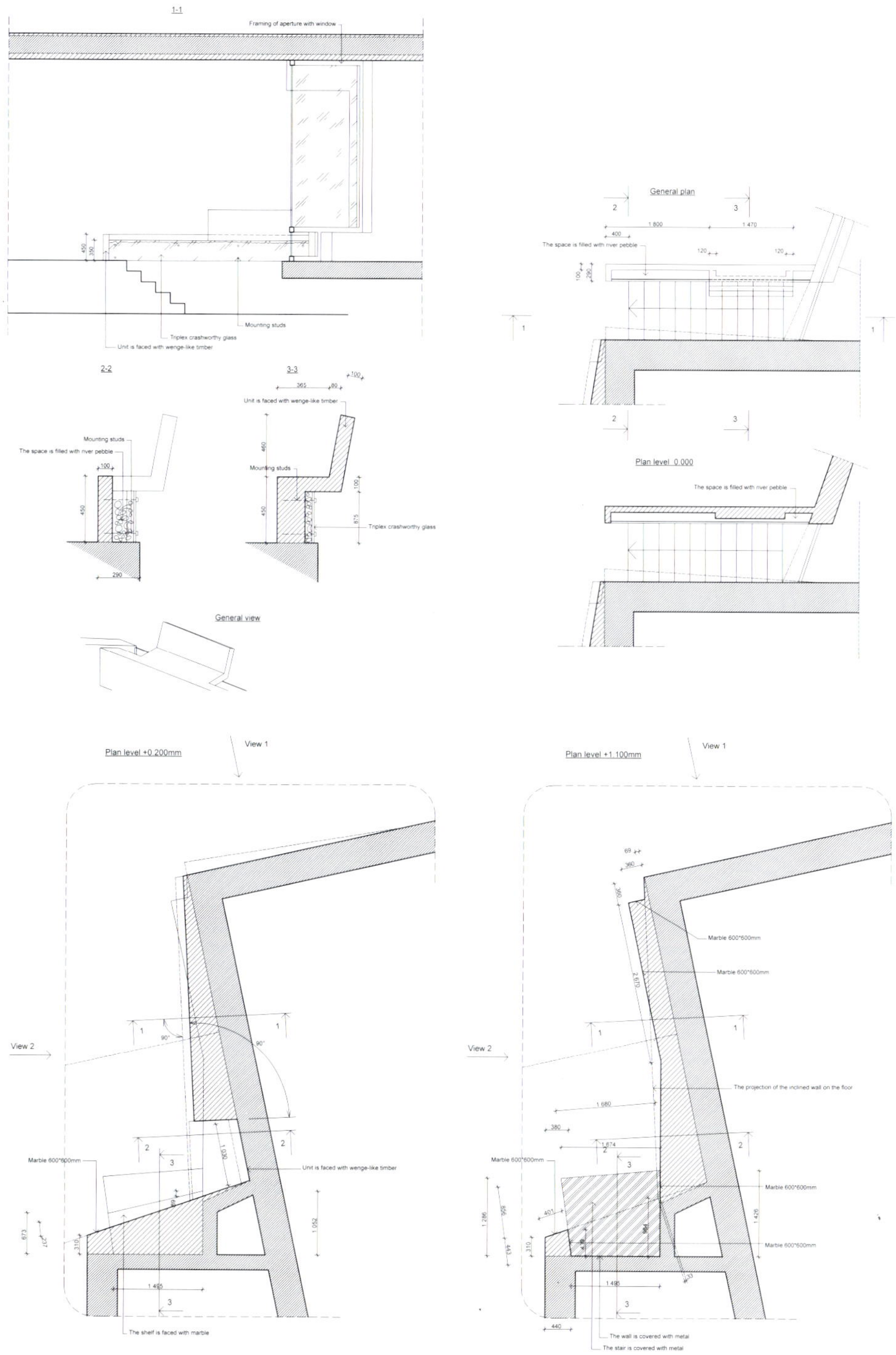

1-1

Framing of aperture with window

Mounting studs

Triplex crashworthy glass

Unit is faced with wenge-like timber

2-2

Mounting studs

The space is filled with river pebble

3-3

Unit is faced with wenge-like timber

Mounting studs

Triplex crashworthy glass

General view

General plan

The space is filled with river pebble

Plan level 0.000

The space is filled with river pebble

Plan level +0.200mm

View 1

View 2

Marble 600*600mm

Unit is faced with wenge-like timber

The shelf is faced with marble

Plan level +1.100mm

View 1

View 2

Marble 600*600mm

The projection of the inclined wall on the floor

The wall is covered with metal

The stair is covered with metal

节点图 DETAILS

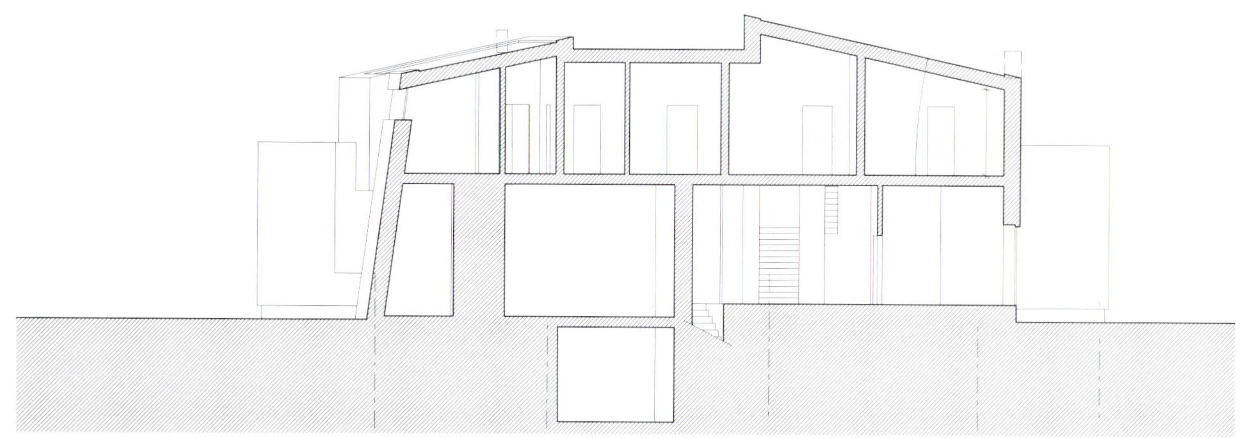

剖面图 SECTION

CONSERVATORY HOUSE

温室别墅

地点：保加利亚瓦尔纳
占地面积：260 m²
建筑面积：780 m²
建筑设计：Ignatov Architects

Location: Varna, Bulgaria
Area: 260 m²
Building Area: 780 m²
Architect: Ignatov Architects

World Green Buildings——Thermal Environmental Engineering Solutions

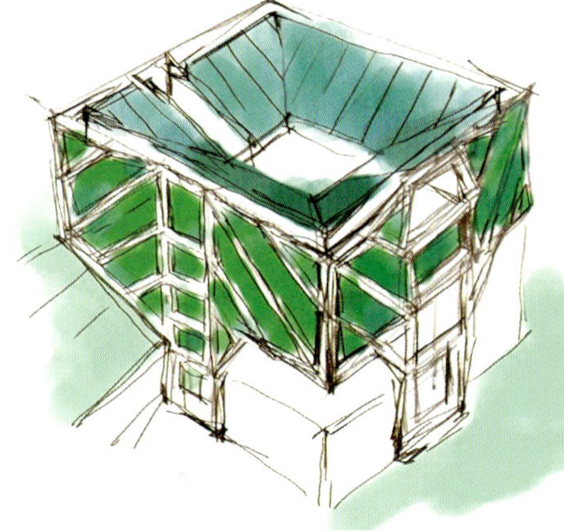

项目概况

位于瓦尔纳附近的温室别墅是一个供主人与亲朋聊天休闲，享受音乐与自然的地方。当地的住宅建设规范要求城郊地区的别墅都要直接或间接地实现某些农业用途，这与房主希望建造大型温室的意愿不谋而合。陡峭而突出的建筑选址原为当地的采石场，后来用作废料堆，这给设计带来了很大挑战。这促使设计师思考如何保持场地原有的斜坡，并恢复当地的生态环境。根据当地的传统和承受能力，选用了钢筋混凝土结构。它内部有中柱以及负荷支撑框架，没有内柱或剪力墙。沿应力线，在南部和东部的外墙有倾斜、垂直或水平的构件，凸显了建筑的科技构造特点。先进的地热系统涵盖了别墅所有的加热和冷却需求，同时得益于温室的绝缘效果，使碳足迹最小化。

建筑热环境解决方案

1. 被动措施

房子被放入一个原有的采石坑中，温室置于起居空间的顶部，最大限度地获取阳光并减少碳足迹。贯穿设计阶段的始终，玻璃温室的受热量及热量损失都经过审慎地处理。因此，温室代表了整个别墅最为主要的功能、美学以及环境特点。它使自然光线照射到别墅的各个楼层，通过土壤的热质 平衡室内微气候，用芬芳的花香调节室内空气，提供氧气和有益微生物，并成为主人最钟爱的休闲场所。在夏季，温室的自然交叉通风结合室内喷泉及快速生长的植物的蒸腾作用使别墅凉爽，减少太阳热能吸收量，并使位于下方的居室降温；冬季通过温室效应提高住户舒适度，获取阳光并通过其液体加热的地板调节空气。西部和北部的热绝缘墙可以防止夏季过度受热、冬季过度降温。由于其保温效果良好，温室下方的起居空间将热量损失降到最低水平。

2. 主动措施

在极端气候条件下的降温和热量损失可以得到补偿，主要通过六个封闭的循环地热探针连接热泵和辐射地板，完成与大地的热量交换。在春秋季节抽取的土壤温度足够保证特定需要。在夏季和冬季进一步通过高效的液－液热泵传输，并通过辐射地板传至各个楼层。

其他可持续设计措施

玻璃屋顶收集雨水作为灌溉之用，并安装集成太阳能集热器提供家用热水。一个500 L的锅炉与一个加热系统及一排集成于玻璃屋顶的太阳能真空管结合，为居室提供热水。用形象的方法比喻这样的组织方式：别墅就像一棵树，温室就是它的绿色树冠，还有主干（中心电梯）以及枝杈（结构），地热探针就如同它的根系。用户居住于此，与之共生。自然景观采用当地植物品种，并保持当地微气候。生物活性废水处理单元（一个分三步处理的净化站）将废水处理成肥料和灌溉用水。通过节能设施和照明，建筑的可持续效能进一步加强。以往各个季节的使用经验证明，温室别墅几乎不需要外部能源。它的设计用一种全新的建筑语言和先进的工程解决方案体现了环境保护的意识。

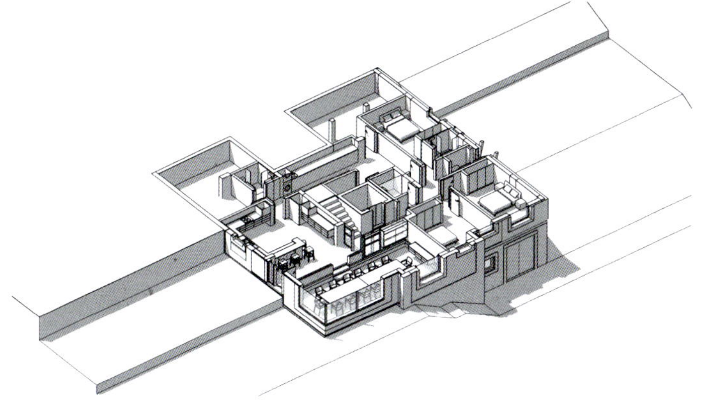

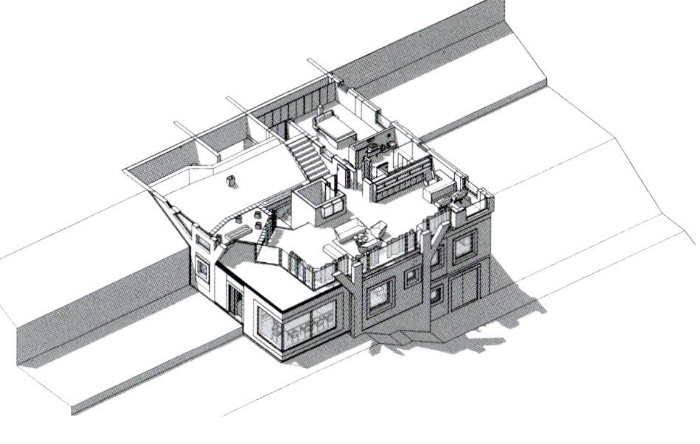

模型图 MODEL DIAGRAM

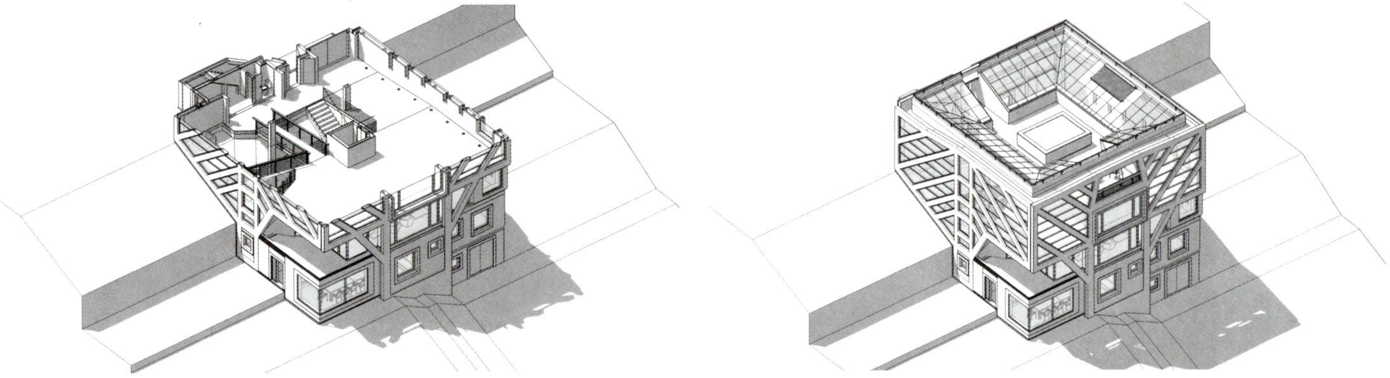

模型图 MODEL DIAGRAM

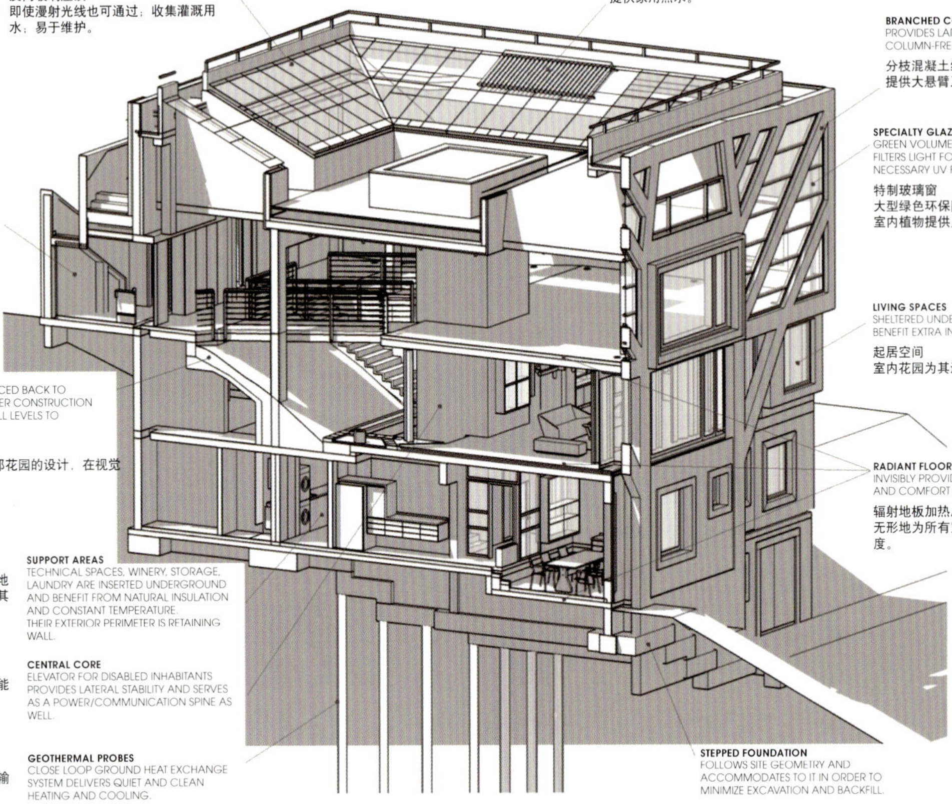

REVERSED GLAZED ROOF
ALLOWS FOR EVENLY DISTRIBUTED LIGHT PENETRATION, COLLECTION OF RAIN WATER FOR IRRIGATION, EASY ACCESS FOR MAINTENANCE.
反向玻璃屋顶
即使漫射光线也可通过；收集灌溉用水；易于维护。

INTEGRATED SOLAR VACUUM TUBES
SUPPLY ALL NEEDED DOMESTIC HOT WATER
集成太阳能真空管
提供家用热水。

BRANCHED CONCRETE STRUTURE
PROVIDES LARGE CANTILEVERS AND COLUMN-FREE INTERIORS.
分枝混凝土结构
提供大悬臂及不用柱子的内部空间。

RAIN WATER TANK
ROOF RAINWATER IS COLLECTED HERE FOR GRAY USE IN TOILETS AND INTERIOR IRRIGATION.
雨水水槽
收集屋顶雨水用于卫生间和灌溉

SPECIALTY GLAZING
GREEN VOLUME-TINTED INSULATED GLASS FILTERS LIGHT FORM HEAT AND ALLOWS NECESSARY UV FOR INTERIOR VEGETATION.
特制玻璃窗
大型绿色环保隔热玻璃透光且隔热，并为室内植物提供足够的紫外线。

LIVING SPACES
SHELTERED UNDER INTERIOR GARDEN BENEFIT EXTRA INSULATION.
起居空间
室内花园为其遮荫，绝缘。

INTERIOR SOIL
TERRAIN IS REINTRODUCED BACK TO INTERIOR GARDEN AFTER CONSTRUCTION AND VISUALLY LINKS ALL LEVELS TO EXTERIOR.
室内土地
地形因素被引入内部花园的设计，在视觉上与外部相连接。

RADIANT FLOOR HEATING/COOLING
INVISIBLY PROVIDES EVEN TEMPERATURE AND COMFORT TO ALL SPACES.
辐射地板加热/冷却
无形地为所有空间提供恒温，提高舒适度。

维护区域
技术区、酒窖、储物室、洗衣房设置于地下，并获益于自然绝热和恒定的温度。其外部为挡土墙。

SUPPORT AREAS
TECHNICAL SPACES, WINERY, STORAGE, LAUNDRY ARE INSERTED UNDERGROUND AND BENEFIT FROM NATURAL INSULATION AND CONSTANT TEMPERATURE. THEIR EXTERIOR PERIMETER IS RETAINING WALL.

中央核心
残疾人电梯提供了横向稳定性，并作为能量/交互中柱。

CENTRAL CORE
ELEVATOR FOR DISABLED INHABITANTS PROVIDES LATERAL STABILITY AND SERVES AS A POWER/COMMUNICATION SPINE AS WELL.

地热探针
封闭的地热循环交换系统安静、清洁的输送冷热能源。

GEOTHERMAL PROBES
CLOSE LOOP GROUND HEAT EXCHANGE SYSTEM DELIVERS QUIET AND CLEAN HEATING AND COOLING.

STEPPED FOUNDATION
FOLLOWS SITE GEOMETRY AND ACCOMMODATES TO IT IN ORDER TO MINIMIZE EXCAVATION AND BACKFILL.
阶形基础
根据地势尽量减少挖掘和回填。

3d剖面图 3D SECTION

Project Description

The Conservatory House near Varna is a place for recreation, relaxed chat with friends and fellow musicians, and contemplation with nature. Local residential building code for rural areas required collateral agricultural use which was in line with the owner's desire for having a large conservatory. Steep, picturesque, but compromised site initially used as local sand quarry and illegal waste dump later presented a good design challenge. It pushed us to think about how to restore the original slope and rehabilitate the local ecology. Reinforced concrete structure is chosen because of its local popularity and affordability. It is designed with a central core and load-bearing facade frames without internal columns and shear walls. Diagonal, vertical and horizontal structural elements on the Southern and Eastern facades follow the stress lines and reveal the building's tectonics. Carbon footprint is minimized by utilizing an advanced geothermal system that covers all heating and cooling needs of the house and benefits from the insulating effect of the conservatory.

Building Thermal Environmental Engineering Solutions

1. Passive Measures

The house was fit into the existing quarry pit and the conservatory was put on top of the living spaces for catching maximum sunlight and minimizing carbon footprint. The typical for every glazed garden heat gains and losses were addressed carefully throughout the design. As a result the conservatory became the main functional, aesthetic and environmental feature of the house. It allows daylight to reach all levels of the house, balances the indoor climate by the thermal mass of its soil, freshens the air, produces oxygen and useful biomass with flower aromas and is a favorite place for relaxation. In summer the house benefits from the conservatory's natural cross ventilation combined with evaporation from interior water fountain and rapid growth of plants dissipating solar heat gain and provides cooling effect to the residential spaces below, while in the winter the residents enjoy its greenhouse effect, catching sun and tempering the air via its liquid-heated floor. Opaque insulated walls to West and North avoid summer overheating and winter overcooling. Living spaces under the conservatory have minimal heat losses due to its insulating effect.

2. Active Measures

Cooling and heating losses in extreme conditions are compensated by a clean exchange with earth via six closed loop geothermal probes connected to a heat pump and radiant floors. In spring and autumn extracted from the ground temperature is sufficient to cover the occupational needs. In summer and winter the energy is further transformed by a highly efficient liquid-liquid heat pump and distributed to all levels via radiant floors.

Other Sustainable Design Features

The glazed roof collects all rainwater for irrigation and features integrated solar collector for domestic hot water. Hot water is supplied by a 500 liter boiler coupled with the heating system and an array of solar vacuum tubes integrated into the glazed roof. Organized in this way, the house resembles a tree with a green crown (conservatory), trunk (elevator core),and branches (structure) and roots (geothermal probes) with the residence accommodated in symbiosis within it. Natural landscaping promotes local plant species and preserves local microclimate. Bio-active wastewater treatment unit (a 3-stage purification station) turns waste into bio- compost and irrigation water. The performance results for the past seasons prove that the Conservatory House consumes very little external power. Sustainable performance of the building is complemented with energy efficient appliances and lighting. Its design actively promotes environmental awareness in new architectural language and advanced engineering solutions.

VILLA "UNDER" EXTENSION

"下沉"别墅

地点：斯洛文尼亚布莱德
占地面积：20 000 m^2
建筑设计：OFIS Architects
设计团队：Mladen Bubalo; Rok Gerbec;
Ivana Sehic; Karla Murovec
摄影：Tomaz Gregoric

Location: Bled, Slovenia
Area: 20,000 m^2
Architect: OFIS Architects
Design Team: Mladen Bubalo; Rok Gerbec;
Ivana Sehic; Karla Murovec
Photography: Tomaz Gregoric

World Green Buildings——Thermal Environmental Engineering Solutions

项目位于阿尔卑斯山麓的Bled湖畔，对19世纪的别墅进行扩建。原有别墅和景观都受到国家自然遗产规定的保护。委托方要求将主要起居室的空间扩大到原来的两倍，而且要求大部分房间都面向湖泊。委托方邀请了多位建筑师提出方案。最主要的任务在于如何将新增的700 m^{2}空间融入建筑中，同时遵守相关环保规定。

本项方案将新增空间设置于既有建筑的地下部分，低于原有建筑一楼地板的高度。延展空间构成了别墅周围“枕垫”一般的圆形基底。如果从对岸来看，新建部分与自然景观融为一体。“枕垫”位于平缓的高地上，俯瞰湖面。地板根据地形分层（± 50 cm）。“枕垫”的屋顶成为别墅高层的露台花园，儿童活动区设置于此。主要的起居空间设在了建筑物新延展出来的部分。衣帽间、主厨房和客人盥洗室位于旧别墅地下室的墙后面。车库和储藏室同样设置于此。

原有别墅成为私密的卧室区域。儿童房设在二楼，家长卧室在顶层。主入口在院子里，在别墅的中轴线转向：一条路从旧房子原来的地下室旁通向新延展部分，另一条越过斜坡，到达楼梯口。这里是一个三层楼高的大厅，形成了旧别墅的核心。它们连接起了新旧区域，成为别墅中的主要通路。所有的房间和开敞空间都通往楼梯，行进路线在大厅交汇。

可持续特色：所有新增部分都以绿色屋顶融入到风景之中。

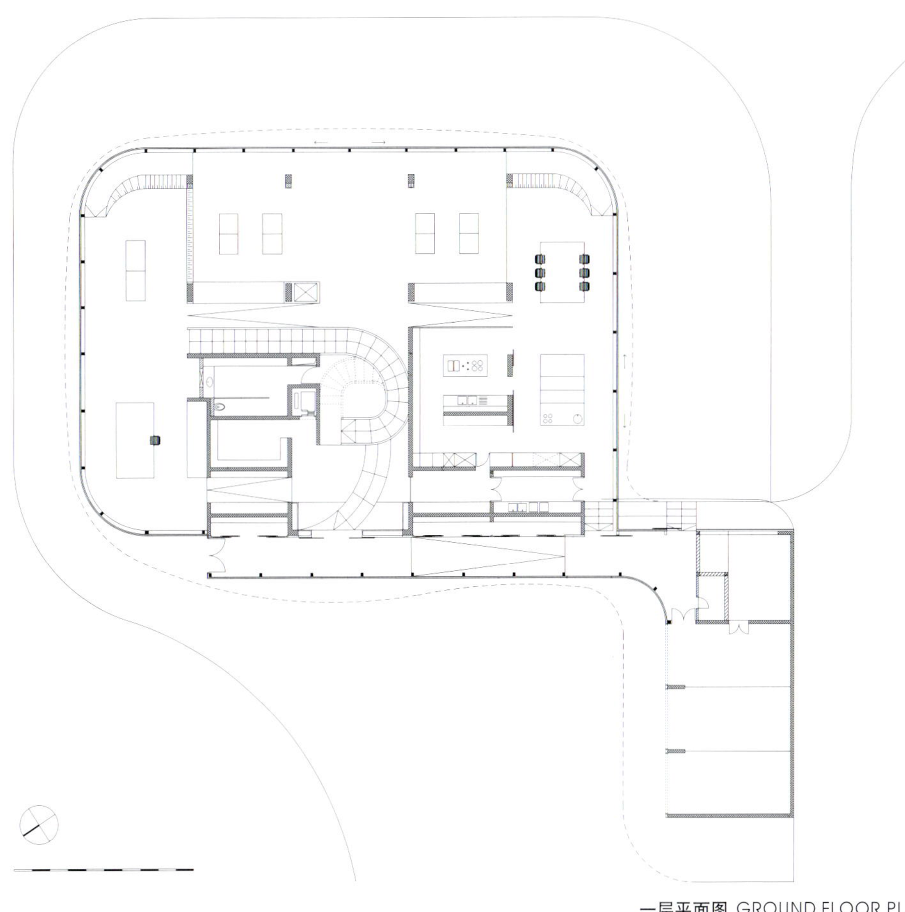

一层平面图 GROUND FLOOR PLAN

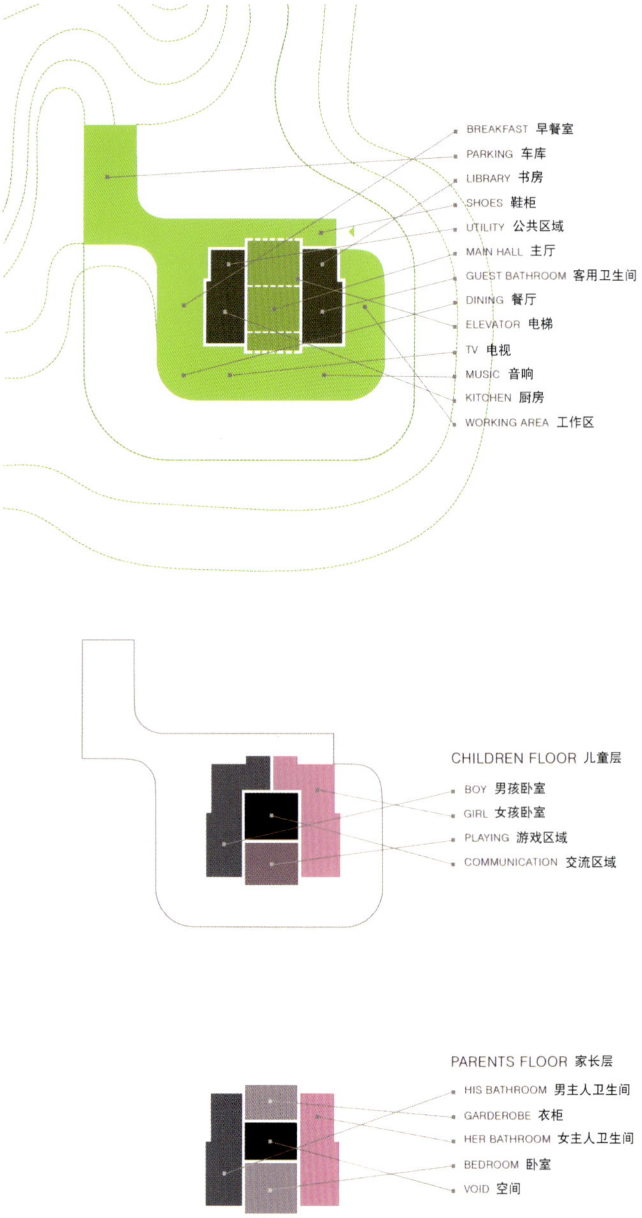

BREAKFAST 早餐室
PARKING 车库
LIBRARY 书房
SHOES 鞋柜
UTILITY 公共区域
MAIN HALL 主厅
GUEST BATHROOM 客用卫生间
DINING 餐厅
ELEVATOR 电梯
TV 电视
MUSIC 音响
KITCHEN 厨房
WORKING AREA 工作区

CHILDREN FLOOR 儿童层
BOY 男孩卧室
GIRL 女孩卧室
PLAYING 游戏区域
COMMUNICATION 交流区域

PARENTS FLOOR 家长层
HIS BATHROOM 男主人卫生间
GARDEROBE 衣柜
HER BATHROOM 女主人卫生间
BEDROOM 卧室
VOID 空间

立面图 ELEVATIONS

分析图 DIAGRAM

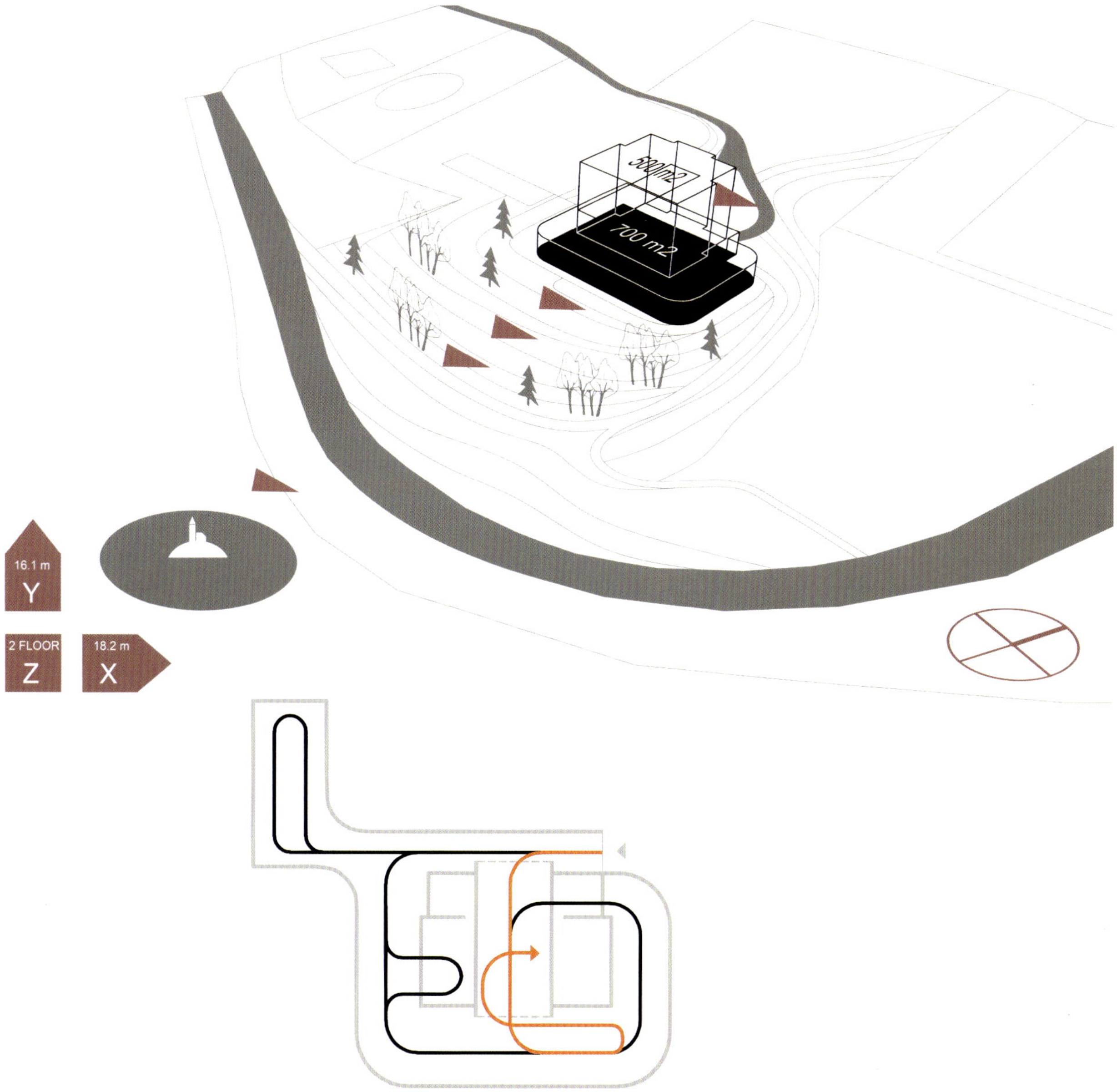

分析图 DIAGRAM

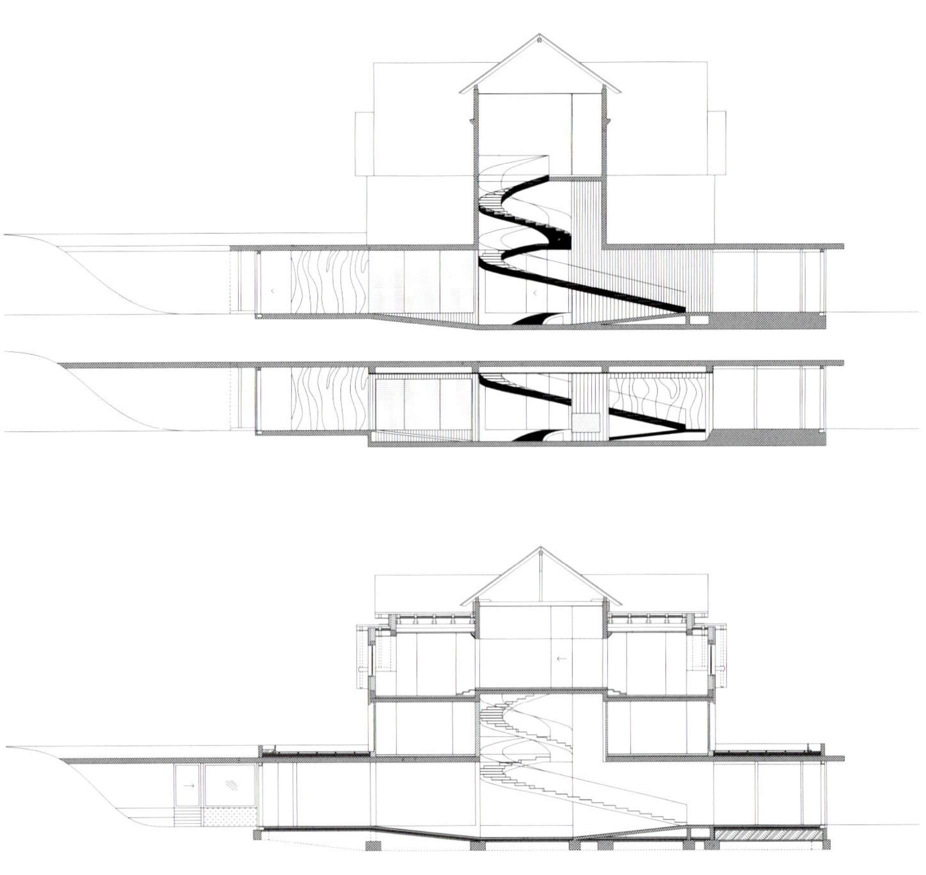

剖面图 SECTIONS

This project involved the extension of a 19th-century villa located in a beautiful Alpine resort next to Lake Bled. Both, the old villa and the landscape were strictly regulated by the National Heritage. The client's request was for the main living area to be twice the size of the old existing villa. On top of that, most of the spaces had to face the lake. The client invited several architects to come up with ideas. The main task was how to incorporate the new 700 m^2 addition while respecting all the restrictions the villa and the surrounding landscape were subject to. Our proposal placed new spaces beneath the ground floor of the existing villa. The extension forms a rounded base around the house - a pillow covered by the landscape. Looking from the other side of the lake, the pillow blurs with the surrounding landscape.

The elevation under the pillow is glazed and overlooks the lake. The floor is organized in levels according to the outside landscape (± 50 cm). They were used as a locator and divider of the opened ground level. The top of the pillow becomes a terrace and garden on the upper floor, where the children's area is located. The main living spaces are placed in the new extension. Services such as wardrobes, the main kitchen and guest bathroom are behind the walls of the old villa cellar. Garages and storage/workshop spaces are also hidden in this volume.

The old villa becomes a private rest area with the children's space on the first floor and the parent's on the top. The main entrance is from the courtyard and is displaced from the house's central axis; one enters the side volume of the old villa cellar to the new part. Then one crosses the curved ramp and passes beneath the staircase. Here, a three-storey hall opens up to form the heart of the old villa, while the visual axis of the lake conducts you to the new living area. The curved stairs are the defining element; they connect the old and the new, and play the role of the main communications core in the house. All rooms and open spaces lead onto the staircase and communicate with the main lobby.

Sustainable factor: all new extension part around the existing villa is integrated into the landscape with the green roof.

绿色建筑室内热环境设计要点

一、公共建筑

1. 采用集中空调的建筑，房间内的温度、湿度、风速等参数符合现行国家标准《公共建筑节能设计标准》（GB 50189—2005）中的设计计算要求。

室内热环境是指影响人体冷热感觉的环境因素。“热舒适”是指人体对热环境的主观热反应，是人们对周围热环境感到满意的一种主观感觉，它是多种因素综合作用的结果。舒适的室内环境有助于人的身心健康，进而提高学习、工作效率；而当人处于过冷、过热环境中，则会引起疾病，影响健康乃至危及生命。一般而言，室内温度、室内湿度和气流速度对人体热舒适感产生的影响最为显著，也最容易被人体所感知和认识；而环境辐射对人体的冷热感产生的影响很容易被人们所忽视；除此之外，围护结构辐射也会对室内空气温度产生直接的影响，因此本标准只引用室内温度、室内湿度、气流速度三个参数评判室内环境的人体热舒适性。根据《公共建筑节能设计标准》（GB 50189—2005）中的设计计算要求，上述参数在冬夏季分别控制在相应区间内。

2. 建筑围护结构内部和表面无结露、发霉现象。

由于围护结构中窗过梁、圈梁、钢筋混凝土抗震柱、钢筋混凝土剪力墙、梁、柱等部位的传热系数远大于主体部位的传热系数，形成热流密集通道，即为热桥。本条规定的目的主要是防止冬季采暖期间热桥内外表面温差小，内表面温度容易低于室内空气露点温度，造成围护结构热桥部位内表面产生结露；同时也避免夏季空调期间这些部位传热过大增加空调能耗。内表面结露，会造成围护结构内表面材料受潮，在通风不畅的情况下易产生霉菌，影响室内人员的身体健康。因此，应采取合理的保温、隔热措施，减少围护结构热桥部位的传热损失，防止外墙和外窗等外围护结构内表面温度过低。另外在室内使用辐射型空调末端时，需密切注意水温的控制，避免表面结露。

3. 采用集中空调的建筑，新风量符合现行国家标准《公共建筑节能设计标准》（GB 50189—2005）的设计要求。

公共建筑所需要的最小新风量应根据室内空气的卫生要求、人员的活动和工作性质，以及在室内停留时间等因素确定。卫生要求的最小新风量，公共建筑主要是对CO_2的浓度要求(可吸入颗粒物的要求可通过过滤等措施达到)。此外，为确保引入室内的为室外新鲜空气，新风采气口的上风向不能有污染源；提倡新风直接入室，缩短新风风管的长度，减少途径污染。公共建筑主要房间人员所需的最小新风量，应根据建筑类型和功能要求，参考国家标准《旅游旅馆建筑热工与空气调节节能设计标准》（GB 50189—1993）、《公共场所卫生标准》（GB 9663—1996～GB 9673—1996）、《饭馆(餐厅)卫生标准》（GB 16153—1996）、《室内空气质量标准》（GB/T 18883—2002）等标准规范文件确定。

4. 建筑设计和构造设计有促进自然通风的措施。

自然通风是在风压或热压推动下的空气流动。自然通风是实现节能和改善室内空气品质的重要手段，提高室内热舒适的重要途径。因此，在建筑设计和构造设计中鼓励采取诱导气流、促进自然通风的主动措施，如导风墙、拔风井等等，以促进室内自然通风的效率。

5. 室内采用调节方便、可提高人员舒适性的空调末端。

公共建筑空调末端是提供室内使用者舒适性的重要保证手段。本条款的目的是杜绝不良的空调末端设计，如未充分考虑除湿的情况下采用辐射吊顶末端、宾馆类建筑采用不可调节的全空气系统等。而个性化送风末端、干式风机盘管、地板采暖等末端，用户可通过手动或自动调节来满足要求，有助于提高使用舒适性。

6. 采用可调节外遮阳，改善室内热环境。

可结合建筑的外立面造型采取合理的外遮阳措施，形成整体有效的外遮阳系统，可以有效地减少建筑因太阳辐射和室外空气温度通过建筑围护结构的传导得热以及通过窗户的辐射得热，对于改善夏季室内热舒适性具有重要作用。

二、住宅建筑

1. 居住空间能自然通风，通风开口面积在夏热冬暖和夏热冬冷地区不小于该房间地板面积的8%，在其他地区不小于5%。

自然通风可以提高居住者的舒适感，有助于健康。在室外气象条件良好的条件下，加强自然通风还有助于缩短空调设备的运行时间，降低空调能耗，绿色建筑应特别强调自然通风。住宅能否获取足够的自然通风与通风开口面积的大小密切相关，本条文规定了住宅居住空间通风开口面积与地板最小面积比。一般情况下，当通风开口面积与地板面积之比不小于5%时，房间可以获得比较好的自然通风。由于气候和生活习惯的不同，南方更注重房间的自然通风，因此本条文规定在夏热冬暖和夏热冬冷地区，通风开口面积与地板面积之比不小于8%。自然通风的效果不仅与开口面积与地板面积之比有关，事实上还与通风开口之间的相对位置密切相关。在设计过程中，应考虑通风开口的位置，尽量使之能有利于形成“穿堂风”。

2. 屋面、地面、外墙和外窗的内表面在室内温、湿度设计条件下无结露现象。

《民用建筑热工设计规范》（GB 50176—1993）对建筑围护结构的热工设计提出了很多基本的要求，其中规定外围护结构的内表面不能结露，绿色住宅应满足此要求。外围护结构的内表面结露会造成居民生活不便，严重时会导致霉菌的滋生，影响室内的卫生条件。绿色建筑应为居住者提供一个良好的室内环境，因此在室内温、湿度设计条件下不应产生结露现象。导致结露除空气过分潮湿外，表面温度过低是直接的原因。一般说来，住宅外围护结构的内表面大面积结露的可能性不大，结露大都出现在金属窗框、窗玻璃表面、墙角、墙面上可能出现的热桥附近，作为绿色建筑在设计和建造过程中，应核算可能结露部位的内表面温度是否高于露点温度，采取措施防止在室内温、湿度设计条件下产生结露现象。

3. 在自然通风条件下，房间的屋顶和东、西外墙内表面的最高温度满足现行国家标准《民用建筑热工设计规范》（GB 50176—1993）的要求。

《民用建筑热工设计规范》（GB 50176—1993）对建筑围护结构的热工设计提出了很多基本的要求，其中规定在自然通风条件下屋顶和东、西外墙内表面的温度不能过高。屋顶和外墙内表面温度的高低直接影响到室内人员的舒适，控制屋顶和外墙内表面温度不至于过高，可使住户少开空调多通风，有利于提高室内的热舒适水平，同时降低空调能耗。《民用建筑热工设计规范》（GB 50176—1993）详细规定了在自然通风条件下计算屋顶和东、西外墙内表面温度的方法。

4. 设采暖或空调系统(设备)的住宅，运行时用户可根据需要对室温进行调控。

从舒适和节能角度，以及收费服务角度，设采暖或空调系统(设备)的住宅，用户应能自主调节室温。

5. 采用可调节外遮阳装置，防止夏季太阳辐射透过窗户玻璃直接进入室内。

夏季强烈的阳光透过窗户玻璃照到室内会引起居住者的不舒适感，同时还会大幅增大空调负荷。窗户的内侧设置窗帘在住宅建筑中是非常普遍的，但内窗帘在遮挡直射阳光的同时常常也遮挡了散射的光线，影响室内的自然采光，而且内窗帘对减小由阳光直接进入室内而产生的空调负荷作用不大。在窗户的外面设置一种可调节的遮阳装置，可以根据需要调节遮阳装置的位置，防止夏季强烈的阳光透过窗户玻璃直接进入室内，提高居住者的舒适感。可调节外遮阳装置对于建筑夏季的节能作用也非常明显。许多住宅在周一至周五工作日的白天室内是没有人的，如果窗户有可靠的可调节外遮阳（例如活动卷帘），白天可以借助外遮阳将绝大部分太阳辐射阻挡在室外，可以大大缩短晚上空调器运行的时间。外遮阳之所以要强调可调节的，是因为无论是从生理还是从心理的角度出发，冬季和夏季居住者对透过窗户进入室内的阳光的需求是截然相反的，而固定的外遮阳(例如窗口上沿的遮阳板)无法很好地适应这种相反的需求。可调节外遮阳应注重可靠、耐久和美观。

摘自国家标准《绿色建筑评价标准》（GB/T 50378—2006）